ADVANCED MASONRY SKILLS

*This book is dedicated to
my wife, Betty,
my son, Ricky,
and my daughters, Pam and Pat,
for all their help and encouragement.*

Copyright ©1978 by Litton Educational Publishing, Inc.

Library of Congress Catalog Card Number 77-29112
ISBN: 0-442-24289-1

All rights reserved. No part of this work covered by the copyright hereon may be reproduced or used in any form or by any means—graphic, electronic, or mechanical, including photocopying, recording, taping, or information storage and retrieval systems—without written permission of the publisher.

Printed in the United States of America.

Published in 1978 by Van Nostrand Reinhold Company
A division of Litton Educational Publishing, Inc.
135 West 50th Street, New York, NY 10020, U. S. A.

Van Nostrand Reinhold Limited
1410 Birchmount Road
Scarborough, Ontario M1P 2E7, Canada

Van Nostrand Reinhold Australia Pty. Ltd.
17 Queen Street
Mitcham, Victoria 3132, Australia

Van Nostrand Reinhold Company Limited
Molly Millars Lane
Wokingham, Berkshire, England

16 15 14 13 12 11 10 9 8 7 6 5 4 3 2 1

Library of Congress Cataloging in Publication Data
Kreh, R. T.
  Advanced masonry skills.

  Includes index.
  1. Masonry. I. Title.
TH5311.K7   693   77-29112
ISBN 0-442-24289-1

PREFACE

Masonry work is a challenging profession, as well as a good means of earning a living. It is a highly skilled trade, requiring not only perfected tool skills but an overall knowledge of building, and an understanding of related technical information. The time spent in apprentice programs is 3 to 4 years, depending on the company or union agreement.

In MASONRY SKILLS, the basics of the trade are covered. ADVANCED MASONRY SKILLS has been developed to cover the application of these basics. The book is divided into 34 units, grouped into 10 sections. The 10 sections present the advanced skills in masonry in logical sequence, as they would be required on the job. These sections deal with the following topics: laying out and constructing footings and foundations; chimneys and fireplaces; building outdoor structures; modular coordination; glazed and prefaced masonry units; arches; stonemasonry; repair and renovation work; advances in masonry construction; and job management practices.

The material in ADVANCED MASONRY SKILLS is appropriate for the masonry student in a high school or community college, as well as for the apprentice working on the job. It is a book that will also serve as a reference to the mason working in the trade.

MASONRY FEATURES OF ADVANCED MASONRY SKILLS

This book contains the following major features:

- Performance objectives begin each unit telling the reader what he should learn through careful study of the unit material.
- All of the material is up to date, presented in a easy-to-follow format.
- The written material and procedures are accompanied by photos of actual, on-the-job masonry practices.
- Safety features integrated throughout the book are highlighted to draw the reader's attention.
- Logical presentation of information as the mason should learn it is based on the author's extensive experience in the trade.
- Related mathematical concepts which apply directly to trade situations are introduced at the appropriate locations in the text.

- Review questions at the end of each unit help the reader evaluate his comprehension of the material. This also allows for a self-pacing, individualized-type program.
- Projects included at the end of many units give the reader practice in applying the techniques discussed in the unit.
- Summary reviews conclude each section, highlighting the important points in the section. These are followed by summary review questions, to evaluate the reader's overall knowledge of the section materials.
- Technical terms are explained as they occur, and defined in an extensive glossary at the end of the book.
- The book has been reviewed by the Brick Institute of America, National Lime Association, National Concrete Masonry Association, the National Bureau of Standards, and other experts in the construction industry.
- An Instructor's Guide includes lesson outlines for each unit, answers to achievement reviews and section reviews, a pretest and final test for evaluating students, suggested activities and field trips, and sources for audio-visual aids.

CONTENTS

SECTION 1 FOUNDATIONS, FOOTINGS, AND WATERPROOFING

Unit 1 Laying Out the Foundation	1
Project 1: Laying Out Batter Boards for a Foundation	11
Unit 2 Constructing Footings	15
Unit 3 The Story Pole	24
Project 2: Laying Out a Story Pole for a Concrete Block Foundation	30
Unit 4 Building a Concrete Block Foundation	33
Project 3: Building a Concrete Block and Brick Foundation on a Changing Grade Line	45
Unit 5 Waterproofing the Foundation Wall	50
Project 4: Parging the Foundation	59

SECTION 2 CHIMNEYS AND FIREPLACES

Unit 6 Cutting with the Masonry Saw	66
Project 5: Cutting Concrete Block with the Masonry Saw	77
Unit 7 History, Theory, and Function of the Fireplace and Chimney	79
Project 6: Two-flue Chimney	87
Unit 8 Design and Construction of a Fireplace	90
Project 7: Building a Brick Fireplace and Chimney	114
Unit 9 Multiple-Opening and Heat-Circulating Fireplaces	119

SECTION 3 BUILDING OUTDOOR STRUCTURES

Unit 10 Patios and Brick Paving	133
Project 8: Laying Brick Paving in a Running Bond	151
Unit 11 Porches and Steps	154
Project 9: Building a Brick Porch and Steps	162
Unit 12 Garden Wall Construction	166
Project 10: Basket Weave Pattern in a Panel Wall	174
Project 11: Diamond Bond Pattern in a Panel Wall	176

Unit 13 Retaining Walls 179
      Project 12: Building a 12-inch Brick and Concrete Block Retaining Wall ... 195
Unit 14 Outdoor Fireplaces 198
      Project 13: Building a Barbecue 209

SECTION 4 MODULAR COORDINATION

Unit 15 Understanding Modular Coordination 215
Unit 16 Estimating Modular Masonry Materials 227

SECTION 5 GLAZED AND PREFACED MASONRY UNITS

Unit 17 Structural Clay Tile and Glazed Bricks 244
Unit 18 Laying Glazed Structural Tile and Glazed Bricks 259
Unit 19 Prefaced Concrete Masonry Units 270

SECTION 6 ARCHES

Unit 20 Development of Arches 281
Unit 21 Construction of a Semicircular Arch 289
      Project 14: Laying Out and Constructing a Two-rowlock
              Semicircular Arch 297
Unit 22 Construction of a Jack Arch 300
      Project 15: Constructing a Common Jack Arch 304
      Project 16: Constructing a Bonded Jack Arch 306

SECTION 7 STONEMASONRY

Unit 23 Development of Stonemasonry 312
Unit 24 Cut Limestone – Composition, Types, and Installation 321
Unit 25 Cutting and Laying Stone 336

SECTION 8 ALTERATION, REPAIR, AND MAINTENANCE OF MASONRY

Unit 26 Cutting Out, Repointing, and Repairing Masonry Work 355
      Project 17: Repointing a Section of Brick Wall 365
Unit 27 Repairing Cracks, Removing Efflorescence, and Dampproofing 366
Unit 28 Brick Veneering Existing Structures 375

SECTION 9 TRENDS AND DEVELOPMENTS IN MASONRY CONSTRUCTION

Unit 29 Constructing Masonry in Cold Weather	388
Unit 30 Engineered Brick Masonry	400
Unit 31 Prefabricated Masonry Panels	410
Unit 32 Advances in Masonry Equipment	423

SECTION 10 PLANNING AND MANAGING THE JOB

Unit 33 Job Planning and Practices to Improve the Mason's Productivity	438
Unit 34 Leadership on the Job — the Mason Foreman	450
Glossary	465
Acknowledgments	473
Index	475

Section 1
Foundations, Footings, and Waterproofing

Unit 1 LAYING OUT THE FOUNDATION

OBJECTIVES

After studying this unit the student will be able to

- describe the important features of a foundation.
- lay out and set up a batter board for the foundation of a small building.
- lay out building lines using the 6-8-10 method.

The foundation of a building consists of walls that are built below grade (surface of the ground), or below the first floor joists. The purpose of the foundation is to support the building.

Two major factors that determine the design and construction of a foundation are the load-bearing strength of the soil, and the weight of the building and contents the foundation must support. When determining wall thickness, an architect must also consider the pressures that will be applied to the foundation walls. Although design problems are not the responsibility of the mason, an understanding of what it takes to construct a strong, safe foundation is essential in today's masonry practices.

MASONRY UNITS FOR CONSTRUCTING FOUNDATION WALLS

Foundation walls can be built from concrete or masonry units, such as brick, stone, and concrete block. The most frequently used masonry unit is the concrete block. Two of the main reasons are that concrete block is low in cost and can be laid faster than other units.

Concrete block is also readily available in almost any area of the country and can be adapted to the requirements of most jobs. Features such as high compressive strength and resistance to fire, moisture (if waterproofed), and deterioration from the elements make the concrete masonry unit foundation a good choice.

LOADS

Loads supported by foundations can be classified as two main types: dead loads and live loads.

Dead load is the weight of the superstructure (floors, roof, and walls, for example) and the foundation itself. Dead load is considered to be constant.

Live load consists of pressures placed on the foundation wall or superstructure of the building. Any moving load applied to the structure is a live load. Some examples of a live load are snow on a roof, movable objects on a floor, furniture, and people in the building. Pressure against the foundation wall from earth, moisture, wind, or impact forces (such as earthquakes) are also live loads and must be considered in designing and building a foundation. Early backfilling of a foundation creates a lot of pressure and is one of the major causes of cracking and leaking foundations. Even though a foundation is well designed, if it is not constructed properly by the mason, it will not support the structure or protect it from the elements.

BUILDING CODES

Local governing agencies in most cities and towns have established ordinances and regulations to protect the owner, builder, and community against substandard construction. The rules governing construction requirements are called *building codes*. These codes must be followed by all those involved in designing and constructing a structure.

Items such as depth of footings, thickness of walls, height of the chimney above the roof, and so forth are included in building codes. A building inspector makes regular inspections to be sure these codes are followed. Note: Before laying out and constructing any masonry foundation, the local building codes should be checked.

THE BUILDING LINE

When laying out a building, one of the first steps is to locate one straight line that will be used to lay out the rest of the structure. This main line is usually the front wall and is called the *building line*.

On large jobs or projects, establishing this line is the responsibility of the engineer or surveyor. In most housing work, however, the building line is taken from the established property line, curbs, or the center of the street or road. This is the responsibility of the contractor or foreman on the job. When the building line has been laid out, the next step is to locate level points.

THE BENCH MARK

A *bench mark* is a definite mark placed in the ground or on some stationary object that is on the job site and near the proposed structure. The stationary object may be a pier, post, or wall of a nearby building. The bench mark must not be moved during the construction of the building since it serves as a permanent reference point for elevations.

The National Oceanographic and Atmospheric Administration establishes key bench marks in all towns and cities throughout the country. These points are based on the elevation either above or below sea level. The bench mark on the job is directly related to these established points.

Bench marks serve as a reference when determining elevations for the structure and specific details of construction parts such as sills, lintels, foundation depth, and finished floors. This is done by either adding or subtracting measurements on the plans from the known bench mark. In residential construction, it is often the finished floor height. In commercial construction, it is usually 4 feet above the finished floor.

The bench mark also aids in the positioning of any objects or materials used in the construction of the building. An example is the setting of a brick windowsill for a house. On the plan, the top of the sill is 30 inches above the finished floor. If there is no floor in place at the time of the sill installation, the bench mark can be used to determine the correct height of the sill. This is done by measuring from the bench mark to the floor level and up the required 30 inches.

On a commercial building, if the windowsill is to be set at 30 inches above the finished floor, the deduction must be made down from the 4-foot level mark. Subtracting 30 inches from 48 inches, the top of the sill is 18 inches. If the floor is not yet in place, a wooden block is nailed at the level where it will be, working from the bench mark. A story pole or gauge rod (see Unit 3) is then made by the mason or foreman and all heights are taken from that point.

BATTER BOARDS

Batter boards, figure 1-1, are erected to hold and preserve the building lines during the period of construction. They are usually constructed of three stakes driven into the ground at right angles on the corner of the building. Two boards are nailed to these stakes, level with each other, forming the right angle.

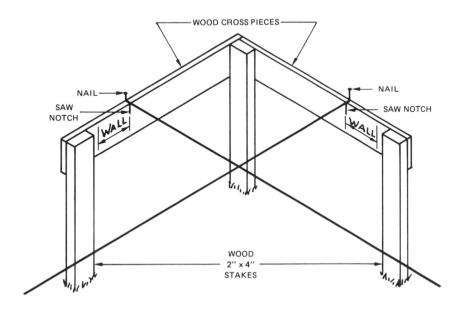

Fig. 1-1 Right-angle corner batter board construction. Nylon line is stretched tightly to the batter board on opposite corners. A saw notch marks the wall line. Notice the area of the wall is marked on the batter boards.

CONSTRUCTING THE BATTER BOARD

Batter boards should be constructed from sound lumber. The boards must be strong enough to stay in the ground without getting knocked out easily or pulled out when the lines are attached. The stakes should be made of 2-inch x 4-inch boards, sharpened on the ends so they can be driven securely into the ground with a sledge hammer. Stakes must be approximately 4 feet long in order to be driven into the ground tightly.

Two 1-inch x 4-inch boards, each 5 feet long, are nailed to these stakes. Strips of form lumber found on most construction projects can be used, but any sound board is acceptable. Make sure the boards nailed on the stakes are level with each other. Batter boards for a single wall require only two stakes, with one board nailed on them, figure 1-2.

LEVELING THE BATTER BOARDS

The batter boards are set back a minimum of 4 feet from the actual building line. This is to allow enough working room and to ensure that the batter boards are not disturbed when excavation is done. Frequently, lines are removed and set up again. In this instance, correct placement of the batter boards ensures that the lines are always in the right place.

After the batter boards are set up, they are leveled with each other on all corners. This is done by setting up a *builder's level* (an instrument that indicates level points). Using the builder's level, a reading is taken for the base height from one corner of the house. All of the other corners are then made level with this base point. This procedure is referred to as shooting points. Establishing level points is usually the responsibility of the contractor or mason foreman and is done by referring to the bench mark. On most homes, the top of the batter board is usually at the same height as the top of the finished floor. The depth of the excavation, and the height of window heads, door heads, and sills can be figured from this elevation.

LAYING OUT THE FOUNDATION USING THE 6-8-10 METHOD

Squaring the foundation is one of the most important steps in the beginning of any structure. On large jobs, the surveyor or building engineer does this by using a transit level. To square the foundation on small jobs or houses, the mason can use the 6-8-10 method (also called the right-triangle method).

Using this method, nylon lines are attached to the batter boards and then squared by applying the 6-8-10 formula. This formula is stated as follows: the base squared, plus the rise squared, equals the hypotenuse squared. The base is some given line (usually the front building line),

Fig. 1-2 Single batter board setup. Notice how lime is used to mark the location of the wall that will have to be dug out on the surface of the ground.

the rise is a line at a right angle to the base, and the hypotenuse is a line drawn connecting the two, forming a triangle.

In the following procedure, the 6-8-10 method is applied to laying out a foundation. Refer also to figure 1-3.

1. Attach a nylon line to batter board No. 1.
2. Attach the other end of the line to batter board No. 2 and tighten securely. This serves as the main building line.
3. Attach a line from batter board No. 1 to batter board No. 3 and tighten.
   Note: The help of another person is needed to perform the remaining steps of this procedure. The task now is to make the No. 1 corner square.
4. Measure out 8 feet from the No. 1 corner (point A to B) using the steel tape. Mark this exact point on the line with a fine pencil point, or by putting a pin through the line.
5. Measure out 6 feet from corner No. 1 in the opposite direction (point A to C). Mark this point in the same way as in step 4.

With one person holding the end of the steel tape at the 8-foot mark (point B), stretch the tape tightly to the 6-foot mark (point C). Detach the line on batter board No. 3 and swing the line either in or out until the 6-foot mark (point C) is exactly in line with the 10-foot reading on the steel tape. Then refasten the line on batter board No. 3. This forms a right angle corner at batter board No. 1. After the procedure is completed, the rest of the building lines can be established by measuring off of the two squared lines according to the dimensions on the plans.

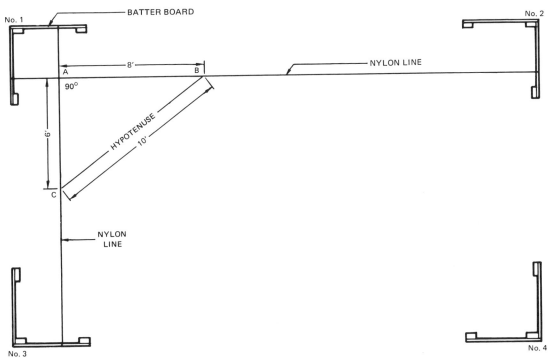

Fig. 1-3 Laying out a batter board corner using the 6-8-10 method.

6 ■ Section 1 Foundations, Footings, and Waterproofing

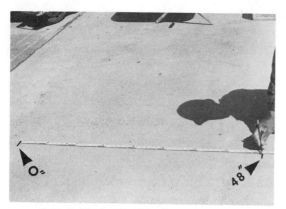

Step 1 Mark off a line 48 inches long.

Step 2 Starting from the original point, measure 36 inches in the direction shown, and swing an arc (moving the ruler to make a curved mark).

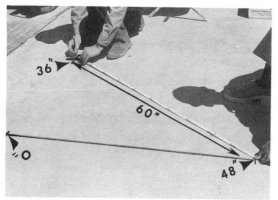

Step 3 Mark from the 48-inch point to the 36-inch point a distance of 60 inches. Arc this mark to bisect the angle. The point where the marks bisect will form a right angle.

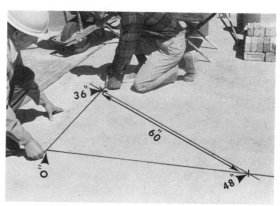

Step 4 Strike a chalk line from the original point passing through the bisected marks of 36 inches and 60 inches. This completes the laying out of the corner square.

Step 5 Check the 3-4-5 layout with the 2-foot framing square. The corner will be square if all of the steps have been correctly followed. The angle formed will be a 90-degree angle.

Fig. 1-4 Laying out a brick corner using the 3-4-5 method

The reason this 6-8-10 method works can be explained mathematically. The 6-foot base squared (multiplied by itself) equals 36 (6 x 6). The 8-foot rise squared equals 64 (8 x 8). Adding 36 to 64 gives a total of 100. The hypotenuse, therefore, is 10 feet because 10 x 10 equals 100.

This formula can be applied on a larger or smaller scale depending upon the size of the project. The larger the triangle, the less chance of error in squaring a foundation. For example, the base and the rise may be doubled (16 feet instead of 8 feet, and 12 feet instead of 6 feet) thus causing the hypotenuse to be doubled (20 feet instead of 10 feet). The formula can also be used to double check the foundation after all lines have been established. Rarely, however, is this formula used where the hypotenuse is greater than 20 feet for a foundation.

The smallest scale generally used for squaring a masonry corner is the 3-4-5 formula. This is very handy when a square is not available. The 3-4-5 formula is useful when laying out a brick corner on a concrete floor or slab. In place of a nylon line attached to batter boards, a chalk line is struck on the slab, as shown in figure 1-4.

LOCATING WALL LINES ON BATTER BOARDS

The exact location of all wall lines is marked on the batter boards with a lasting mark, using a permanent ink marker or lumber crayon. A saw notch is made on the top of the board to further ensure that the location of the line is not lost or confused. On the top of the batter board be sure to mark on which side of the saw cut the wall goes. (This is to ensure that the wall is built on the correct side of the line during the construction process).

As a final check, measure the diagonals of the foundation, figure 1-5. They should be equal in length (if the structure is to be rectangular). Be careful that as the diagonal measurement is made, the tape does not sag in the middle and give a false measurement.

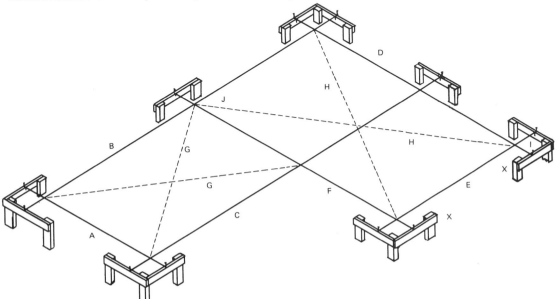

Fig. 1-5 Batter board layout with an offset. Note how the diagonals are checked. Diagonal measurements should be equal (G = G, H = H).

This can be avoided by having another person hold the tape level in the center during the measurement. Be sure to pull all layout lines tightly and securely by wrapping and knotting them around the nail driven into the top of the batter board even with the saw notch.

ESTABLISHING DEPTH OF THE FOUNDATION

The most important consideration in determining the depth of a foundation is the frost line. The *frost line* is the depth to which the ground freezes. When water freezes, its volume increases. This also happens when moist soil freezes. When soil freezes, it rises upward. When it thaws, it returns to its original position. Masonry work, including foundations, should be built on a base below the frost line to prevent the frost from causing it to shift or crack. The depth of the frost line depends upon the temperature in each locality. The frost line in a particular area should always be checked before digging a foundation or base for masonry walls.

The depth of the foundation is determined by measuring down from the lines attached to the batter board. When laying concrete block, a multiple of 8 inches is used to determine the depth. A good method to follow as the foundation is being pushed out (excavated) by the bulldozer is to cut a wood pole of the proper length. As the excavating progresses, the depth can be easily checked by stretching a line diagonally from the batter boards and holding the pole vertically in the excavated foundation. Earth is removed until the top of the pole is even with the line. This saves much work when the base is leveled up for the concrete floor. It also makes digging the footings easier since a minimum of earth has to be removed.

BASEMENT FOUNDATION

A basement is a practical and economical use for the foundation area. The distance from the basement floor to the basement ceiling should be a minimum of 6'-10". If the basement is to be finished and used as a living space, a distance of 7'-6" to 8'-0" is recommended.

Having a basement constructed is an inexpensive means of providing added space in a home. Since a building must have some type of foundation, just an additional two to six feet depth allows for a basement. However, conditions such as too much rock or moisture can prevent the building of a full basement.

Cutting costs in the construction process, if it is going to affect the strength or durability of the foundation, is false economy. Since the foundation is the base of the structure, the building can be no stronger than the foundation it rests on.

CRAWL SPACE FOUNDATION

When no basement space is specified, a crawl space type of foundation can be used. The term *crawl space* comes from the fact that only enough space is provided for a person to crawl around under the floor joists for access to plumbing or electrical wiring. A 24-inch minimum space is recommended. Crawl space can also be used for limited storage space.

The crawl space foundation, figure 1-6, including the footing, is dug deep enough in the ground to eliminate the danger of frost or freezing temperatures getting under the wall

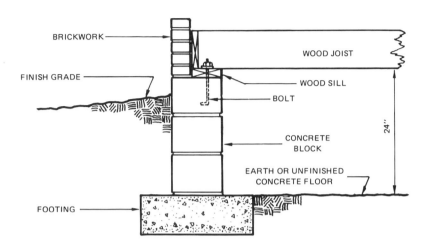

Fig. 1-6 Crawl space foundation. The depth of the foundation below grade is determined by the frost line in a specific area. Note that there should be a minimum space of 24 inches from the top of the rough floor to the underside of the wood joists. This is to allow workers to move around, and to provide ventilation.

causing the foundation to crack. This depth is known as the frost line. A crawl space foundation should never be closed up tightly because the dampness can cause mold and rot to form on wood floors. Vents must be installed in the masonry walls in such a manner that air can circulate.

EXCAVATING THE FOUNDATION

A bulldozer or front-end loader can be used to excavate for the foundation. Mason's lime can be poured along the building line giving an approximate digging line to follow. This helps the equipment operator since the nylon lines that usually mark the building line must be removed temporarily during initial ground breaking.

Topsoil should be piled separately outside of the building area so it can be used later for the finish grading. After the topsoil is removed from the foundation area, the excavation can be continued until the proper depth has been reached.

> In areas where heavy equipment such as bulldozers or front-end loaders are being operated, take care to keep out of the path of the machines.

ACCESS TO FOUNDATION

Another important consideration that affects the mason is the job of getting construction materials into the foundation area. If necessary, excavation should also provide a driveway for trucks, an area for storing materials, and a space for erecting scaffolding.

Block trucks can lower concrete block units into the foundation site with a hydraulic lift. Access must be made, however, for the concrete mixer truck when the footings are placed.

SUMMARY

In conclusion, the reasons for building a foundation are to

- provide support for the structure.
- hold backfill and moisture out.
- provide protection against frost and freezing temperatures.
- provide a basement or crawl space.

ACHIEVEMENT REVIEW

Select the best answer from the choices offered to complete each statement. List your choice by letter identification.

1. One of the following is not a dead load:

    a. roof of the structure.
    b. partition wall.
    c. snow on the roof.
    d. steel beam supporting the floor joists.

2. Most foundations that are built of masonry units are constructed of

    a. stone.
    b. concrete block.
    c. tile.
    d. brick.

3. The building line of a structure is usually the

    a. back of the building.
    b. right side of the building.
    c. left side of the building.
    d. front of the building.

4. Bench marks are for

    a. laying out the wall lines.
    b. determining heights.
    c. finding the property boundaries.
    d. determining the load to be placed on the foundation walls.

5. Batter boards are used for

    a. mixing mortar.
    b. bracing the foundation walls.
    c. laying out the foundation square.
    d. holding the earth back out of the foundation.

6. To allow working space and ensure that batter boards are not dug out during the excavation, the batter boards should be set back a minimum of

    a. 4 feet.
    b. 3 feet.
    c. 2 feet.
    d. 1 foot.

7. The best known formula for squaring a foundation if a transit level is not used is the
   a. 2-3-4 method.
   b. 6-8-10 method.
   c. 9-10-20 method.
   d. 12-16-20 method.

8. The most lasting method of marking the wall lines on a batter board is with a
   a. fine pencil.
   b. saw cut.
   c. scratch mark.
   d. paint mark.

9. The square of a foundation can be most accurately checked by measuring
   a. length and width.
   b. with a steel square.
   c. diagonally.
   d. with a plumb bob.

10. To aid the bulldozer operator in the initial ground breaking, a line should be marked with
    a. a length of old garden hose.
    b. sand.
    c. lime.
    d. nylon line laid on the ground.

11. The most important consideration in determining the depth of a foundation is
    a. settling.
    b. frost and freezing.
    c. providing a crawl space area for access to electrical wiring and plumbing.
    d. basement space.

12. To prevent the problem of dampness from occurring in a crawl space foundation, one of the following should be installed:
    a. windows.
    b. fans built into the wall.
    c. vents built into the wall.
    d. vapor barriers.

PROJECT 1: LAYING OUT BATTER BOARDS FOR A FOUNDATION

OBJECTIVE

- The student will lay out a four-corner batter board for a small house.

EQUIPMENT, TOOLS, AND SUPPLIES

builder's level
6-foot mason's rule
50-foot steel tape
hatchet

sledgehammer
mason's hammer
handsaw
pencil and indelible pen

ball of nylon line
square
fourteen 2-inch x 4-inch wood stakes about 4 feet long
eight 3/4-inch by 4-inch boards, about 5 feet long, relatively straight
supply of 8d common nails

SUGGESTIONS

- When sharpening the stakes, cut against another board to prevent damaging the blade.
- Follow good safety practices at all times.
- When measuring, keep the steel tape out of water or wet grass.
- Drive all nails securely in the batter boards.
- Stretch the nylon lines tightly to prevent sagging.

PROCEDURE

1. Stretch 50 feet of nylon line level between two stakes driven securely in the ground. This serves as the main building line.

2. Using this line as a guide, drive three sharpened 2-inch x 4-inch stakes in the ground to form a right angle as shown in figure 1-7. This will be corner post A.

3. Nail the 3/4-inch x 4-inch boards even with the top of the stakes and level with each other as shown in the figure at corner post A.

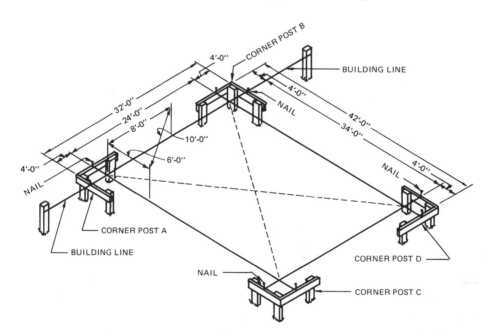

Fig. 1-7 Project 1: Laying out batter boards for a foundation

4. Using steel tape, measure from outside of corner post A 32 feet to establish the location of corner post B. This requires the help of another person to hold the tape.
   NOTE: The actual width of the foundation is 24 feet. The additional 8 feet is for setting back from the actual edge of the excavation to prevent the stakes from being dug out during excavation. Stakes are set back 4 feet on each side for a total of 8 feet.
5. Drive the stakes for corner post B as shown. Level the 3/4-inch x 4-inch boards and attach them to the stakes with nails. These boards should be level with the boards on corner post A. Use a builder's level to level corner post B with corner post A. If one is not available, a long straightedge board and a mason's level can be used.
6. Return to corner post A and using the steel tape, measure 42 feet in the opposite direction. This locates corner post C.
7. Drive the stakes for corner post C and attach batter boards to the stakes level with corner post A. Use the builder's level to do this.
8. At corner post A drive a small nail in about the middle of the top of the batter board to hold the line.
9. Attach the line to the nail and pull the line to corner post C. Fasten the line to the batter board on corner post C, even with the top of the board, and about in the middle of the board.
10. Using the 6-8-10 method, measure from where the lines cross at corner A exactly 8 feet toward corner B. Mark this point on the line with a sharp pencil. (The aid of another person is required to hold the end of tape.) This is line AB.
11. Next, using the steel tape, measure 6 feet along the line from corner post A toward corner post C. Mark this point exactly on the line with a sharp pencil. This is line AC.
12. With one person holding the end of the steel tape at the 8-foot mark (line AB), adjust line AC by moving either in or out until the 6-foot mark is exactly even with the 10-foot reading on the tape. This forms a right angle, indicating that the corner is square. Pull line AC to corner post C and attach it permanently to a small nail driven in the top of the batter board.
    NOTE: Line AB, which is the main building line, must not be changed or altered during the above procedure, to make sure the lines remain square on the batter boards.
13. To establish the location of corner post D, measure 32 feet from corner post C and 42 feet from corner post B. See figure 1-7.
14. Drive the stakes at this point and attach the batter boards to the stakes. Make this corner post level with corner posts A, B, and C, using the builder's level.

15. To establish the remaining wall lines on the batter boards, measure 24 feet from the nail driven in corner post A to corner post B. Drive a nail in the top of the batter board to mark the location of the wall line. Measure 24 feet from corner post C to corner post D and drive a nail in the top of the batter board. This establishes the width of the building.

16. To find the length of the building, measure 34 feet from the nail driven in corner post B to corner post D and drive a nail in the top of that batter board.

17. Attach the nylon line to the nails at all corners. Draw tightly and fasten. The line indicates the outside wall line of the foundation.

18. As a final check, measure from one corner to the corner diagonally across. All diagonal measurements should agree if the corners are square.

19. Wall lines should be marked permanently by notching with a saw. This is necessary when lines have to be removed for excavation of the foundation.

20. Have the instructor check the project for accuracy.

Unit 2 CONSTRUCTING FOOTINGS

OBJECTIVES

After studying this unit the student will be able to

- determine the size of footings for various widths of concrete block walls.
- describe the correct concrete for the footing for a house or small commercial building.
- define the two types of footings used for a foundation.
- describe how a footing is placed.
- clean concrete tools properly.

The footings are the first actual structural work done on a building. A *footing* is the base on which a foundation is built. Footings are usually concrete, which is placed in a trench or form at the bottom of the foundation. Their function is to support the weight of the structure and distribute it over a greater area.

The two key factors that influence the size of a footing are the weight of the structure and the load-bearing strength of the soil it rests upon. The engineer determines the load-bearing strength of the soil. This is considered in writing the building codes which govern the size of footings. The architect shows the size of the footings on the plans for the building.

Good footings are important because without them a building can settle and cracks may develop in the walls. Since the footings are never seen after the floor is in place, their importance is often overlooked.

SIZE OF THE FOOTINGS

The earth the footings rest on should be firm and undisturbed to give maximum support. Concrete is the material used most often for footings, because it can be made to conform to the requirements necessary for strong footings.

As a general rule, for a residential or small commercial building, the footing should be twice as wide as the masonry wall foundation that will rest on it. The depth of the footing should be one-half to three-fourths of its width. For example, the footing for an 8-inch concrete block wall should be 16 inches wide and 12 inches deep. It must be remembered that this rule does not fit all conditions; it is only suggested as a general rule when there are no special problems. Special cases are dealt with by the engineer.

Two important rules to observe in placing any footing are

- Always try to place the entire footing at one time, with no breaks, to ensure full bonding of the concrete.

- Regardless of the size of the masonry wall, never make the footing less than 6 inches deep.

15

Footings for chimneys, columns, or piers should be larger than for an ordinary wall, because they must support a more concentrated load. This load must be supported without settling, because not only will the chimney settle, but the wall it is bonded to will be damaged. The footings for chimneys with fireplaces should be at least 18 inches deep. The concrete should extend 18 inches beyond the actual masonry work on all sides to provide better weight distribution, figure 2-1. The footing for a chimney is commonly called a *chimney pad*.

STEEL REINFORCEMENT

Steel reinforcement in the footing is not required for houses or small commercial buildings. However, in the case of chimney and fireplace footings, it is a good practice to place a few steel rods or concrete reinforcement wire in the footing for extra strength. If steel is to be used, about 4 inches of concrete is placed, the steel is laid on the concrete, and then the remainder of the concrete is placed. Never lay the steel on the ground and place concrete on top of it. The steel must be encased in the concrete to be of any value. Half-inch steel rods are the size selected for most small jobs. When piers which carry a heavy load are to be built on the footing, the reinforcing rods can be left projecting out of the footing, figure 2-2. In this manner, masonry or concrete piers can be tied to the footing.

CONCRETE FOR FOOTINGS

Concrete for use in footings for homes or small commercial buildings should support at least 2500 psi (pounds per square inch). Adding more portland cement to the mix makes the concrete stronger. However, the mix for residential construction usually requires no more than six bags per cubic yard. Concrete for a footing mix can be ordered either by

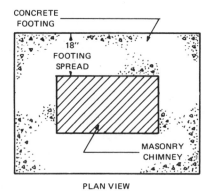

PLAN VIEW

Fig. 2-1 Masonry chimney foundation showing the footing spread for weight distribution. Consult local building codes for exact footing size in a specific area. This is an average footing for a chimney.

specifying a 5-bag mix, called a *prescription method* (5 bags of portland cement to the yard); or by a psi rating of 2500. Both mean the same. In certain areas, building codes should be checked for unusual conditions, such as earthquakes or high winds.

TYPES OF FOOTINGS

When laying out footings the building lines are first established using nylon line, as discussed in Unit 1. A *plumb bob* (a small weight attached to a line) is suspended from the point where the building lines cross, figure 2-3. A stake is driven in the ground at this point to locate the exact outside corner of the building.

There are two major methods of placing the footings for a structure. *Trench footings* are constructed by digging out the soil and placing the concrete in the trench. *Formed footings* are constructed in forms which are built before the concrete is placed. Either method is an acceptable way of placing footings, depending upon the job conditions.

Fig. 2-2 Steel rods projecting out of the footings to reinforce piers that will be built at a later time

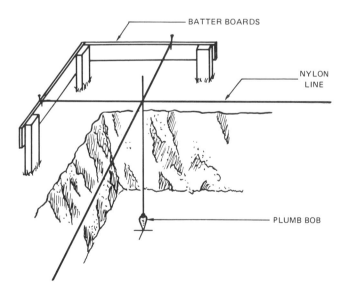

Fig. 2-3 Locating the building lines with a plumb bob. The point located by the plumb bob is the exact corner of the building. This method is used to transfer the corner of the building from batter board level to the bottom of the excavation, determining the location of the footings.

Trench Footings

One of the most important items to consider in digging the trench for a footing is the shape of the sides and bottom of the trench, figure 2-4. Often the trenches are dug carelessly and the footings do not have the strength which is expected of them. The trenches should have straight sides and should be dug to the full thickness and width indicated on the plans. The soil should be shoveled back far enough out of the way to provide working room. This also helps prevent soil or rock from sliding back into the trench before the concrete is placed.

After the trench is dug, level stakes are placed in the center of the footing. These are stakes which show where the top surface of the footing is to be. Usually wood stakes or pointed lengths of steel rod are used, depending upon the hardness of the ground. These stakes or rods are placed about 5 feet apart in the trench and are leveled with each other. The concrete is then placed to the top of the stakes.

If the soil in the trench is very dry, it should be dampened with a garden hose before placing the concrete. This prevents too rapid drying of the concrete.

Formed Footings

When the foundation is being dug rock is sometimes found. If rock is present, it may be necessary to form the footing with wood, figure 2-5, and place the concrete into the forms. This is sometimes done because digging out the rock is expensive and delays the job.

On large buildings the footing forms are constructed by carpenters. However, it is common practice on small buildings, such as houses and garages, for the mason to form and place the footings. Lumber used to form footings must be rigid and braced securely to prevent the concrete from pushing the forms out of position. Once this happens it is nearly impossible to correct.

An economical way of forming footings on a small job is to order the first floor joists (the framing lumber for the first floor) as soon as the foundation is excavated. This lumber is used to construct the forms for the footings so that no special lumber must be ordered. After the concrete has set, remove the forms, pull out the nails, and clean the lumber with a steel brush. The same lumber can then be used for floor joists when needed.

When setting the forms, it is very important that the top of the boards or forms are level with each other and at the height specified on the plans. The footing must be placed exactly as specified.

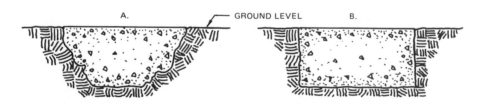

Fig. 2-4 Trench footings. Improperly shaped footing (A) will not support the load of the structure as the sides are not straight. Properly shaped footing (B) will support the load of the structure as it completely fills the footing area. Note: Steel rods should not be used with this footing.

Stepped Footings

When a building is constructed on sloping ground it is not always possible to place the footing around the foundation at the same level. In this case the footings must conform to the existing grade and be placed in steps, figure 2-6. Each step must be level and positioned in increments of eight inches from the top of the next step. (Eight inches is the height of a course of concrete blocks). The steps should be no more than 2 feet high and no less than 2 feet long, to help prevent settling and cracking. Where there is a sharp drop in the grade of the excavation, it may be necessary to use several steps. This is done by using form boards cut to the size of the steps, figure 2-7. The boards are braced securely in position. Steel rods can be inserted across the steps to increase the strength of the footing. The width and thickness of a stepped footing are the same as for a regular footing.

Fig. 2-5 Formed footing. Notice the bracing that holds the forms in place.

Fig. 2-6 Stepped footing that has been placed, and the forms removed. Steps are 8 inches in height so they will work the full height of the concrete block without any cutting required.

Fig. 2-7 Stepped footing forms. Boards are in place, ready to receive the concrete. Notice the steel rods laid from the level top of the footing into the step for strength.

PLACING CONCRETE FOR FOOTINGS

NOTE: Recheck all forms before placing the concrete.

Position the truck carrying the concrete relatively close to the trench or form (inside the building lines if possible), so the concrete can be transferred directly from the truck by means of the metal chute, figure 2-8. This saves time and labor that would be wasted in wheeling the concrete in a wheelbarrow from the truck to the footing.

The tools used in placing concrete for a footing are very simple, but it is important to have them ready when the concrete is delivered to the job site. Shovels, rakes, mortar hoes, and a few brick trowels are usually adequate.

Part of the water should be left out of the concrete mix until the truck arrives on the job site. The builder or mason can then tell the driver of the truck what consistency mix is desired. This is very important because just enough water is needed to allow the concrete to pour from the truck and to allow the mason to work it. Too much water in the mix greatly decreases the strength of the concrete.

Fig. 2-8 Placing the concrete into the formed footings

Immediately after placing the concrete in the forms, it should be compacted so that it fills out completely against the forms. This causes the footing to have a neat, smooth finish, free from holes or voids when the forms are removed or stripped.

Compacting can be done easily on footings for a house by tapping the sides of the forms with a hammer. It also helps to insert a trowel blade in the concrete along the side of the form and work it up and down with a slicing motion. The tapping and slicing motion causes the concrete to compact and the cement paste in the mix to fill out the form again, closing all voids and holes. On large jobs, gasoline-powered vibrators are used.

The top of the footing should be left relatively rough. This provides for good bonding to the mortar joint when the first course of concrete blocks is laid. Leveling with an ordinary garden rake and tamping with the rake tines leaves an excellent surface for the first course. Remember, it is very important to place the entire footing in one day, with no breaks, so that it cures as one mass.

CARE OF TOOLS

As soon as tools and equipment are no longer being used, they should be washed off with water to remove all concrete. If left to harden, the concrete is difficult to remove and can rust or pit the metal parts.

It is advisable to have a supply of water, preferably a hose with a nozzle, to clean the tools. If a water supply is not available, have the driver of the concrete truck rinse the tools when washing out the truck before leaving the job. All concrete trucks are supplied with a water tank and hose.

CURING TIME

In warm or moderate weather, concrete blocks should not be laid on the footing until three days have passed. The forms should not be removed until four days have passed. In cold weather, the footing should not be loaded or forms removed for at least eight days. Freezing can be prevented by covering the footing with straw. The strength rating for concrete is usually based on twenty-eight days curing time. However, concrete actually takes longer to reach full strength. The best temperature for curing concrete is about 72 degrees Fahrenheit. Common sense should always be used when starting concrete block work on a green (not cured) footing. If masonry materials are being stocked in the foundation area (figure 2-9) before the footing has cured completely, care should be taken not to damage the footing.

Fig. 2-9 Concrete block being placed in the foundation with a hydraulic arm mounted on the truck. Care should be taken not to damage the footing.

SUMMARY

Although the construction of footings is considered one of the simpler tasks done by the mason, it is one of the most important parts of the structure. The entire building rests on the footings. On residential and small commercial buildings, the mason places the footings. For this reason, it is important to be familiar with all aspects of footings and how to properly construct them.

Good footings make it easier to erect the walls properly. Poor footings not only threaten the stability and strength of the building, but require many extra hours of labor to level and get the job started properly. A large part of the contractor's or mason's profit can be lost in the beginning of the job if the footings are not properly constructed.

ACHIEVEMENT REVIEW

Column A contains statements associated with footings. Column B lists terms. Select the correct term from Column B and match it with the proper statement in Column A.

Column A

1. two key factors that influence the size of footings
2. primary function of a footing
3. the width of the footing for a ten-inch concrete block foundation
4. the minimum depth of the footing for a fireplace and chimney
5. a specification for concrete for footings in light construction
6. the exact location of the footing is determined from the batter boards by using this
7. the best type of footing to use when rock is encountered in the excavation of a basement
8. footing used when it is not possible to place the footing around the foundation at the same height, due to a drop in the grade

Column B

a. stepped footing
b. steel reinforcement
c. rough
d. support the building
e. 3 days
f. 8 days
g. 2,500 psi
h. water
i. the job site
j. a plumb bob
k. weight of structure and load-bearing strength of soil
l. 18 inches
m. materials and finish desired
n. formed footing
o. tapping with a hammer
p. 20 inches

Column A

9. the top of the footing should be in this condition to provide for good bonding to the mortar joint

10. greatly improves the finished sides of a formed footing, and fills voids or holes

11. the final amount of water should be added to the concrete after the concrete is on this

12. added to increase the strength of a concrete footing

13. in moderate to warm weather, the mason should wait this number of days before laying concrete blocks on the footing

14. used for cleaning tools and equipment immediately after placing concrete

15. in cold weather, the mason should wait this number of days before laying concrete blocks on the footing

Unit 3 THE STORY POLE

OBJECTIVES

After studying this unit the student will be able to

- explain why a story pole is needed for a masonry building.
- describe how a story pole is laid out for the average job.
- lay out and mark off a story pole from a set of plans for a foundation of a building.

PREPARING TO MAKE THE STORY POLE

Before any masonry units can be laid, a story pole must be made. A *story pole* is a wood pole approximately 1 inch x 2 inches x 10 feet long. This is a suggested size; any wood pole about this size is adequate. It should be remembered, however, that the pole must last throughout the whole job, so care should be taken to select a straight pole free from loose knots or other defects. A length of exterior construction grade plywood this size is a good choice because it withstands the weather.

The various heights of all openings, lintels, vents, sills, and similar features must be marked on the pole at the proper place according to the plans. The individual courses of units are then laid out with the aid of the mason's rule and marked so they are level with the top of these objects. Try to eliminate the need to cut masonry units.

The plans should be studied thoroughly in advance rather than waiting until the day the job is to start. The main reason for this is so that the other workers will not have to wait while the pole is being made. In this way, there is also time to contact the builder or architect if a mistake is found in the plans. Specifications should be carefully checked to obtain a full picture of the job. The story pole should not be made until all of these things are done.

The following description of making a story pole is for a brick wall. Due to a low grade, some foundations have brick exposed on several sides, from the footing to the top of the wall. The project at the end of this unit is for a story pole for an average foundation. A student should be able to make a story pole for a brick or block foundation wall.

MAKING THE STORY POLE

Select a straight pole, and cut it to a height that will be a little longer than the actual height of the story (as previously mentioned, this is about 10 feet). Use a small square and a sharpened pencil to mark where the pole is to be cut off. Be sure to cut even with the line so the edges are square.

Next, using a mason's rule, measure up from the bottom of the pole and mark all of the overall heights of the important features with the point of a pencil. Using the square, draw lines across the pole. Be sure to indicate what the feature is by printing underneath the line an abbreviated name for the feature such as W. H. (window head) or D. H. (door head), as shown in figure 3-1. Recheck all of the measurements with the plans. When satisfied that they are correct, mark over the lines with a red pencil or indelible ink to make a permanent mark.

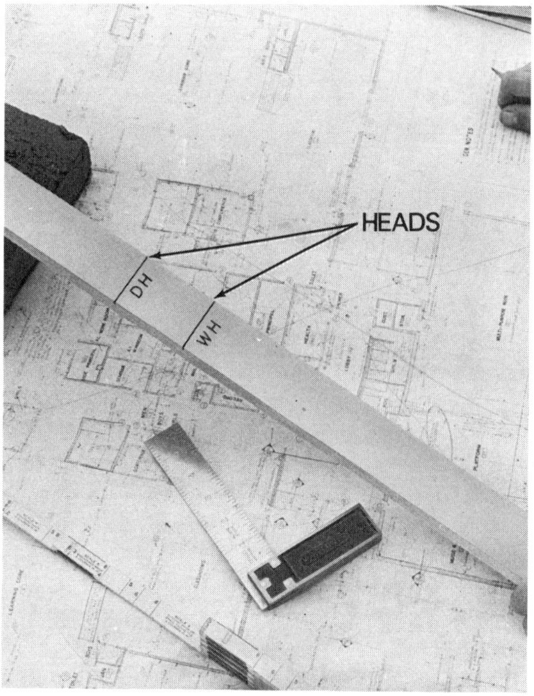

LAYING OUT STANDARD BRICK COURSES ON THE POLE

Start measuring at the bottom of the pole and mark upward using number six on the mason's modular rule. Usually, the first feature reached is the bottom of the windowsill. If this height does not work out even on the modular scale, the mason's spacing rule can be used. The mason's spacing rule allows adjustments of the mortar joints with as little deviation as possible.

Fig. 3-1 Marking off the heads of the door and windows on the story pole. Brick courses are marked off to work even with these heights.

The thickness of a brick windowsill including the mortar joint and the pitch is approximately 4 1/2 inches. This should be marked with an "X" so it is not confused with the regular courses of brick.

Continue marking off the courses, figure 3-2, until the top of the pole is reached. Mark the top of the pole with the word "top" so there is no possibility of using the pole upside down.

The heights of windows and door heads do not always match perfectly with the brick courses on a modular scale. Adjust the difference with the mason's spacing rule. The purpose of this is to make all mortar joints about the same thickness rather than having a large joint and then a number of small joints.

NUMBERING THE COURSES

After the courses are marked off, they are numbered. Number each course from the bottom to the top of the pole, beginning with number 1, as shown in figure 3-3. This is very important because it prevents the building of a *hog in the wall* (a wall that has more courses on one end than the other, but built to the same level).

Course numbering is also helpful when installing objects in the masonry work such as fans or vents that have to be built in at specific heights. As a final step, mark over the numbers either in black or red permanent marker or ink so they will not fade out.

SAWING IN THE MARKS

When the story pole has been rechecked for accuracy, the coursing marks can be made permanent by sawing in with a carpenter's handsaw. This is an excellent practice to follow

Fig. 3-2 Laying off the individual courses on the story pole from bottom to top with a modular spacing rule, a pencil, and a square.

Fig. 3-3 Numbering the courses of brick from the bottom to the top of the pole prevents a hog in the wall.

since damp weather, moisture, or bright sunlight can cause the lines to fade away. Using the handsaw, cut in approximately 1/16 inch deep for each course mark, figure 3-4.

GAUGE ROD

The gauge rod, more commonly called a *sill stick,* is a shorter version of the story pole that is used when building to the sill height. It helps speed work, being smaller and easier to handle than the story pole. Once the sill height is reached, the mason again uses the story pole until the story height is reached. The sill stick is usually about 4 feet high and made of the same material as the story pole. See figures 3-5 and 3-6.

HOLDING POLE CORRECTLY WHEN CHECKING HEIGHT

Once the story pole is made, the bottom is set on the bench mark to check the height of the masonry work. Make sure there

Fig. 3-4 Sawing in the course marks with a handsaw establishes a permanent mark that will not fade out. The depth of cut is approximately 1/16 inch.

Fig. 3-5 Gauge rod (or sill stick) used in building the brick wall to sill height. The height of the rod is about 4 feet.

Fig. 3-6 Student mason shown with the completed story pole and gauge rod. Gauge rod must match the master story pole for unit coursing.

are no obstructions such as hardened mortar at the bottom of the story pole. Level marks are located by shooting level points on each of the outside corners of the building. These are referenced from the base bench mark on the job. A concrete nail or small wooden block (usually plywood) is nailed into the mortar bed joint nearest the finished floor, level with this point. The bottom of the pole is then set on this level point.

To obtain an accurate reading with the story pole, hold the pole in an upright plumb position on the level point. Lay the trowel blade flat on the wall and note where it levels across to the story pole marks. If the level points are not on the wall itself but are on a stake near the wall, lay a plumb rule on the stake and level across from the wall to the pole.

> NOTE: Check the height of the wall frequently, figure 3-7. A good rule of thumb to follow is to check the height every other course. Do not exceed three courses before checking the height.

RESPONSIBILITY OF THE MASON TO BUILD TO THE MARKS ON THE POLE

The corner mason is responsible for building the corner exactly to the marks on the pole. This is especially important for the heights of the features such as the heads of windows and doors. Remember that if the corner is built incorrectly, the entire wall will be wrong.

Using the story pole for measuring eliminates the errors made when measuring with a rule. Building and determining heights with the rule is very inaccurate because the rule can slip out of position when being used. The mason may not hold the rule at the same exact position each time the wall is checked. Once the corner is too low or too high, the error is repeated since there is no base point to set the rule on. There is, however, a base point when a story pole is used. The difference may not be noticed until the wall is built but then it is usually too late to correct the error. The wall would have to be torn down and rebuilt.

Therefore, the story pole is the most efficient and accurate method of gauging masonry work (with the exception of the manufactured corner pole, which is fastened to the structure of the building). On many jobs, laying out the story pole is the job of the mason foreman, but any competent mason should be able to perform this task. Masons who can work ahead on their own without constant supervision are highly respected and often rewarded by contractors.

Fig. 3-7 Mason checking the height of a concrete block foundation with the story pole. Notice how the trowel is laid across the wall to serve as a straightedge when checking the marks.

The student mason should learn, and be able to use, the following common terms associated with the story pole.

- *sill high*—height of the masonry wall where the mason installs the windowsill.
- *topping out*—process of laying the last section of the wall to the finished height.
- *story high*—finished height of the masonry story on a building from the floor line to the level where the wood plates will be attached.

CARE OF THE STORY POLE

To ensure that the story pole lasts for the entire job, it must be properly cared for. Store it in a dry place when not in use to prevent warping and bowing out of shape. Do not lay it on the floor since it can be run over by equipment or thrown away by mistake.

When the job closes down at the end of the workday, the story pole should be stored in the mason's toolshed or trailer. As the story pole is the guide for much of the masonry work, the mason should know where it is at all times. It is a lot of trouble and costly to make a replacement story pole for one that has been lost or broken. The contractor will not appreciate this added expense. If there are many masons working, or if several identical houses are being built at one time, several story poles can be made off the master pole. This helps to keep the job running efficiently.

ACHIEVEMENT REVIEW

Part A

Select the best answer from the choices offered to complete each statement. List your choice by letter identification.

1. The length of a story pole is usually

   a. 4 feet.
   b. 6 feet.
   c. 8 feet.
   d. 10 feet.

2. The most accurate method of marking the individual courses of masonry units on the pole is with

   a. ink.
   b. a pencil.
   c. a saw cut.
   d. paint.

3. To lay out standard size bricks on the story pole, the number used on the modular scale is

   a. 6.
   b. 4.
   c. 8.
   d. 9.

4. A standard brick windowsill including the mortar joint and pitch measures

   a. 2 3/4 inches
   b. 4 1/2 inches.
   c. 6 1/2 inches.
   d. 8 inches.

5. A hog in the wall can be prevented by
   a. measuring with the spacing rule.
   b. increasing the height of the story.
   c. marking off each course of masonry units and numbering them.
   d. moving the bench mark to match the height of the wall after it has been built.

6. The shorter pole used to build the corner to sill height is called a
   a. leveling rod.   c. transit level.
   b. modular rule.   d. gauge rod.

7. The height of the masonry corner should be checked
   a. every course.   c. every four courses.
   b. every other course.   d. every six courses.

8. On most jobs the task of laying out and making the story pole is the responsibility of the
   a. masons.   c. contractor.
   b. mason foreman.   d. architect.

Part B

Describe the various steps in laying out and constructing a story pole for an average job. List the steps in order.

PROJECT 2: LAYING OUT A STORY POLE FOR A CONCRETE BLOCK FOUNDATION

OBJECTIVE

- The student will select a pole, and using the heights stated on the elevation, mark off all the features and courses of masonry units with a mason's rule.
  NOTE: This foundation elevation is typical of an average house with a basement. A door and window are included.

EQUIPMENT, TOOLS, AND SUPPLIES

mason's modular rule   small square
sharp pencil   handsaw
red pencil   wood story pole about 10 feet long

SUGGESTIONS

- Use a pencil with an eraser on it.
- Mark all features and courses with a light pencil. Recheck to be sure measurements are correct, then darken the marks.
- Lay the pole out flat when marking on a table or flat surface.
- Always observe safety practices when using the handsaw.

PROCEDURE

1. Select a straight wood pole about 10 feet long.
2. Lay the pole out flat on a worktable or flat surface.
3. Study the plans.
4. Measure a distance of 88 inches from the bottom of the pole to the top. That is the total height of the finished foundation. This will work courses of masonry units with no cutting. Cut the pole off square at this height.
5. Measuring up from the bottom, mark the top of the door and window head with the letters DH and WH on the pole.
   NOTE: The top of all features are always marked on the pole before any courses are marked.
6. From the top of the door head or window head (which is the same in this case) measure down 24 inches plus 4 1/2 inches for the brick sill. This includes the pitch of the sill. Square the line across the pole and mark an X in the 4 1/2 inch area, indicating this is where the sill is installed.
   NOTE: It is important to mark the top of the sill with the letter T as sills must be set exactly on the correct height.
7. Mark off each course of concrete blocks from the bottom of the pole to the top of the door. Each course should be 8 inches (this can be done using number 2 on modular rule). The courses work out evenly. There are 10 courses of blocks as shown in figure 3-8.

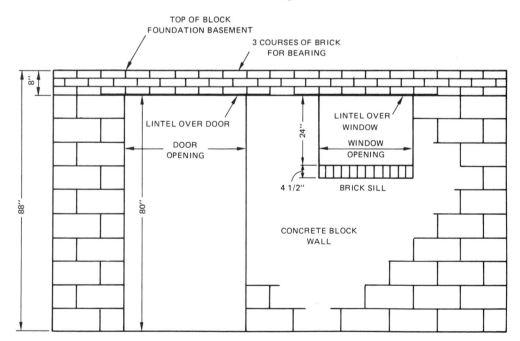

Fig. 3-8 Project 2: **Elevations for a concrete block foundation**

8. From the top of the door to the top of the wall, mark 3 courses of bricks using number 6 on the modular rule. These courses also work out evenly.

9. Mark the top of the wall on the pole with the word *top* so the pole will not be used upside down.

10. Recheck all measurements and courses that have been marked on the story pole.

11. Using a red pencil, number the courses of blocks and bricks from the bottom of the pole to the top, starting with the number 1.

12. With the handsaw, saw in the coursing marks about 1/16 inch deep.

13. Have the instructor check the pole for accuracy.

Unit 4 BUILDING A CONCRETE BLOCK FOUNDATION

OBJECTIVES

After studying this unit the student will be able to

- describe how to locate the corners of the foundation wall on the footings, from the batter boards.
- explain how to lay out the bond and first course.
- discuss how to stock materials and locate traffic patterns.
- build the foundation project at end of the unit.

IMPORTANCE OF BEGINNING CORRECTLY

After the foundation has been excavated, footings placed, and the story pole made, then the block walls can be laid up. Before attempting to locate the wall lines in the footings, the plans should be studied carefully. The mason must understand exactly how the job is to be built before starting.

The footings should be brushed clean of any loose dirt so that the mortar will bond well with the base. In hot, dry weather, it is a good idea to dampen the footings by using a brush and a bucket of water. Do not soak the footings, only dampen them. This will cut down on moisture being pulled from the mortar into the footing concrete, and improve bonding. In freezing weather, if ice or snow is present on the footings, no blocks should be laid until it is removed or melted and the excess water is removed. A good bond with the base cannot be established if there is ice on the footing.

Use a solid bed of mortar under the first course. Never furrow with the trowel. This helps to prevent moisture from penetrating through the bed joint. Lay the first course to the height obtained from the bench mark. Make sure the course is level and plumb. Recheck all measurements and make sure the foundation is laid out square before proceeding to build the walls. Tool the mortar joints when they are thumbprint hard for a neater, more waterproof finish.

Good working practices must be observed when building the foundation. The foundation is a very important part of a structure. If a mistake is made in the foundation and goes undetected, it will affect the rest of the structure and all of the workers who must build the house.

LAYING OUT CORNERS ON FOOTING WITH A PLUMB BOB

Attach lines to the batter boards as previously described and pull them tightly. Fasten a line to a plumb bob and drop it into the foundation at the point where the lines intersect at the corners. The plumb bob works on the principle of gravity and will hang absolutely

Fig. 4-1 Checking the diagonal corners of foundation with a steel tape

vertical (plumb) unless the wind is blowing strongly. Hold the line with the plumb bob attached so that the point is just slightly off the footing. After it has stopped swinging and is perfectly still, mark this spot with a pencil or crayon dot.

Drive a hardened nail into the footing to mark the corner point. This serves as a reference point until the first course of concrete block is laid around the foundation. Repeat this procedure for each corner. As a check to make sure that the foundation is square, measure with a steel tape diagonally from one corner nail to the other, figure 4-1. The measurements should be the same.

A chalk line can then be struck on the footing from corner point to corner point where the wall will be built. This helps the mason to lay the first course of blocks plumb; the line on the footing is used as a guide in addition to the nylon line at the top of the course.

LAYING OUT CORNERS ON FOOTING WITH A TRANSIT LEVEL

Another method of laying out the corners on the footing of a foundation is with the aid of a transit level. After one corner nail is located from the batter board, the transit is set up directly plumb over the nail in the footing, figure 4-2. To be sure it is set up plumb, a plumb bob is dropped from the transit to line up with the center of the nail head.

The telescope of the transit is then sighted on a downward angle to the far corner of the foundation. Another nail is driven into the footing in line with the sighting. The transit is then swung towards the opposite corner 90 degrees on the transit scale. A nail is driven into the footing at this point. This forms a right angle that is square. The length and width of the building is then taped off and chalk lines are struck.

Either method can be used to lay out the foundation corners. Whichever method is used, the batter board is needed to establish the first corner point.

LAYING OUT THE FIRST COURSE OF BLOCK

As mentioned before, first brush off the footing. Then, using a steel tape, measure from one corner to the other to establish the bond. Each concrete block lays out to 16 inches, including the mortar head joint. If the wall works out perfectly to multiples of 16 inches, use a pencil or crayon and mark off every third block (48 inches) on the footing. These checkpoints are useful when laying out the first course.

If the blocks do not work out perfectly, there are two corrections that can be made: (1) reverse the corner block on one end of the wall and open the head joints to make up the difference, or (2) tighten the mortar head joints and if necessary put a piece in the center of the wall. Pieces of blocks smaller than 12 inches should not be put in the center of the wall. It may be necessary to cut 2 blocks so they are not very noticeable.

Fig. 4-2 Setting up the transit over the corner nail in the footing

LAYING THE CORNER BLOCKS IN MORTAR

After the bond is marked off on the footing, spread a full mortar joint at one corner and lay the corner block. Do not furrow the mortar joint with the trowel because water can leak through the bed joint. Be sure to lay the block level with the first course mark on the story pole and make sure it is plumb.

Next, go to the opposite corner and repeat this procedure. Fasten a line from one corner block to the other and line up the tail end of the blocks. Make sure the line is stretched tight.

Fig. 4-3 Laying the first course of concrete blocks to the line on the footing

Lay the first course of blocks to the line, figure 4-3, using the checkpoints as a guide. Work from the ends of the line to the middle, because any adjustment should be made there. If the footing is too high to allow the blocks to lay level with the line, cut the first course of blocks to fit. If the footing is too low in places, build up a thicker bed joint (no more than 3/4 inch) to compensate for the difference. Differences of more than 3/4 inch should be corrected by laying split pieces of block or tile in mortar. When laying out the first course, be sure to locate all openings, pilasters, and so forth that are located on the drawings. Remember that it is very important that the first course is laid level with the story pole mark, and plumb.

PLACING MATERIALS

After the first course is laid out completely around the foundation, the wall should be stocked with blocks. The blocks should be located in such a position that the mason can take the least time and expend the least amount of effort in laying them in the wall.

Stock blocks in piles about 2 feet back from the inside of the wall line and approximately 6 feet apart. Mortar pans or boards should be spaced between the stacks of blocks. On every other course, alternate the direction of the individual courses of blocks stacked on the piles, to prevent them from falling over and injuring someone.

Stock only enough blocks along a wall to build that wall. This can be done by counting how many blocks are on the first course and multiply the number of courses to scaffold height (approximately 4 to 5 feet). Rehandling the blocks several times tends to chip and mar the finish of the face and increases the cost of the job.

Consider the space available for building the scaffold in the foundation so it is close to where it will be used. The scaffold builder can assemble the frames and move them into position as soon as the wall reaches scaffold height. This saves time, because the mason does not have to wait for the scaffolding to be moved from outside the foundation to the work area.

Locate the mortar mixing area as close to the foundation as possible without interfering with the building of the masonry work. Keep the bagged cement covered with plastic or canvas to keep it dry. The cement bags should also be stacked off the ground on a wood pallet, figure 4-4.

Fig. 4-4 Bags of masonry cement are stacked on wood pallets off the ground to prevent moisture from entering the cement.

Fig. 4-5 Materials are stocked in piles with traffic lanes left between the piles for passage of wheelbarrows.

ESTABLISHING TRAFFIC LANES FOR MATERIAL MOVEMENT

Materials must be brought from outside of the foundation to where the wall is to be built. Some type of traffic lanes, figure 4-5, must be established for movement of the materials. The shortest direct route that is free from obstructions is usually best. Conditions on each individual job determine where traffic lanes should be.

Muddy conditions or soft ground may require the building of *runways* (planks laid down for wheelbarrows to run over when carrying materials).

> When building runways for the movement of materials, watch for nails sticking up through boards and scraps of building materials under foot that may cause accidents. Make sure that all boards are supported evenly with each other where they butt together, so the wheelbarrow does not run off them.

Arrange the bulk of materials needed to build the wall before laying up the wall. This is particularly necessary if power equipment is being used to transport the materials to the work area. A mason must not be delayed by poorly stacked materials. Work must progress without interruptions if a profit is to be made on the job.

BUILDING THE WALLS

Build the corners to scaffold height. Attach the lines to the corners and fill in between the lines with concrete blocks. From time to time, check the height of the center of the wall with the story pole to make sure that the wall is being built level with the marks on the pole.

Instead of building the masonry corner, a metal corner pole, figure 4-6, can be used. It consists of a metal pole marked with individual masonry unit courses. Metal braces hold it in position and it is fastened to the ground with stakes. Either method is an acceptable way of building the corners and walls.

STARTING THE BRICKWORK AT GRADE

On a brick building, the finish grade level of the ground may be at a lower point than the top of the foundation wall or there may be a rising and falling grade. If so, the brickwork must be started at that grade. Most builders try to take advantage of the natural grade of the land when building a structure because of the high cost of fill and topsoil. To accomplish this, the foundation wall must be changed to another size block at the grade line so that there is a 4-inch shelf on the wall from the outside face of the foundation, figure 4-7. The brickwork is laid on this 4-inch shelf.

For example, a building may have a 10-inch block foundation wall. The blocks are laid up to grade height, and then 6-inch blocks are laid flush with the inside of the foundation wall. This leaves a 4-inch shelf on which to lay the brickwork. Usually the 6-inch block wall is laid to the top of the foundation wall, figure 4-8, and then the brickwork is done after the earth has been filled in around the foundation.

When laying the brickwork on the shelf, project the bricks out from the wall 5/8 inch, figure 4-9, or the thickness of a mason's rule

Fig. 4-6 Corner pole setup on a concrete block foundation. Line blocks, fastened to the poles, eliminate the need for building the masonry corner.

Fig. 4-7 Blockwork is set back 4 inches at the grade line to receive the brickwork.

Unit 4 Building a Concrete Block Foundation ■ 39

Fig. 4-8 View of a foundation wall that is ready to receive the brickwork.

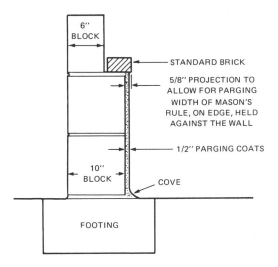

Fig. 4-9 Setting the brick out 5/8 inch for parging on the foundation wall. The mason's rule turned on its side is often used since it is 5/8 inch wide.

held on its side against the blockwork (both are about the same distance). This is done to allow for parging. Also, in this way, tar applied to the wall does not project out past the brickwork and a neat job results.

STEPPING UP A RISING GRADE

When there is a drop in the grade of the earth next to the foundation wall, the brickwork must be built to match the grade, because the concrete blockwork should not be exposed to view. The grade line is determined by holding a line from the lowest point of exposure to the height where all of the foundation wall can be seen after the wall has been backfilled with earth.

The brickwork is stepped up over the blockwork to match the finish grade, figure 4-10. It is important to lay the brickwork about 12 inches lower than the actual finish grade line to allow for settling of the earth from rain and moisture. The filled earth around a foundation will settle at least 8 inches (depending on weather conditions) over a period of years. It is unsightly and a sign of poor planning to have a waterproofed concrete block foundation wall projecting out of the ground. Laying a few extra courses of bricks allows for this settling. Remember, always start brickwork at least three courses below the finish grade line to allow for settling of the earth.

Fig. 4-10 Falling grade line shown by the brick being stepped up to match the fall of the land. This is rough grade; topsoil will be added to bring the soil up to the finish grade.

LINTELS AND BEAM POCKETS

Concrete block lintels or angle irons are used over openings such as windows and doors. Allow a minimum of 8 inches of bearing on each side of the openings. For more support, fill mortar in the last course of blocks where the lintel will be, or use solid block. Concrete lintels are the same height as standard concrete blocks. They should be set to the line when the course that covers the opening is laid. Never rest the lintel directly on the door or window frame. Allow approximately 1/2 inch for a caulking joint. This prevents the weight of the lintel from interfering with the opening of a window or door.

Beam pockets, figure 4-11, are sometimes left in the last course of concrete blocks where the main beam that supports the floor joists rests. As a rule, the block cells that are 8 inches on each side of the beam pocket should be completely filled with mortar down to the bottom of the foundation wall, for load distribution. If this procedure is not followed, bricks can be used instead. The bricks should be bonded back and forth into the blockwork. No stack joints are used. Either method is an acceptable way of building beam pockets in a foundation wall.

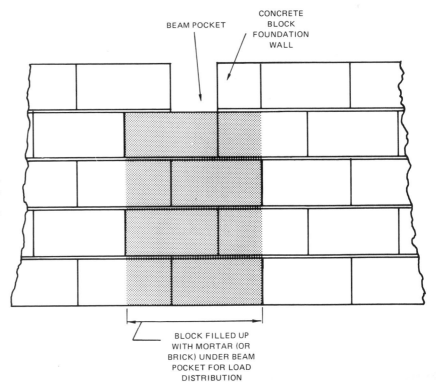

Fig. 4-11 Beam pocket in foundation wall. The block is filled with mortar underneath and to the sides of the pocket for load distribution.

AREAWAYS

Areaways (outside masonry entrances to the basement) built of concrete blocks should be bonded into the foundation wall by interlocking every other course into the main wall. Usually, the inside face of the areaway wall is exposed to view because steps will be there. The mortar joints should be tooled and blocks should be free from chips or cracks.

Metal wire reinforcement can be laid into the mortar bed joints every 2 courses to add extra strength. This is done because there will be pressure from the fill dirt against the wall. A footing for the areaway should be placed at the same time and depth as the foundation footing. It is best to build the areaway wall at the same time as the foundation walls, figure 4-12. If this is not done, dirt can fall in which must be cleaned out before the areaway wall can be built.

Fig. 4-12 Concrete block areaway wall return. Note how the areaway wall is tied into the main foundation wall.

PORCHES

Porch foundations, figure 4-13, are almost always buried under the grade lines, so it may not be necessary to *strike* (tool) the mortar joints. Waste pieces of blocks left from the foundation walls can be used to build the porch walls. The same is true for the foundation wall of the chimney if it is on the outside of the wall. Be sure to form solid mortar joints and bond the pieces so the vertical head joints are not stacked one over the other. The sign of an efficiently built masonry foundation is to be able to finish the walls with a minimum of pieces of blocks or waste masonry units remaining. Any materials left over have to be hauled away or discarded. It is important to be economical by using as many pieces as possible in areas that will not be visible. Make certain they are laid or bonded correctly.

SOLID JOIST BEARING

If the job specifications require solid joist bearing, there are three ways of doing this. In the first method, the last course of blocks laid on the foundation are solid blocks. The second method uses three courses of either concrete bricks or common (building) bricks to accomplish the same results. The cost is greater if bricks are used. This is because bricks cost more than blocks and take longer to lay. In the third method, the cells of the last course of concrete blocks laid are completely filled with mortar. Filling with mortar is the least desirable method, because the courses are weaker than the solid courses.

Regardless of which method is used, the goal is to distribute the weight of the joists over a greater area. Another goal is to eliminate a place for insects to collect.

Fig. 4-13 Foundation supporting wall for the porch and bay window. Notice the use of lintels for tying the porch wall into the main foundation wall.

ANCHOR BOLTS

Anchor bolts should be set in the wall in the last course of blocks and spaced according to the plans. Spacing is usually indicated on the plans by the architect but the normal distance between bolts for the sill is approximately 4 to 6 feet.

The bolts should be mortared in securely. Care should be taken not to damage the threaded area that projects above the top of the foundation wall, and upon which the nut will fit.

On some jobs the wood sill is bedded in mortar, however, if the foundation wall is level, as it should be, bedding sills in mortar should not be necessary. Good working habits usually make this practice unnecessary.

TERMITE SHIELDS

Termites cause problems because they eat the woodwork of a building. They are found in many parts of the country. The warmer areas have the worst problem with termites, while the colder areas are least affected.

A metal shield, figure 4-14, is placed on top of the foundation wall and the wood sill rests on the shield. The shield is extended slightly under the bed joint of the outside masonry wall and into the inside of the foundation. It should project approximately 2 inches on a downward angle of about 45 degrees on the interior.

The purpose of the shield is to stop termites from boring into the wood. Termites only survive in tunnels they have made to get to wood materials. They cannot tunnel through the metal shield.

Metal shields can be made of copper, copper-coated kraft paper, or aluminum sheets. If splices must be made between the sheets, the splice should be soldered. Termite shields should *not* be installed completely through the bed joints, from the outside to the inside of the wall. If they are, the mortar will not bond with the course below.

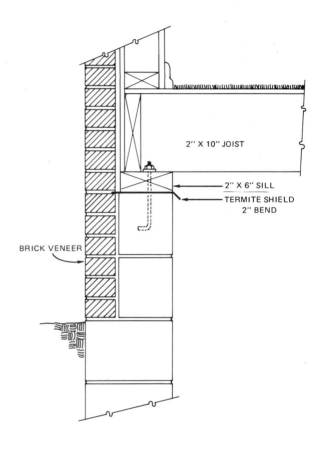

Fig. 4-14 Termite shield built into a masonry veneer wall.

CENTER PARTITION MASONRY BEARING WALLS

If a steel beam is not used to carry the joists, a concrete block partition wall can be built instead, as long as the basement is meant to be divided in half. The advantage of a concrete block partition wall to carry the joists is that it is strong and not likely to sag. It also gives extra protection to the furnace room (if there is one), in the event of a fire or an explosion. As a rule, 8-inch blocks are used for center partition bearing walls in basements.

The disadvantage of this type of bearing wall is that it is permanent and cannot be moved. If the basement is used as a living space it must be designed around the partition wall. A steel beam with steel columns can be boxed in with wood, allowing the full, open use of the basement. Either method of supporting the floor joists is acceptable and is the decision of the architect, owner, or builder.

Good working practices should be used in the building of a foundation as well as for work built above the finished grade line. A well-built foundation provides for a better building and simplifies the tasks of all other trades that work together in the completion of the structure.

ACHIEVEMENT REVIEW

Part A

Select the best answer from the choices offered to complete each statement. List your choice by letter identification.

1. The corners of the foundation wall that are built on the footings are determined from the batter boards by using a
    a. transit level.
    b. straightedge.
    c. plumb bob.
    d. chalk line.

2. To permanently mark the corner on the footing, it is best to use a
    a. pencil.
    b. nail.
    c. lumber crayon.
    d. piece of chalk.

3. When laying out the corners of the foundation on the footing with a transit level, the transit is swung
    a. 45 degrees.
    b. 60 degrees.
    c. 90 degrees.
    d. 360 degrees.

4. Stacks of concrete blocks near the wall in the foundation should be placed
    a. 2 feet apart.
    b. 4 feet apart.
    c. 6 feet apart.
    d. 8 feet apart.

5. The finished surface of the ground, in relation to the building, is known as the
    a. slope.
    b. subsoil.
    c. elevation.
    d. grade.

6. To allow for parging, the proper distance for setting bricks out over the concrete block foundation wall is
    a. 1/4 inch.
    b. 1/2 inch.
    c. 5/8 inch.
    d. 3/4 inch.

7. The caulking joint that is left from the top of a frame to the bottom of a concrete block lintel is
    a. 1/8 inch thick.
    b. 3/8 inch thick.
    c. 1/2 inch thick.
    d. 3/4 inch thick.

8. The outside entrance with steps to a basement is called
    a. a sill.
    b. a well.
    c. an areaway.
    d. an apron.

9. Concrete lintels used to cover openings are the same height as a standard concrete block. This height, including the mortar bed joint, is
   a. 6 inches.
   b. 8 inches.
   c. 10 inches.
   d. 12 inches.

10. Termite shields installed in foundation walls are made of
    a. wood.
    b. plastic.
    c. tar.
    d. metal.

Part B

Draw a sketch of a concrete block foundation wall with the brickwork on steps to accommodate the grade of the land.

PROJECT 3: BUILDING A CONCRETE BLOCK AND BRICK FOUNDATION ON A CHANGING GRADE LINE

OBJECTIVE

- The students will build, from the plans, a small concrete block foundation with brickwork starting on a grade line and built to the top of the wall. This project also teaches bonding of brick and blocks together at varying levels. It gives the students practice in leveling and plumbing, and working together with other students as a member of a team, figure 4-15. The building of pilasters and a center pier in the foundation provides practical experience in building a small project within a larger one. This requires mastery of the basic skills.

Fig. 4-15 Project 3: Foundation being built in a shop situation. Note that a group of students can work together on this project.

EQUIPMENT, TOOLS, AND SUPPLIES

set of mason's basic tools
chalk box
large 2-foot framing square
6 to 8 mortar pans or boards
mortar box and mixing tools
two wheelbarrows—one for mortar and one for brick
brick carriers (tongs)
water buckets
pencil or lumber crayon
builder's level or straightedge and plumb rule
hydrated lime and sand (washed)
supply of water and hose with a nozzle
standard bricks, 8-inch x 8-inch x 16-inch concrete blocks and 8-inch x 4-inch x 16-inch concrete blocks estimated by students from the plan
wood pole approximately 3/4 inch x 2 inches x 10 feet long for making a story pole

SUGGESTIONS

- The 8-inch x 8-inch x 16-inch concrete blocks are used to build the wall up to the grade height shown on the plans. The brick courses must be set out past the block wall 5/8 inch, to allow for the parging. Use the thickness of the mason's rule held against the wall as a gauge.
- The 4-inch x 8-inch x 16-inch concrete blocks are used to back up the brickwork forming an 8-inch thick wall.
- When laying out the brickwork on top of the block, bond from the corner to the point where the brickwork meets the block at grade line. If there is a piece left due to large and small bricks, put it at that point because the fill will hide it from view.
- Use solid mortar joints under the first course of blocks and with all brickwork.
- Keep all materials and mortar pans or boards spaced far enough away from the wall to allow for working room (approximately 2 feet).
- Mix only enough mortar to use in a one hour period.
- Follow good safety practices. Cut away from other students and do not leave tools in the path of work.
- Try not to smear the masonry work.

PROCEDURE

1. Sweep the area clean where the project is to be built.
2. Study the plans, figures 4-16, page 47, and 4-17, page 48.
3. Make a story pole according to the plans.
4. Mark off all courses of concrete blocks to number 2 on the modular mason's rule (or 8 inches). This includes the mortar bed joints. Mark

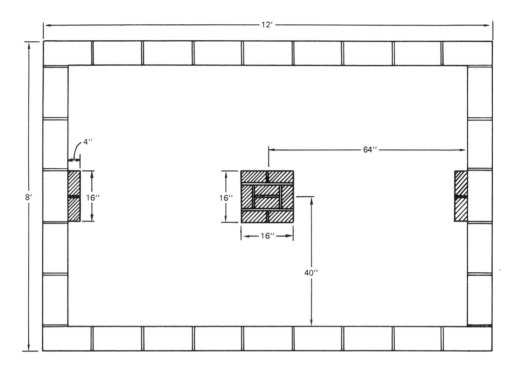

Fig. 4-16 Project 3: **Plan view for a block foundation with brick above grade. Eight-inch concrete block foundation with 2 brick pilasters and 1 brick pier.**

    off all courses of standard size bricks to number 6 on the mason's modular rule.

5. Lay out the foundation with the help of another student on the base. Strike the wall lines with a chalk line.
6. Assemble the blocks and bricks in the work area and stack them near the wall. Be sure to allow for working space.
7. Mix the mortar and load the pans or boards. Wear safety goggles when mixing mortar.
8. Lay out the first course of 8-inch blocks as shown on the plans. Level up the first course either with a builder's level or wood straightedge and plumb rule.
9. Racking back as shown in the plans, lay all of the 8-inch blocks in the foundation before laying any bricks.
10. Bond and lay out the brickwork starting from the lowest point, as shown on the plan.
11. Align the first course of bricks with the opposite corner and to the same projection using a nylon line.
12. Build up the corners on the low end.
13. Fasten the line to the opposite corner by marking off the courses of bricks on the blockwork. Fill in the wall.

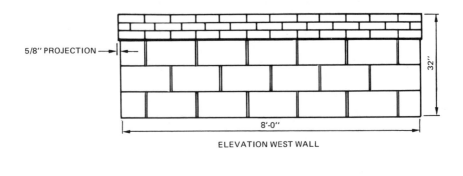
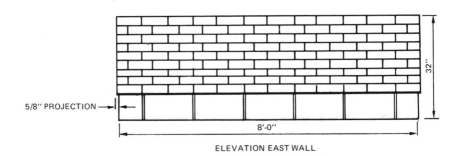
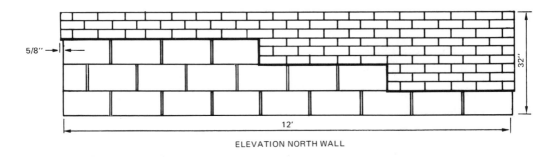
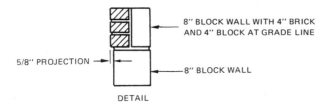

Fig. 4-17 Project 3: Elevation of concrete block and brick foundation on a changing grade line. Notice that the block and brick are to be laid in a running bond (half over the unit under it).

14. Back up the brickwork with 4-inch x 8-inch x 16-inch blocks as needed.
15. Continue to step up the brickwork over the blockwork until the brickwork is level all the way through, from corner to corner, as shown on the plan.
16. Check with the story pole as often as needed.
17. While the outside walls are being built, three students can be working inside the foundation, building the 2 pilasters and center pier.
18. Strike the brickwork with either a round (concave) or V joint when the mortar becomes thumbprint hard.
19. Build the foundation to completion, according to the height and grade lines on the plans.
20. Brush off the work and have the instructor check it.
    NOTE: Concrete blocks on the outside of the wall do not have to be struck (joints are not tooled). Only fill any holes in the mortar joints. Mortar joints on the inside of the foundation wall should be struck with a convex jointer and brushed so this can be used as a finished wall.

Unit 5 WATERPROOFING THE FOUNDATION WALL

OBJECTIVES

After studying this unit the student will be able to

- explain why foundations have to be waterproofed.
- discuss the different types of waterproofing used.
- describe how portland cement mortar is applied and cared for until cured.

NECESSITY FOR WATERPROOFING CONCRETE BLOCK FOUNDATIONS

In any soil in which a foundation is being built, there is going to be some amount of moisture present. To prevent this moisture from entering through the walls, some type of waterproofing finish must be applied to the outside of the foundation walls below ground level.

The amount of moisture in the soil varies in different parts of the country and at different times of year. Factors which cause dampness are rainfall, natural springs in the ground, poor drainage of the soil, closeness to rivers and streams, and elevations below sea level. Land that is low to surrounding terrain is usually wetter than high ground. This results in the need for more widespread waterproofing in order to keep the basement dry.

This unit deals with foundations built with concrete block rather than cast concrete. This is because the concrete block foundation is part of the mason's work.

In very wet soils, the places most likely to have water seeping through are the inside of the foundation walls and where the concrete floor meets the walls. In a case such as this, the basement is of no value to the homeowner as a recreation room, laundry room, boiler room, or storage facility. It is important for the homeowner to have the full use of the structure.

Water pressure that builds up against the outside of the foundation wall can cause cracking and pushing in of the wall. Engineers call this type of pressure *hydrostatic*. To relieve the water pressure, it may be necessary to install drain tile or plastic drain pipe to carry the water away from the foundation.

Excess humidity and water in the basement of a structure is damaging not only to the immediate basement area but to the floors, woodwork, and other parts of the building. Many lending institutions, banks, and government agencies require (as part of the contract between the builder and owner) a guarantee that the basement of a new house be free from moisture for a period of one year from the time of completion.

It is a waste of time and money to have to return to a job and make repairs to prevent the entrance of moisture in a basement. This usually indicates poor construction and can ruin the contractor's reputation. If each step in the construction of a foundation (including the waterproofing) is done correctly, the basement of the house should remain dry.

A very important part of the mason's job is the parging of the foundation wall with a portland cement mortar. The application of tar and other materials is not usually the responsibility of the mason. However, in some instances, the mason may have to perform these tasks. The mason should, therefore, be familiar with the total process of waterproofing the foundation. Many times, on small jobs, the mason is called upon to apply all of the waterproofing.

KINDS OF WATERPROOFING

For a very mild condition of dampness in the soil, a hot or cold type of tar waterproofing can be applied. The wall should be free of mortar drippings or obstructions that can affect the ability of the tar to stick to the surface of the block. Use a scrap of concrete block or scraper to remove the particles.

One coat applied with a mop or brush is usually adequate. An asphalt base material can be applied in a cold condition. This is easier to work with because there is no danger of being burned by hot tar. There are some types of cold waterproofing that never completely harden. This makes them more flexible and less subject to shrinkage from drying.

APPLYING IN LAYERS

Waterproofing compounds can be applied with layers of tar paper felt that form what is called a *membrane*. This method is recommended for walls that are subject to excessive moisture.

The block wall is first coated with a bituminous primer. Bituminous refers to materials derived from coal tar, wood tar, petroleum, or pitch. The hot or cold tar is mopped onto the wall, then a layer of the felt membrane is placed on the hot, tarred surface. This procedure is repeated until two to six layers of felt are applied. The number of layers depends upon the water pressure to be resisted. A final coat is then applied to seal the layers.

> NOTE: Special care should be taken not to damage the waterproofing when backfilling with earth.

PORTLAND CEMENT MORTAR AND TAR

When soil conditions are normal, parging the wall with portland cement mortar followed by applying a coat of tar or bituminous compound provides a dry basement.

It is recommended to use either mortar that consists of 1 part portland cement and 2 1/2 parts washed building sand, proportioned by volume, or type M mortar. Type M portland cement mortar consists of the following: 1 part portland cement to 1/4 part hydrated lime to 2 1/4 times the sum of the volumes of cement and lime used in the mix. The addition of lime to the mix allows more workability when applying with the trowel.

Type M mortar is somewhat more durable than other mortar types. It is specifically recommended for masonry below grade in contact with earth such as foundations, retaining walls, walks, sewers, and manholes.

PREPARING THE SURFACE FOR PARGING WITH PORTLAND CEMENT MORTAR

When applying any type of waterproofing, the surface must be clean and free of obstructions. Dirt or dust can be removed by scrubbing or hosing off with a spray of water. Any grease should be removed by washing with a solvent or highly alkaline solution such as lye. Efflorescence (powdery crust) can be removed by dry brushing or using appropriate commercial cleaners. Usually, though, all that is necessary is to rub the wall with a scrap piece of concrete block and brush it off with a stiff-bristle brush.

Particular attention should be given to the point where the first course of concrete block is laid on the footing. If the wall is going to develop a leak, this is the most likely spot for water to settle. The application of a mortar cove base greatly lessens the chances for leakage.

DAMPENING THE WALL WITH WATER

To reduce the absorption of water used in the mortar mix and to keep the mortar from drying too fast, spray the wall with a fine mist of water before parging. This can be done best by using a garden tank type pressure sprayer. A hose with a fine spray nozzle can also be used.

> NOTE: It is important not to soak the wall with water because the mortar parging will not bond to the concrete block. In addition, excessive shrinkage takes place in the drying process, resulting in cracking.

APPLYING THE PARGING TO THE WALL

Parging is the portland cement mortar troweled on the foundation wall to waterproof it. For best results, parging should be plastered on in two separate coats, allowing drying time between coats. Each coat should be about 1/4 inch to 3/8 inch thick, resulting in a finished coat of 1/2 to 3/4 inch thick.

A mortar cove is plastered on at the point where the footing and exterior wall meet. This is to ensure that the water is turned away. The mortar cove is made by building up the mortar on an angle. The mortar is thicker here than at any other place on the wall.

The trowel (plasterer's trowel) is held on an angle. Mortar is plastered upwards with a sweeping movement of the wrist forming a beveled cove at the base of the wall, figure 5-1. Keep the finish as smooth as possible. The smoother the finish, the more waterproof the wall.

The rest of the wall should be plastered upwards with long strokes of the trowel, keeping it as smooth as possible. The parging line should go from the cove to where the brickwork starts or to the grade line. Mortar should be mixed with enough water so that it trowels on easily with a minimum of effort, figure 5-2.

After the mortar has partially hardened, its surface should be scratched with a rough brush, straw broom, or a piece of metal lath. This provides a better bond for the second coat when it is applied. The second coat should not be applied until a period of 24 hours has passed to allow the first mortar coat to cure properly. In very hot climates it is helpful to keep the parging damp by covering with moist burlap.

Unit 5 Waterproofing the Foundation Wall ■ 53

Fig. 5-1 Portland cement mortar cove applied with the trowel at the bottom of the foundation wall. This is to ensure that the water is turned away.

Fig. 5-2 Parging the foundation wall with portland cement mortar

Before applying the second coat, repeat the same procedure of spraying the wall with a fine mist of water. Parge with another 1/4-inch coat of mortar for the finished job. Remember, the smoother the finish the more water-repellent it is.

To ensure proper curing and to keep it from drying out too fast, the finish coat should be continually dampened with a fine spray of water. Existing weather conditions determine how often the parging has to be dampened. The parging must be dampened more on hot, sunny days than on cool, cloudy days. For proper curing, a good rule of thumb to follow under average conditions is to keep the second coat damp for at least 48 hours after hardening.

APPLYING THE TAR COMPOUND OVER THE PARGING

After the parging coats have cured, a more waterproof job can be made by brushing or rolling on one or two coats of bituminous waterproofing. One heavy troweled-on coat of cold, fiber-reinforced, asphaltic mastic can be substituted for the brushed-on bituminous coatings. Either one makes the parging more effective on the foundation.

Most building supply dealers stock tar-type waterproofing in 5-gallon containers. All that is needed to apply the tar is a long-handled brush or a paint roller. To use the roller, the tar must either be heated or in a warm state. Old work clothes that can be discarded after use should be worn because tar will not come out of the clothes.

> Long gloves should be worn to protect sensitive skin.

Any type of petroleum solvent will remove the tar from the skin. Care should be taken if a person has sensitive skin or allergies. Test the solvent on only a small spot of skin, and observe if there is any reaction before wiping it over a large area.

As was the case in applying the parging, be sure to coat all of the surface with the waterproofing for best results. There must not be any loose soil between the tar coats or it will not stay on the wall. Figure 5-3 shows a completed, waterproofed foundation.

Fig. 5-3 A completed, waterproofed foundation wall with the tar showing

WATERPROOFING THE FIRST THREE COURSES BEFORE COMPLETING THE ENTIRE FOUNDATION

Often when a foundation is being built, rain occurs before the basement is waterproofed, causing some of the surrounding soil to fill in. In this case, the dirt must be removed before any parging can be done.

A good method of eliminating this problem is to lay up the first three courses of concrete blocks and then parge and tar them before continuing to build the foundation. The proper drying time must still be followed between applications of the waterproofing. If this is done while building the walls, the work should not be delayed.

Fig. 5-4 A mason is parging the wall after the first three courses of concrete block have been waterproofed.

This assures a better job of waterproofing because the waterproofing seals to the masonry work better if no dirt is present. Figure 5-4 shows a mason parging the wall after the first three courses have been completed.

DRAIN TILES LAID AROUND THE FOOTINGS

Drain tiles are used around the outside of the footings to trap and remove excess water. The design of the system is very important if it is going to work properly. Drain tiles are laid in a bed of crushed stone or gravel around the outside perimeter of the foundation, figure 5-5. The bottoms of the drain tiles should never be lower than the bottom of the footing. The tops of the tiles should never be higher than the top of the footing. Strips of tar paper are laid over the joints to prevent dirt from entering and clogging the tiles.

Drain tiles should be laid in such a manner that there is an even slope at all times. No flat places should be left where water can lay. The water must flow away from the footing. This can best be done by using a builder's level and a line to determine the correct fall.

Two methods of draining the water are used. One is to run the water from the drain tiles to a natural drain or sewer away from the foundation wall, as shown in method A, figure 5-6, page 56. The other method is to drain the tiles to a sump pit that is located inside the basement. In the pit, a pump forces the water out of the house through the sewer system, as shown in method B,

Fig. 5-5 Drain tile installed around the foundation wall in a bed of stone. The strips of tar paper stop the earth from entering and clogging the pipes.

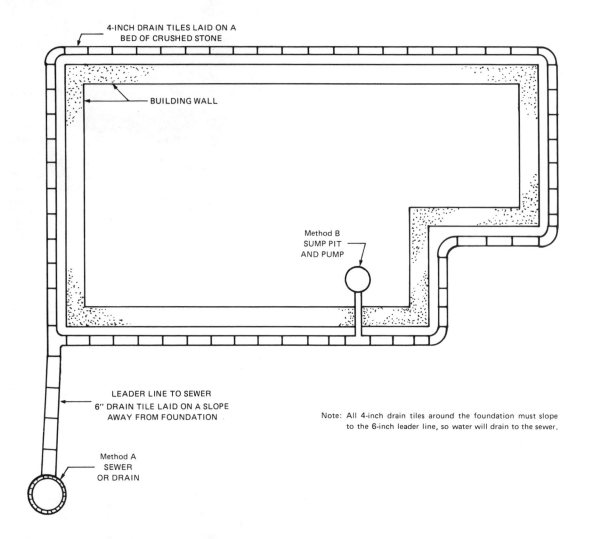

Fig. 5-6 This drawing shows the proper way of installing drain tile around the outside of the foundation wall. Method (A) is draining to a sewer and method (B) is draining to a sump pit.

figure 5-6. The first method is preferred because it does not require the use of a pump to get rid of the water.

The installation of drain tile around a foundation is often done carelessly, and this causes problems. If the drain tiles are not sloped correctly, the water is trapped and circulates around the house, creating a more damp situation than if the tiles were not used. This condition can build up pressure and create a major problem.

After the tiles are laid out and covered with the tar paper, they should then be covered with crushed stone or gravel to a minimum depth of 18 inches. The stone must be placed carefully since the paper joints should not be disturbed nor the tile broken. Figure 5-7 shows a section view of drain tiles and the waterproofed wall of a foundation. Lengths of flexible plastic pipe with small holes are also available in place of drain tile.

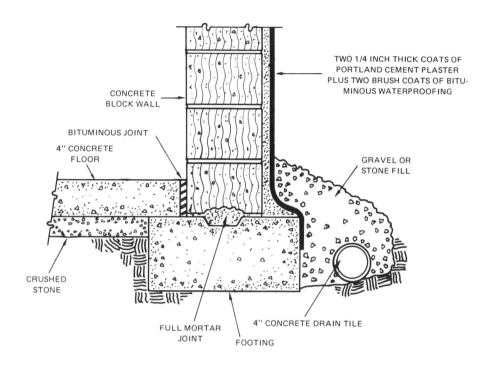

Fig. 5-7 Section of a concrete block foundation wall waterproofed below grade

BACKFILLING WITH EARTH AROUND THE FOUNDATION

The foundation must be safeguarded against cracking or pushing in of the walls. Therefore, backfilling should only be done when the walls are adequately braced and some of the weight of the first story is in place to help hold the walls. Frozen earth, trash, boulders, or anything that can rot in the ground should not be used as fill around foundations. Backfilling must be done carefully so that pressure on the walls is built up evenly. Heavy earth-moving equipment should not be operated too close to the walls. The fill should be pushed in parallel to the wall in 1-foot levels, and compacted by hand. A careless bulldozer operator can crack the walls or push them in. This would require rebuilding of the basement walls at a great expense to the builder.

The backfill should be built up on a slope away from the wall to allow water drainage and expected settlement of approximately 8 inches to 1 foot. Soil should not be soaked with water, because too much hydrostatic pressure can develop.

The bracing should remain until enough weight is on the foundation of the structure to prevent the shifting or cracking of any walls, figure 5-8, page 58. Figure 5-9, page 58, shows a front-end loader being used to fill dirt in around a house. Remember, no matter how well the foundation is waterproofed, if the backfilling damages the walls, they will probably leak.

Fig. 5-8 The foundation wall is braced on the inside while the backfilling is being done. This prevents the wall from being pushed by pressure from the earth or a bulldozer.

Fig. 5-9 Using a front-end loader to fill dirt in around a foundation.

ACHIEVEMENT REVIEW

Select the best answer from the choices offered to complete each statement. List your choice by letter identification.

1. The length of time required to guarantee a dry basement in a house is usually
   a. 6 months.
   b. 1 year.
   c. 18 months.
   d. 2 years.

2. Pressure caused by water building up against the foundation wall is called
   a. membrane.
   b. capillary.
   c. hydrostatic.
   d. bituminous.

3. If there is only a small amount of moisture in the soil around a foundation wall, the type of waterproofing that should be used is
   a. membrane.
   b. parging.
   c. parging and tarring.
   d. hot or cold tar.

4. The correct type of portland cement mortar to use for the parging of a foundation wall below grade line is
   a. type O.
   b. type K.
   c. type M.
   d. type S.

5. The total thickness of two coats of parging on a foundation wall should be
   a. 1/4 inch.
   b. 1/2 inch.
   c. 3/4 inch.
   d. 1 inch.

6. The thickened, angled mortar that is plastered on the foundation wall at the point where the footing and first course of block is laid is called a
   a. cove.
   b. wash.
   c. mastic.
   d. primer.

7. To ensure better bonding of mortar between coats, it is a good practice to
   a. spray with a hardener.
   b. add metal reinforcement wire.
   c. scratch the surface.
   d. trowel the surface smooth.

8. When parging two coats of mortar on a foundation wall, the second coat should not be applied until the first has cured for a period of
   a. 8 hours.
   b. 16 hours.
   c. 24 hours.
   d. 48 hours.

9. After both coats of parging have been troweled on the wall, the mortar should be cured, before the application of any tar waterproofing, for a period of
   a. 8 hours.
   b. 16 hours.
   c. 24 hours.
   d. 48 hours.

10. After drain tiles are laid in place on a bed of crushed stone, the joints are covered with
    a. strips of aluminum.
    b. strips of tar paper.
    c. strips of plastic.
    d. portland cement mortar.

11. To permit water to seep through to the tile, the tile should be covered with a layer of crushed stone to a thickness of
    a. 12 inches.
    b. 18 inches.
    c. 24 inches.
    d. 36 inches.

12. Backfilling of the foundation should be done only after the
    a. foundation walls are built.
    b. entire building is finished.
    c. weight of the first floor is in place.
    d. weight of the first floor is in place and the walls are braced.

PROJECT 4: PARGING THE FOUNDATION

To familiarize the student with the process of parging, the foundation project that was constructed in Unit 4 can be parged up to the brick line. If this is not possible, a concrete block wall can be built approximately 4 feet long and 2 feet high. This will work with full-size block. The student can be graded on the building of the wall; then the wall can be parged with 2 coats of mortar. Use mortar made of hydrated lime and sand in the proportions of 1:2 for the parging. The mortar can be screened and reused in a training situation. Be sure to have the student apply a mortar cove at the bottom of the wall.

SUMMARY – SECTION 1

- Two major factors that determine the design and construction of any foundation are the weight the soil will carry (load-bearing strength) and the load the foundation will have to support.
- Loads can be classified into two groups: dead loads and live loads.
- Rules and regulations govern the construction of buildings in most areas. These rules are known as building codes.
- Basement-type foundations serve not only as foundations but as living and storage space.
- The main line from which the building is laid off is known as the building line.
- Bench marks are key reference points established on the job site from the known elevations above sea level.
- All height or construction points are derived from the bench marks during the period of construction.
- Wood batter boards are erected at each corner of the foundation to maintain and preserve the lines of the building during the excavation and construction of the foundation.
- Batter boards can be laid out with the transit for level, and squared by the 6-8-10 method.
- The 6-8-10 method is based on the formula that the base squared, plus the rise squared, equals the hypotenuse squared.
- The depth to be excavated for the foundation can be found by measuring down from the stretched lines on the batter boards (with a rule or story pole) as the foundation is being dug.
- After batter boards have been established, wall lines should be marked with saw cuts on top of the boards.
- Crawl space foundations are built when no basement area is needed or when it is not practical to construct a basement.
- Foundations for any type of masonry work, including crawl space, should be started below the frost level of the ground.
- Topsoil should be stockpiled for finish grading.
- A footing is the base that supports and distributes the weight of the structure.
- Concrete is the material used for the installation of footings because it conforms to any type of form and can be designed to support any load.
- Concrete footings for masonry foundations should be twice the width of the unit being laid and one-half to three-fourths the width of the footing in depth.

- For the best bonding, footings should be placed continuously with no breaks.
- Regardless of the size of the wall being built, footings should be placed no less than six inches in depth.
- Steel reinforcement is not required for footings in houses unless there is a special problem. Large structures, however, do require reinforcement in the footings.
- Generally a 5-bag mix (5 bags portland cement to the cubic yard) is strong enough for footing mixes.
- The two general types of footings are classified as trench and formed.
- It is important that trench footings are dug with straight sides and bottoms for maximum strength when concrete is placed.
- Formed footings must be braced securely to support the pressure exerted by the concrete when placed in the forms.
- Footings can be formed and placed on a rising grade by building forms in a stepped pattern.
- Concrete should be settled in the forms by tamping, by tapping on the forms, or by vibration.
- The top of the footing should not be troweled smooth.
- All tools and equipment should be cleaned immediately after use to remove concrete.
- Installation of footings is considered part of the mason's job.
- Poor footing installation results in added cost (leveling the first course of blocks laid), and a weakened structure.
- A story pole is so-named because it is made and used for the construction of one story of a building.
- The purpose of the story pole is to lay out masonry work. All of the heights and individual courses are marked on the pole.
- All the plans and specifications should be studied before making the story pole.
- Select a straight length of lumber for the story pole.
- Mark all important heights on the story pole before marking individual courses.
- Use a square to mark off all unit courses on the pole.
- Number the courses from the bottom to the top of the story pole to eliminate any chance of getting a "hog in the wall."
- Lightly saw in all individual masonry courses on the story pole.
- Hold the story pole plumb (vertical) when checking height.

- Sill sticks or gauge rods (both are the same) are made like the story pole, but are only used as a guide to build the wall to sill height.
- Brush all dirt from the footings before laying any block.
- A solid, unfurrowed mortar bed joint should be used under the first course of block to prevent water penetration.
- Locate the corners of the walls by using the plumb bob suspended from the batter board lines.
- The transit can be used to lay out the foundation after the first corner point is located.
- Lay out one course of concrete blocks around the foundation to determine the bond before building the corners.
- Use safe work practices in the stacking of materials.
- Masonry materials should be stacked in the foundation in the best position to increase the mason's work output.
- Establish traffic lanes for the movement of materials in the work area.
- Do not stack more materials in the work area than are needed to build the walls.
- Brickwork should be started 3 courses below the actual finished grade to allow for the settling of the soil.
- To allow for parging, the brickwork should be set out over the concrete blockwork the thickness of the mason's rule held sideways against the block wall. This is 5/8 of an inch.
- Areaways, porches, chimneys, etc., should always be connected to the main walls of the foundation.
- To conserve materials, leftover pieces of concrete blocks from the main walls can be used in areas of the foundation that will not be seen.
- On a joist bearing wall of a foundation the last course of masonry units should be made either of solid block or brick.
- Allow proper bearing on each side of an opening for all lintels.
- Metal shields should be built into the foundation wall to protect the woodwork from termites.
- Center partition concrete block walls can be used instead of steel beams to carry the floor joists.
- Some type of waterproofing must be applied to the foundation to ensure a dry basement.
- Parging the wall is the job of the mason.
- Tar waterproofing increases the effectiveness of a portland cement mortar parging.

- Membrane layers of waterproofing are used for very wet conditions.
- Type M portland cement mortar is the best type for parging below grade.
- Any loose dirt should be removed before waterproofing.
- Walls should be dampened with water before parging.
- Enough drying time must be allowed between coats of parging if the parging is to cure properly.
- Good working practices should be followed in waterproofing procedures.
- It is a good practice to waterproof the first several courses as soon as possible.
- Drain tiles help remove water from the foundation area only if they are installed with the correct slope and toward a drain area.
- Backfilling should only be done after the waterproofing has cured and after part of the weight of the first story is in place to prevent the wall from pushing in.
- Rubbish and large stones should not be included in the backfill soil.
- Backfilling should be done very carefully.
- Waterproofing of the foundation is one of the most important tasks in the construction of a structure.

SUMMARY ACHIEVEMENT REVIEW, SECTION 1

Complete each of the following statements referring to material found in Section 1.

1. The weight of the structure that rests on the footing is known as _____ load.

2. Rules and regulations established by local governments to control the construction of buildings in a township or city are called _____ _____.

3. The majority of masonry foundations are constructed of _____ _____.

4. A permanent reference point is established on the job site from sea level elevations. All building points can be taken from this point, which is known as a _____.

5. Wood stakes and boards erected at the corners of the foundation for the purpose of maintaining and preserving the lines of a structure during the period of construction are called _____ _____.

6. The most frequently used method of squaring a foundation when a transit level is not available is the _____ _____.

7. A foundation that does not have a full basement but only room for pipes and electrical utilities is known as a _____.

8. The base of concrete that the walls and building rest on is called the _____.

9. The width of the footing for a concrete block wall should be _____.

10. Footings are installed in two ways. These two ways are _____ and _____.

11. When installing a stepped footing on a falling grade, the step height should not exceed _____.

12. When concrete is delivered to the job in a regular ready mix truck, the final amount of water should be added _____.

13. Immediately after use, concrete should be removed from the tools and equipment with _____.

14. Marking off the courses of masonry units on the story pole should be done accurately by using the _____ and _____.

15. Each course of masonry units on a story pole is numbered from the bottom to the top to prevent the possibility of _____.

16. The gauge rod is used as a guide when building the wall up to _____ height.

17. After marking off all of the courses on the story pole, the best way to permanently mark them is with _____.

18. The corner points of a foundation can be transferred from the batter board lines to the footing by dropping a _____.

19. A good method of checking the foundation to see if it is square is to measure with a steel tape on a _____.

20. To help keep water from leaking through the bed joint under the first course of blocks laid on the footing, the bed joint should not be _____.

21. If the ground is soft or irregular in the work area where wheelbarrows must be pushed, it may be necessary to build _____.

22. The height at which the blockwork stops and the brickwork starts is known as the _____ line, due to the level of the ground.

23. The correct distance to set the brickwork out over the blockwork to allow for parging is _____.

24. The concrete block wall that is built for the purpose of having an outside entrance to the basement is called an _____.

25. The last course of block on a foundation is sometimes filled solid with mortar or built of solid masonry units for the purpose of _____.

26. The upward tunneling of termites inside a foundation can be stopped by installing a _____.

27. Portland cement mortar parging can be made more waterproof by applying _____.

28. The bonding of the parging to the concrete block wall can be improved by first _____.

29. The angled, thickened coat of mortar that is applied at the bottom of the foundation wall to help drain the water away is called a _____.

30. Excess water can be drained away from the foundation wall by installing a bed of crushed stone and _____.

Section 2
Chimneys and Fireplaces

Unit 6 CUTTING WITH THE MASONRY SAW

OBJECTIVES

After studying this unit the student will be able to

- explain the two main methods of cutting: wet and dry.
- list the important safety practices to be followed when operating the masonry saw.
- make some of the more common cuts described in the unit.

The masonry saw is introduced at this point in the text because it will be needed for cutting masonry units around electrical and mechanical fixtures and when building fireplaces, arches, and complicated brick panel walls. Operating the masonry saw safely and correctly is one of the skills that all masons have to perform as part of their trade.

When laying masonry work around electrical and mechanical fixtures, usually the masonry unit must be cut rather thin to enclose it in the wall. This cannot always be done with a hammer or chisel because the shocking power may break the unit. To prevent breaking the masonry unit, architects on many jobs require that all masonry cuts be made with a saw.

Although it is true on some jobs laborers operate the saw, the mason who is laying the unit in the wall should operate it. The mason has a better understanding of the cutting needs and therefore can be more exact in making the cuts.

Cutting with the masonry saw is faster, neater, and less wasteful than cutting with a chisel or hammer. On a well run job, cuts are made ahead of the masons and stocked in the work area. This greatly speeds up the work because the masons do not have to wait for wet blocks or bricks to dry, or make cuts with the hammer or chisel. It also decreases the possibility of the bond shifting back and forth on the wall since all of the starting pieces are the same size. The mason must adjust the mortar joints so they work bond in the wall. Cutting with the masonry saw, therefore, is economical and is an essential part of the masonry business.

DESCRIPTION OF THE MASONRY SAW

All brands of masonry saws have the same basic construction. If they are different, it is mainly in the quality of materials and devices built into the saws to improve performance. The parts of a masonry saw are shown in figure 6-1.

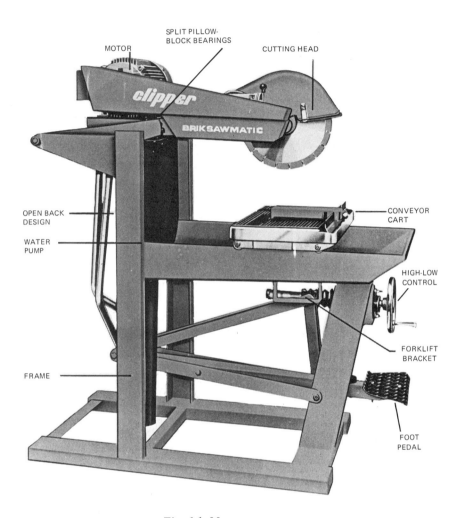

Fig. 6-1 Masonry saw

A full-size stationary or portable saw can be purchased. The full-size saw can be made more movable by buying a wheel arrangement that can be lowered if necessary so the saw can be rolled from place to place. This requires the services of a truck and several people to move the saw from one job location to another. The portable saw, however, can be loaded into the trunk of a car and transported more easily.

Select the type of saw that fits the needs of the job. If much cutting is required, the full-size saw should be used. One of the major improvements incorporated in new models is the electric pump. The pump forces water up from the pan onto the blade for use in wet cutting with a diamond blade. The adjustable cutting head operates by turning a crank to either raise or lower the blade.

Masonry saws can be run on either 115/120 volts or 220/440 volts. They will operate more efficiently on the higher voltage. Running on higher voltage, they have less chance to bog down or overheat, and will probably last longer.

The motor is protected with a thermal overload switch for use if the machine does overheat. Once the motor cools, the reset button can be pushed and the motor will run again, unless it has been damaged.

METHODS OF CUTTING WITH MASONRY SAWS

There are two ways of cutting with masonry saws; with water to cool the blade *(wet cutting)*, or with a dry shatter-resistant blade *(dry cutting)*.

Wet Cutting

A wet saw blade has industrial diamond chips fused to the blade. As the blade cuts through the masonry unit, water is continuously pumped against the cutting edges and sides. This prevents the diamonds from overheating and burning up. If a diamond blade is run without water to cool it, the blade will be ruined in a very short time.

> Eye protection must be worn at all times when operating the masonry saw.

At the close of the workday, the saw operator should remove the blade from the saw and lock it in the toolshed or trailer. Be very careful not to bump or ram the cutting edge against a hard object since it can be damaged, making the blade useless for cutting.

With proper care, the diamond blade withstands long, hard wear and lasts a long time. Diamond blades are, however, expensive. They can be purchased in different diameters depending on the depth of cut desired. The average all-around size used is the 14-inch blade.

Diamond blades can also be purchased to suit the hardness of the materials being cut. When ordering blades be sure to specify the type of materials to be cut and the diameter of the blade. The edge of the diamond blade is made in segments (see figure 6-2 A) to help it run cooler and prolong its life.

Rapid advances have been made recently in diamond blade technology. A diamond blade properly used on well designed and maintained saws can provide the lowest cost per cut of all the methods now used for cutting masonry materials. However, if the same blade is put on a poorly maintained piece of equipment, and used by a poorly trained operator, this results in a much greater cost of cutting.

Fig. 6-2 (A) Diamond blade for wet cutting
(B) Abrasive blade for dry cutting

Dry Cutting

Cutting without the aid of water requires an abrasive type of blade, figure 6-2 B. Shatterproof abrasive materials such as silicon carbide are bonded with a reinforcing mesh built into the blade to prevent it from breaking. Resins are added to help bond the blade and cutting abrasives together.

Dry cutting blades are also available to cut different hardnesses of masonry materials. If the blades are given proper care, they should not break.

> Dry cutting blades that are not reinforced should not be used; they can shatter, injuring the operator. Read the instructions carefully when ordering saw blades.

ADVANTAGES AND DISADVANTAGES OF CUTTING WITH WET BLADES

Wet cutting allows thinner, more delicate cuts with the saw because the diamond blade is sharper than the abrasive blade and less heat builds up. Wet cutting is also much faster than cutting with a dry blade. The diamond blade, although it does cost more, lasts longer than any other type of masonry saw blade with proper care and cutting practices. Therefore, the diamond blade is actually more economical to use.

The major disadvantage of cutting with the wet saw is that the masonry unit becomes soaked with water and is difficult to lay in the mortar bed. The wet cut does not set up as does a dry cut. The proper practice is to cut the units far enough in advance that they can dry out before being laid. This of course requires some planning ahead.

There are sometimes problems with the wet saw in cold weather. If it remains outside, the saw can freeze up. It is also uncomfortable for the person operating the saw and this can cause reduced work output. The water can be heated or the saw set up in a heated enclosure.

ADVANTAGES AND DISADVANTAGES OF CUTTING WITH A DRY BLADE

There are two important advantages of the dry saw. The masonry units set up more quickly in the mortar bed joint and there is no weeping or smearing of the joints.

The disadvantage of dry cutting is that the units cannot be cut as thin as with the wet blade. Also, dry cutting is a much slower process. The heat generated from the dry blade can cause the masonry unit to blow out on the side or crack if the cut is too thin.

A major problem for the mason is the large amount of dust caused by the sawing process. A dust mask or respirator must always be worn when operating a dry saw. A strong exhaust fan can be placed in front of the saw to draw the dust away from the operator. The dust should be blown into a duct or tube and vented to the outside. This is for the safety of all workers inside the building.

SAFETY PRACTICES TO FOLLOW WHEN OPERATING THE MASONRY SAW

- No horseplay.
- Wear an approved hard hat to protect the head from any flying particles.

- Have the electrician ground the saw frame to a steel rod in the floor or ground to prevent electrical shock.
- Stand on a wooden platform or board to prevent grounding of the operator.
- Never stand in water when operating the saw.
- Make sure the blade is on the saw correctly and tightened before starting the saw.
- Do not wear loose clothing.
- Do not wear rings that project out from the fingers.
- Wear shatterproof eye protection.
- Never ground or touch a metal object to another nearby object when running the saw because an electrical shock can result.
- Do not operate the saw if wires are frayed (worn) or bare.
- Never start the saw unless the masonry unit to be cut is clear of the blade.
- Do not cut close to the fingers. Back up the cut with a waste piece of masonry unit.
- Do not use the saw blade to dress chisel edges or points.
- Wear rubber outer wear to prevent getting wet and to reduce the chance of getting an electrical shock.
- Never force the saw blade to cut faster than it will freely cut.
- Make sure the unit to be cut is laid square and flat, free of chips, on the saw table so it cannot bind or kick back at the operator.
- Use the miter gauge on the saw table when cutting angles.
- When operating a dry saw, face the saw so the wind carries the dust away from you.
- Secure an exhaust fan and place it in front of the saw to help carry away the dust.
- Wear a dust mask or respirator when running the dry saw.
- Never mark a unit on the saw table while the blade is running.
- The conveyor cart that the masonry unit lays on should be rolled back and forth as the blade is applied to the cut. This lessens the possibility of binding the blade in the cut.
- If the saw blade binds in a cut, do not attempt to hold or grab it. Let it go.
- Do not cut a cracked masonry unit.
- Do not operate the saw when you are ill.

> Accidents happen to experienced operators, too, if they become careless after doing the cutting for a long period of time. Observe all safety rules and regulations and report any accidents immediately to the instructor, foreman, or employer.

MAKING CUTS WITH THE SAW

Before actually cutting a masonry unit, review the previous safety rules. A safely dressed mason is shown in figure 6-3.

Unit 6 Cutting with the Masonry Saw ■ 71

Although there are many cuts that can be made with the saw, there are certain basic cuts that are required on almost all masonry jobs. Masonry cuts are usually the same whether they are block, brick, tile, or other materials. The following cuts are made using concrete blocks.

One Cell Out Sometimes there is a pipe or wire in the wall that the block cannot be slipped over by using the hollow cell. If this is the case, a web has to be cut out with the saw.

If only the end web of the block is to be cut out, it is called one cell out. It is confusing to use the term cell when ordering cuts from the saw operator. Therefore, to save time, some masons have devised a set of terms to describe the number of cells to be cut out of a block.

If the end of one cell is to be cut out, it is called a quarterback. A quarterback cut is shown in figure 6-4.

Two Cells Out When it is necessary to cut out the cross webs of two cells, it is called a halfback, figure 6-5.

Fig. 6-3 The mason operating the saw is properly dressed to keep dry, and is protected with a hard hat and eye protection.

Fig. 6-4 Quarterback cut block. Notice that the end cell is cut out of the block.

Fig. 6-5 Halfback cut block. Notice that two cells are cut out.

72 ■ Section 2 Chimneys and Fireplaces

Fig. 6-6 Fullback cut block. Notice that all of the webs have been cut out but the last one.

Fig. 6-7 Comparison of all 3 cut blocks described. Six-inch blocks are shown.

Three Cells Out If all of the cross webs must be cut leaving only the end of the block holding it together, it is called a fullback, figure 6-6.

A comparison of the three cuts is shown in figure 6-7. If a concrete block is being used that has only two cells, the cuts are called halfbacks and fullbacks; there is no quarterback due to the cell arrangement.

CUTTING AROUND ELECTRICAL BOXES

Electrical boxes installed flush with the face of the masonry wall must be cut back neatly into the block. Only the wall plates that the electrician fastens on later will hide the edges.

First, mark the area to be cut out with a crayon or soapstone. Then, using the saw, cut out as close as possible to the marked area. As a last step, chip out the remaining particles with a small tile hammer to allow the block to fit over the electrical box neatly. Figure 6-8 shows a standard receptacle box installed in a block lead.

Fig. 6-8 Electric receptacle box cut into a concrete block corner. Notice the entire box is enclosed in the wall.

SLABS

Standard size concrete block slabs can be ordered from the concrete block company. These slabs measure the same height and length of a standard block and are 1 5/8 inches thick. However, often when building around fixtures on the job, a slab thinner than normal is needed. A cut concrete block slab is shown in figure 6-9. If cut wet, be sure to allow the slab time to dry before laying if possible. If time is not allowed, it will be difficult to hold the block in place.

RIPS

Concrete blocks that are cut across the face side lengthwise (horizontally) are called *rips*. They can be of different sizes

Fig. 6-9 Cut concrete block slab

depending on the need. If the block is cut exactly in half lengthwise, it is sometimes called a *half-high* rip. Rip blocks are often used as the starting course off the floor. They are also used when there is a structural tile base, and a filler block piece is needed to reach a height of eight inches for normal coursing. Other places where rip blocks are used are against the bottom of steel and concrete beams when sealing off a masonry wall, and under and over window frames.

In figure 6-10 the mason is shown cutting a rip block that would be very difficult to cut with a mason's hammer. Note the proper dress and eye protection worn by the operator.

Fig. 6-10 Mason cutting a rip block on the saw (wet saw with diamond blade). Notice the proper clothing and eye protection.

BOND BEAM

Sometimes the specifications require a course of concrete blocks to be laid across an opening or around the top of the wall to be filled with concrete and reinforced with steel rods. This is called a *bond beam*, figure 6-11. To accomplish this, the inside webs of the blocks are cut out approximately 3/4 of the way down. The cut block is then laid across a wooden form, if it is to serve as a lintel for an opening. Concrete and steel rods are installed in the center of the blocks. The bond beam can also be used in a structure to give added strength when necessary for special conditions. After the concrete or grout has cured, the form can be removed.

Fig. 6-11 Bond-beam block cut out of a full block. Notice that the lower part of the block is left in for the placement of the steel rods and concrete.

CUTTING ANGLES

It is possible to cut masonry units on angles on the masonry saw. This is done by either turning the block to match the line of the blade on the table, or by using the miter attachment which is graduated in degrees. Two angle cuts are shown in figure 6-12.

The cuts discussed here are only a few of the many that are needed on a masonry job. Although concrete blocks are shown in the illustrations, any masonry unit can be cut on the saw, if the proper blade is used.

Fig. 6-12 Concrete blocks cut on an angle or miter

> Good safety practices should be followed at all times regardless of the type of materials and the nature of the cut.

CARE OF THE SAW

A saw operates more efficiently and for a much longer period of time if given good care and maintenance. This list of good work practices should be closely followed at all times:

- Use the recommended blade for the materials being cut.
- Never operate the diamond blade dry, even for a short period of time. This causes permanent damage to the blade.
- Make sure that the electric pump is operating and a good flow of water is on the blade at all times.
- Keep the water tray full of water.
- Change the water as often as needed to keep it clean and free from sludge buildup.
- Check the tension and wear of the belts and adjust or replace as needed.
- When there is an electrical problem, have an electrician fix it.
- Any operating problems that affect the use of the saw should be reported to the foreman.
- At the end of the workday, drain and flush the saw with clean water.
- Roll up the saw cords, and lock the blade and cord in the toolshed at quitting time.

USING THE SAW TO THE BEST ADVANTAGE

The saw can be used to its best advantage by following these recommendations:

- Set up the saw as close to the work area as possible, but not so close that it interferes with the work of the mason. This eliminates unnecessary waiting for cuts.
- On a large job, have one person place the orders for cuts with the saw operator. Having the same person placing the orders will make the work more accurate.
- When using the wet saw, let the cuts dry out as much as possible before laying them in the wall. This is done by laying out the walls in advance and determining all of the cuts needed. The saw operator can then make the cuts and stack them near the wall to dry.
- If cutting with the wet saw, mark all cuts for electrical boxes or fixtures with a crayon or waterproof marker.
- Replace a dull saw blade with a sharp blade.
- Follow good safety practices and report any accidents to the foreman immediately.

ACHIEVEMENT REVIEW

Part A

Select the best answer from the choices offered to complete each statement. List your choice by letter identification.

1. When wet cutting, the best type of blade to use is
   a. silicon carbide.
   b. fiberglass.
   c. diamond.
   d. carbon.

2. The most popular all-around size blade to use on a masonry saw is
   a. 10 inches.
   b. 12 inches.
   c. 14 inches.
   d. 20 inches.

3. Diamond blades that are used on the saw must be lubricated with
   a. oil.
   b. graphite.
   c. alcohol.
   d. water.

4. The biggest disadvantage of a dry blade is that it
   a. cuts slower.
   b. has dust problems.
   c. will break.
   d. is high in cost.

5. The main disadvantage of the wet saw is that
   a. the blade cuts slower.
   b. the blade is high in cost.
   c. the unit being cut becomes wet.
   d. there is a great danger when using a diamond blade.

6. To prevent the operator from getting an electrical shock, the saw should
   a. be grounded.
   b. never be filled with water.
   c. be placed on a wood platform.
   d. only be operated on 115/230 volts.

7. A block that has the end of one cell cut out is called a
   a. halfback.
   b. quarterback.
   c. fullback.
   d. bond beam.

8. Concrete blocks that are cut lengthwise and used when a full-size block is not needed are called
   a. angle cuts.
   b. bond beams.
   c. lintel blocks.
   d. rips.

9. When cutting a concrete block, if the block is pinched or bound in the saw blade, the best practice to follow is to
   a. call the foreman.
   b. turn the saw off.
   c. release your hand from the block and let it go.
   d. try to free the block from the saw blade as quickly as possible.

10. A block that has the inside web cut out to receive concrete and steel rods is called a
    a. rip.
    b. lintel.
    c. half high.
    d. bond beam.

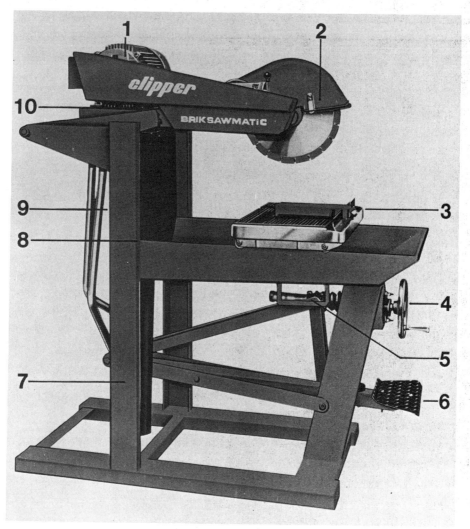

Fig. 6-13 Parts of the masonry saw

Part B

Identify the names of the parts numbered on the saw in figure 6-13.

PROJECT 5: CUTTING CONCRETE BLOCK WITH THE MASONRY SAW

OBJECTIVE

- The student will perform some of the more common cuts of concrete blocks required on the job using the masonry saw. The safety practices outlined in the unit must be followed. A wet saw equipped with a diamond blade should be used. If a wet saw is not available, a dry saw can be used.

EQUIPMENT, TOOLS, AND SUPPLIES

1 masonry saw equipped with a diamond or abrasive blade
rubberized clothing and gloves
1 brick hammer
mason's rule
yellow crayon or soapstone, to mark the masonry units
small square

safety goggles or glasses
standard electrical receptacle box
brush, to keep the saw table clean
supply of water close to the saw
supply of 6-inch or 8-inch concrete blocks (to cut) and a few bricks

SUGGESTIONS

- In addition to having an old brush and a supply of water near the saw, also have handy a garden hose with a nozzle to flush the saw tray.
- Long-armed gloves are better than short gloves.
- Make sure that the saw frame is grounded to prevent electrical shock.
- Secure a wooden pallet or boards to stand on.
- Locate the saw close to the stockpile of materials.
- Keep a clean cloth handy to wipe the safety goggles off as needed.
- Do not force the saw when it is cutting; let it cut freely.
- Drain the saw tray when it is very dirty and when sludge has built up. Fill the tray with clean water.

PROCEDURE

1. Fill the saw tray with water.
2. Install a blade on the saw, making sure that it is tightened securely.
3. Check the conveyor cart to make sure it rolls smoothly back and forth.
4. Set a 6-inch block on the saw table.
5. Mark the webs with the soapstone or crayon to make a quarterback.
6. While rolling the saw table (tray) back and forth and pushing down with a foot on the foot pedal, cut the cell out.
7. Chip out the corners of the cut with the mason's hammer if the cut must be neat.
8. Next, cut a halfback with the masonry saw in the same manner.
9. Then cut a fullback with the saw.
10. To become more familiar with the masonry saw, cut a rip block, slab, bond beam block, and a few bricks.
11. Holding the electrical box even with the bottom and in the center of the block, mark the outline of the box on the block with the soapstone or crayon as shown.
12. Cut out the box opening with the saw by making a series of vertical cuts where the box is marked on the block.
13. With the hammer, gently chip out the block so the box fits in the cut section and is flush with the face of the block.
14. Wear safety glasses at all times when cutting.
15. Have the instructor evaluate the work at the completion of the cuts.
16. Wash out and drain the saw when finished.

Unit 7 HISTORY, THEORY, AND FUNCTION OF THE FIREPLACE AND CHIMNEY

OBJECTIVES

After studying this unit the student will be able to

- briefly describe the history of the fireplace and chimney up to modern times.
- define the meaning of draft and explain how it works in the operation of the fireplace.
- list the parts of the two- or three-flue chimney and state how each part contributes to chimney operation.

HISTORY OF THE FIREPLACE

Fireplaces and chimneys have been used for many centuries for heating and cooking purposes. The earliest fireplaces in the United States can be traced back to about the year 1600 in the eastern part of the country.

Early fireplaces were not very efficient because the damper was not yet developed. Without a damper, there was no control of the draft or burning process and much of the heat escaped up the chimney.

After the American Revolution, information was published on fireplaces by several people including Benjamin Franklin. The curved back, common on modern fireplaces, is still called the *Franklin back* after its founder.

Fireplaces of Franklin's time used heavy wood beams to support the masonry work over the firebox opening. After a period of years the beams often weakened or burned out from the constant heat of the fire. The beams had to be carefully checked and repairs made when necessary. Today, iron or steel lintels are used, thus eliminating the problem. Figure 7-1 shows a modern colonial fireplace which uses a steel lintel over the opening.

Fig. 7-1 Stone fireplace built to resemble the old colonial or early American fireplace. The hearth is recessed from the floor level.

MANTELS

Mantelpieces were first built on chimneys between 1725 and 1750 when plastering replaced the use of wood paneling on the other walls of the room. During this period, it was common to make the fireplace masonry facing smaller on the front and to make the mantel quite fancy.

For today's fireplaces, it is common to use a projected masonry mantel or no mantel at all instead of the elaborate wooden and marble mantels. Figure 7-2 shows a modern version of the colonial "ring around" fireplace in which only one stacked course of bricks is seen around the opening of the fireplace.

BASEMENT FIREPLACES

Basement fireplaces, figure 7-3, were rare until about 1750. They were large and also served as a support for part of the home. Pockets were built into the base to store the wood ashes for making soap, potash, and lye. Lime mortar was used because portland cement mortar was not in use at this time. Large stone slabs were laid level with the finished first floor. These slabs served as the inside and outside hearth for the fireplace on the first floor. The first-floor fireplace was then started off this base.

The basement fireplace of today serves a different purpose. In clubrooms, recreation areas, or enlarged living spaces for the family, a basement fireplace is used to provide a pleasant atmosphere as well as heat. Many of today's homes have a fireplace on the basement level as well as the first floor. The additional fireplace increases the value of the home when it is to be sold.

Fig. 7-2 Modern version of the colonial "ring around" fireplace. It is made by stacking one course of bricks up the jambs and around the opening.

Fig. 7-3 Early American basement fireplace built about 1775.

MODERN FIREPLACES

Fireplaces today are basically built the same as years ago, but with many improvements. Some improvements include the use of dampers to control the draft, fireproof fired clay flue linings (terra-cotta), highly improved portland cement mortars, and special firebrick for lining the firebox. In design, the outer hearth on many fireplaces is raised off the floor to provide a bench seat in the room and a higher level of heat radiation from the fireplace. On old fireplaces, the hearth was usually built level with the finished floor line.

The fireplace is growing in popularity today because of the energy situation and the high cost of fuel. Past traditions influence many people to include a fireplace in the construction of their homes.

It is true that a fireplace is not as efficient as a stove or heating device located in the middle of a room. However, if the dampers are regulated properly, it will radiate much heat. In colonial times, the fireplace was used primarily for heating and cooking purposes. In more recent years the fireplace has served more of a decorative use rather than as a functional source of heat. However, since the cost of energy is rising, there is a growing consideration of fireplaces as an additional source of heat.

WORKMANSHIP

It is necessary that a fireplace operate in a safe manner to contain the open fire in the firebox and to carry the smoke and gases to the outside. To achieve this, it must be correctly designed and constructed.

The workmanship involved in building a fireplace has to be of the highest quality. Not only must the fireplace work correctly, it must also show that neat, careful work was used in the laying of the masonry units. As it is usually the focal point of the room, any inferior work is noticeable and reflects poorly on the mason's craft.

Importance Of Draft In The Burning Process Of The Fireplace

Draft is the current of air created by the variation in pressure resulting from differences in weight between the hot gases in the flue and the cooler air outside the chimney. The movement of draft or intensity depends on the height of the chimney and this difference in temperature. Draft is stronger in the winter than in the summer because of this fact. In addition, a fireplace and chimney built in the center of the home rather than on an outside wall provide a better draft because it remains warmer.

High trees near the chimney or the rise of a nearby hill can influence the proper draft and should be considered when constructing a chimney. Trees can cause downdrafts which result in puffs of smoke coming from the firebox into the room. They may also cause the fireplace to smoke and not draw at all. The logical solution is to build the chimney top higher above the roof than is regularly done. Under usual conditions, however, the chimney is built 2 to 3 feet above the top of the ridge of the roof, figure 7-4. Metal extensions to increase the chimney's height should not be used because they rust and burn out in a relatively short period of time. The chimney should be built high enough the first time so that extensions are not necessary.

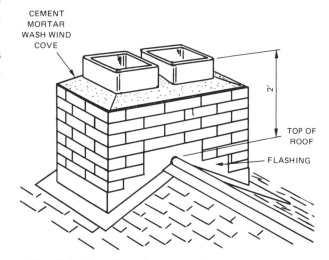

Fig. 7-4 Two-flue chimney built above the roof a minimum of 2 feet to prevent downdrafts in the chimney. Notice the mortar wash wind cove on top of the chimney to deflect the air currents upward.

82 ■ Section 2 Chimneys and Fireplaces

Two common mistakes are often encountered in chimney design. Sometimes the flue size is not large enough, or else the flue is not high enough to produce a good draft. Either one or both of these mistakes will result in a chimney that does not draw properly, forcing smoke into the house.

Draft Operation In The Fireplace

The actual operation of draft in the fireplace is shown in figure 7-5. Cool air is drawn down one side of the flue. The warm air, smoke, and gases return up the other side. When the air coming into the chimney reaches the smoke shelf directly behind the damper, the air is deflected up the chimney on the opposite side of the flue. Smoke and gases are taken up the chimney in the air. Air is sucked from the room into the firebox as a result of this action. As the air enters the firebox it quickly becomes heated and rises through the throat of the damper. It then meets the colder air bouncing off the smoke shelf.

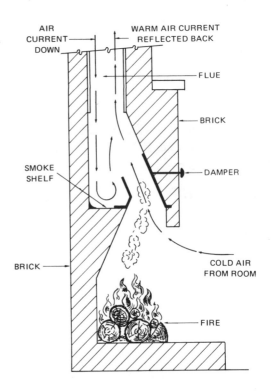

Fig. 7-5 Draft operation in the fireplace. Notice that air comes down one side of the flue, bounces off the smoke shelf, and goes up the other side of the flue. Cold air is sucked in from the room, completing the draft.

NOTE: The *smoke shelf* is a brick ledge laid level with the bottom of the damper. It is built by extending the back firebrick lining forward until it meets the bottom rear flange of the fireplace damper. The mixture of hot and cold air causes the column of air to be drawn rapidly up the chimney. If the fireplace is not built correctly this action and reaction does not occur.

The damper of the fireplace must always be checked to make sure it is completely open before building a fire in the fireplace.

The air drawn from the room is replaced by outside air which enters the room through cracks around the windows, doors, or vents. This is one of the major problems with electrically heated homes; they are insulated so tightly that there is no draft from inside the house. This problem can be cured by opening a window slightly to allow the passage of air.

As the fire burns, the damper is regulated slowly to the point where it does not smoke but still burns freely. The maximum amount of heat possible is radiated into the room, discounting the heat that naturally goes up the chimney. If the damper is left entirely open when the fireplace is in use, a great amount of the heat escapes up the chimney with little benefit to the room. After the fire is completely out, the damper should be closed. This keeps out insects, squirrels, and rain, and protects against the loss of heat. An open damper sucks out a large amount of heat when not in use, increasing the cost of heating the home.

> **When building a fire in a fireplace, keep the fire to the back. Start with a small fire until the flue heats up and then increase the size of the fire.**

Radiation Of Heat From The Fireplace

Heat is generated from the fireplace by the burning of the fire, and by the heat radiating outwards into the room from the hot back and sides of the firebox. In an ordinary fireplace, almost no heating effect is produced by a moving air current, such as a warm air duct in a room. As previously described, the cold air drawn into the room from cracks around windows or doors is lower in temperature and is especially noticeable to someone in the room.

Heat radiation, like light, travels in straight lines. Therefore, unless a person is in the direct line of radiation, little heat is going to be felt. This is why warmth is felt on the front of the body when facing a fire, while the back of the body may remain cold.

As a result, a fireplace cannot cause the room temperature to rise greatly unless the fire is very hot and burns for a number of hours. The heat must completely fill the space in the room.

Tests have shown that about five times the amount of air required for good ventilation can be drawn into the room by the operation of a fireplace. Further tests also showed that a fireplace is only about one-third as efficient as a good circulating-type heater. Nevertheless, fireplaces still provide a source of heat, add to the cheerful atmosphere of the home, and are highly desirable by the homeowner.

It is very important that the above theory and function of fireplace and chimney operation be understood before the mason attempts to build a fireplace.

TWO-(OR MORE) FLUE CHIMNEYS

Solid Masonry

Only masonry materials should be used in the construction of a chimney that serves a fireplace. Hard fired brick is a good choice. If brick is used as the facing material, concrete block or stone can be used inside as the fill unit, except against any area that will be subjected to direct heat. It is recommended to always use a hard fired brick for the roughing in of the firebox and the smoke chamber of the chimney. Flue liners should also be surrounded by at least 4 inches of hard fired brick for fire protection.

Do not build up the outside walls of the chimney and then fill with trash laying around the job. Using leftover mortar particles, scraps of nonmasonry materials, or sand results in an inferior chimney. Scraps of wood should never be used to fill in around the chimney.

Masonry units must be laid in a good portland cement mortar, such as type M or N. All of the masonry units should be bonded together into the main outside walls either by using metal wall ties or by interlocking the units into each other.

Footings

The footings must be of a high-grade concrete as discussed in Unit 2. The footing should also be wider on all sides than the actual chimney, for load distribution. Reinforcement steel can be added to the concrete for extra strength.

If the chimney is part of the outside wall, make sure it is built below the frost line to prevent freezing and thawing. The chimney can be located on one of the exterior walls or in the center of the home. It can be built as a separate part completely removed from the outside structural walls. This allows the architect more freedom in designing the structure. Do not locate the chimney and fireplace too close to a doorway or closet.

Installing Flue Linings

As previously mentioned, there must be a separate flue for every fireplace or heat source such as a furnace. Start the flue lining for furnaces approximately 12 inches below the thimble intake, or where the furnace pipe is cut into the chimney. (The *thimble* is a round tile ring used to put furnace pipe through.) The lowest flue should be supported on at least three sides with corbeled brick projecting from the inside wall of the chimney, and flush with the inside wall of the flue lining. Set in a bed of mortar and cut off flush with the trowel. Wipe the joint with cloth to make sure it is smooth and full of mortar. Corbeling support eliminates the chance of the flue slipping down into the chimney as the construction proceeds.

Set any additional flues in a bed of mortar not more than 3/8 inch. Finish off in the same manner with a smooth joint. If possible, set the flues ahead of the brickwork, level and plumb. This results in a better mortar bed joint.

Every flue should be divided by a 4-inch brick partition wall to prevent air in one flue from leaking over and affecting the draft in the flue next to it. Mortar should not, however, be slushed tightly down between the flue and brick because there must be room for expansion due to heat. Full mortar joints are important when building chimneys or fireplaces. There should be a soot pocket below the flue lining and a cleanout door to remove the soot.

When more than one flue is needed in a chimney, it is often necessary to offset one of the flues for proper fit in the chimney. These offsets or bends should not vary more than 30 degrees from a vertical position to eliminate the chance that the flue might be choked off. The flue lining should be cut neatly on a miter to fit the angle of the slope. This can best be done by first filling the flue with sand, tamped tightly. A sharp chisel and light hammer are used to make the cut along the marked line. After working all the way around the flue, it breaks cleanly. Thimble holes are cut in the same way, except the circle size is drawn on the flue and a point chisel is used to cut a hole in the center. The cut is then made from the center out to the edges of the circle by chipping with the hammer head. Figure 7-6 shows a mason cutting a flue lining that has been packed full of sand.

Clearance Around Wood

No wood should be in contact with the chimney at any time. Leave a 2-inch space between the outside face of the chimney and all woodwork. This rule also applies at the point where the chimney passes through the roof.

Unit 7 History, Theory, and Function of the Fireplace and Chimney ■ 85

Fig. 7-6 The mason is using a chisel to cut a thimble hole in the furnace pipe. The flue is packed tightly with sand to support the cutting without breaking.

Fig. 7-7 A typical 3-flue chimney.

Flashing

The mortar bed joints should be raked out to receive a weatherproof flashing slightly above the roof line, or the flashing can be built into the masonry work if it is available. Installing flashing is usually the job of the carpenter.

Finishing The Chimney Above The Roof

As previously mentioned, the chimney should be built at least 2 feet above the highest point of the roof to prevent downdraft. The chimney is capped off with a portland cement mortar wash coat on an angle to form a wind cove. Use type M mortar because of its strength. To prevent the mortar from drying out too fast, cover it with wet burlap and dampen periodically until it has cured. This prevents cracking that sometimes takes place from sun and wind.

The flue linings should extend at least 4 inches above the top of the finished cove. They are laid level with each other if more than one flue is in the chimney. Figure 7-7 shows a completed 3-flue chimney.

DUMMY FLUES

If the chimney is rather large and only two flues are being used, sometimes a *dummy flue* is added in between the two flues to balance out the top of the chimney. This is done by cutting off a section of flue and building it into the top of the chimney. To prevent moisture or water from laying in the dummy flue, it should be filled to the top with mortar.

The chimney is a column of masonry work. If the work is not plumb and level, it is very obvious to any person viewing it from the ground. Therefore, it is very important that good workmanship practices be followed at all times. Building a chimney tests the skill of the mason.

Section 2 Chimneys and Fireplaces

ACHIEVEMENT REVIEW

Select the best answer from the choices offered to complete each statement. List your choice by letter identification.

1. Fireplaces and chimneys in the United States can be traced back to about the year
    a. 1550.
    b. 1600.
    c. 1776.
    d. 1800.

2. The early fireplaces lost most of their heat up the chimney due to the absence of
    a. a flue lining.
    b. an ash dump.
    c. a wind shelf.
    d. a damper.

3. The main purpose of the fireplace in colonial days was to
    a. provide light for the room.
    b. give warmth and cook food.
    c. provide a cheerful atmosphere.
    d. increase the value of the home.

4. The main reason flue linings are added to chimneys is to
    a. add beauty and design to the chimney top.
    b. provide better draft.
    c. prevent the walls from burning out.
    d. make the chimney stronger.

5. The draft in a chimney is stronger in the winter than in the summer because
    a. the prevailing winds are stronger.
    b. the fireplace is used more.
    c. there is a difference in temperature between the air outside the flue and the air inside the flue.
    d. air currents change in the atmosphere.

6. Air coming down the flue is deflected back up by bouncing off of the
    a. damper.
    b. smoke shelf.
    c. hearth.
    d. mantel.

7. Heat from the fireplace is forced into the room by
    a. radiation.
    b. air currents coming down the flue.
    c. forced air.
    d. the draft inside the room sucking the air.

8. The dividing wall that is built around the flue in a two- or three-flue chimney is made of
    a. stone.
    b. block.
    c. tile.
    d. brick.

9. In order to prevent downdrafts, the chimney top is built above the roof
    a. 12 inches.
    b. 2 feet.
    c. 4 feet.
    d. 6 feet.

10. Offsets or bends in a chimney flue should not vary from vertical more than
    a. 30 degrees.
    b. 40 degrees.
    c. 45 degrees.
    d. 60 degrees.

11. All woodwork surrounding a chimney should be kept back
    a. 1 inch.
    b. 2 inches.
    c. 3 inches.
    d. 4 inches.

12. The top of the flue lining projects above the mortar wash a distance of
    a. 4 inches.
    b. 6 inches.
    c. 8 inches.
    d. 12 inches.

PROJECT 6: TWO-FLUE CHIMNEY

OBJECTIVE

- The student will build a two-flue chimney, the type that has a fireplace and furnace flue. The basic skills of leveling and plumbing can be developed by the student in this project. A column of masonry, such as a chimney, requires good workmanship.

EQUIPMENT, TOOLS, AND SUPPLIES

mortar pan or board
mixing tools
mortar box or wheelbarrow
mason's trowel
brick hammer
plumb rule
chalk box
2-foot framing square
convex sled runner jointer
mason's modular rule
brush

supply of mortar (lime and sand if to be reused)
two 13-inch x 13-inch flue liners (wood plywood form can be used instead of flue liners)
two 9-inch x 13-inch flue liners
standard bricks for project are estimated by counting the number on one course and multiplying it by the number of courses shown on the plan

SUGGESTIONS

- After the first course is laid out according to the plan, set the flue liner in place to make sure it fits correctly.
- All mortar protruding to the inside cutoff should be kept clean to prevent it from building up inside the chimney.
- Use well-filled head and bed joints.
- Plumb and level each course as it is laid.
- Strike the mortar joints as soon as they are thumbprint hard.
- Make sure the first course is laid square since it is the guide for the entire project.

- Lay up one side of the chimney 3 courses high. Rack back and then build the opposite side.
- Check the height of the chimney every other course with a modular rule.
- Keep all scraps and tools out of the way.
- Set the flue liner in place immediately after the first course is laid out on the base. This prevents the chimney from being shoved out of position by dropping in the flue after the work is up.
- Follow good safety practices at all times.

PROCEDURE

1. Mix the mortar.
2. Prepare the work area by stocking the bricks and mortar back away from the project so the mason is not disturbed when working.
3. Lay out the project with the 2-foot framing square and pencil.
4. Dry bond the brick with a 3/8-inch head joint as shown on the plan in figure 7-8.

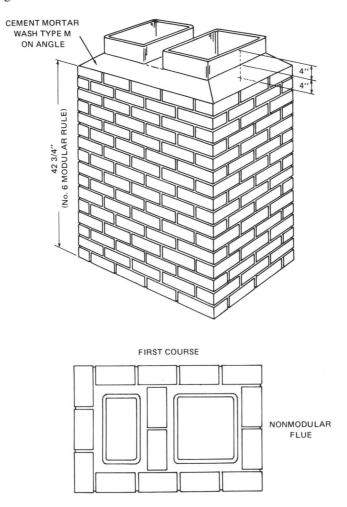

Fig. 7-8 Project 6: Two-flue chimney

5. Lay each course to No. 6 on the modular rule.
6. Lay the first course in mortar.
7. Set the first two flue liners in place as shown on the plan.
8. Lay the 4-inch brick partition between the flues.
9. Build the chimney to the height of the first flue liner.
10. Set the second flue in place with mortar.
11. Continue to build the chimney up to the finished height shown on the plan.
12. Strike the mortar joints with the convex jointer.
13. Apply the mortar wash coat to the top of the chimney on a slope as shown on the plan. The flue liner should project 4 inches above the top of the wash.
14. Brush the project when dry enough not to smear the joints.
15. Recheck the project to be sure it is plumb.
16. Have the instructor inspect the work.

Unit 8 DESIGN AND CONSTRUCTION OF A FIREPLACE

OBJECTIVES

After studying this unit the student will be able to

- list key factors to consider when planning and building a fireplace.
- identify the parts of a fireplace.
- describe how the flue size is determined for different sizes of fireplaces.
- explain the construction techniques for building a fireplace step by step to the point where the flue lining is set.

FACTORS TO CONSIDER WHEN BUILDING A FIREPLACE

Location

When deciding the location of the fireplace, consider that it is a permanent structure. It is not a good idea to build the fireplace near a doorway because it would be in the way of household traffic. The door could also affect the draft pattern of the fireplace. The fireplace is usually the center of attraction in a room. Therefore, thought should be given whether to build it on an outside wall or in the center of the room.

Traditionally, the fireplace was built as part of the outside wall. Now this is not always done because the fireplace and chimney may be a separate unit within the home. They also serve as a column of masonry that strengthens the structure.

The fireplace can be built close to a window if it does not block the light entering the room. Many architects design a window or bookcase on either side of the fireplace using it as a centerpiece. All of these factors are usually considered by the architect when the plans are developed.

Size

The size of the fireplace depends on the size of the room in which it will be located. This is not only for appearance sake, but also to ensure successful operation.

A fireplace that is too small, even if it burns correctly, will not supply enough heat to the room. If the fireplace is too large, a fire that fills the combustion area (firebox) would be too hot for the room. In addition, a larger fireplace requires a larger chimney flue which draws more air through doors and windows to use for combustion.

A room with 300 square feet of floor area is well served by a fireplace with an opening 30 inches to 36 inches wide. For larger rooms, the width of the fireplace opening is increased. The height of the opening is just as important as the width. The height and width must be in correct proportion for the fireplace to function properly. (See the table in figure 8-5, page 96, for fireplace dimensions.)

A fireplace that is poorly designed and incorrectly built has many disadvantages: it is unattractive; it may allow leakage of water into the room; it creates constant problems from soot, dirt, or downdrafts; and it will not draw properly.

Offsetting The Fireplace Into The Room Past The Wall Line

The front of the fireplace can be built either flush with the wall or offset into the room. Most are built flush with the wall because it is more economical. However, a fireplace that extends into the room adds a feel of depth and character of design.

The major problem to consider when offsetting the fireplace into the room is that the flue lining must be laid on more of an angle than usual at the top of the smoke chamber to attain the proper position. Also, the damper handle may be too short and have to be custom-made. Therefore, the fireplace should not be offset more than eight inches into the room; four inches is suggested.

RECOMMENDED METHOD OF BUILDING A FIREPLACE

A fireplace can be built two different ways. One way is to complete the fireplace as the chimney is built. This includes the firebox and all of the facing work.

Most masons, however, use the second method of building a fireplace, in which the chimney is built completely to the top, leaving a rough opening for the later installation of the finished fireplace. Flue linings that will be installed above the firebox for the fireplace are set at the top of the smoke chamber at the prescribed height on brick corbeling. This corbeling must be built to this height, regardless of which method is used.

It is *strongly* recommended to use the second method of building a fireplace. If the finished fireplace is built as the chimney is constructed, mortar, chips of masonry units, and other objects may be dropped down the open flue lining causing an obstruction or possibly breaking the brittle cast iron damper. These objects are very difficult to remove or clean out. Also, if the fireplace is installed as the chimney is built, the face can be smeared or coated with mortar from building the chimney. It is possible, too, that the damper rod in the front of the fireplace may be struck with a fallen object breaking the control. This is expensive to repair and may require tearing out the front of the fireplace.

Another reason for using the second method is that most contractors have the masons do the finish work of building a fireplace on bad or rainy days. This also provides a good incentive for the masons.

LAYING OUT THE FOUNDATION

Once the footing is in place the chimney base is laid out to accommodate the fireplace and any other heating units that are needed, such as a furnace flue. The size of the chimney is determined by the number of fireplaces and flues needed for furnace connections. Usually, no more than two fireplaces and one furnace flue are used in one chimney.

The first step in laying out the foundation is to study the plans and mark with a pencil and square or strike a chalk line on the footing to correspond with the size indicated on the plans. The location of the chimney in relationship to the windows or openings can be determined by checking the plans. The opening for the chimney must not interfere with the layout of windows or doors. The chimney may be shifted without hindering any features of the building.

Fig. 8-1 (A) **Dry bonding bricks for proper mortar head joint**

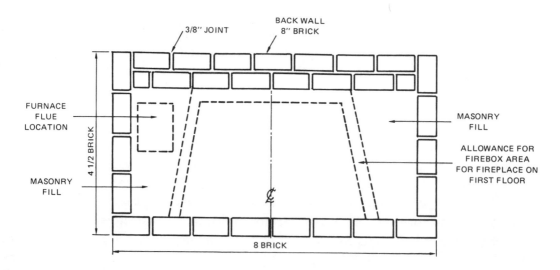

Fig. 8-1 (B) **Plan view of dry bonding the first course layout of a chimney and fireplace. This plan is for a fireplace to be installed on the first floor and shows the foundation layout.**

Brick or concrete block, whichever material is being used in the base, should be dry bonded to work full units without any cutting. If the base is to be concrete block (due to being below grade), be sure to use brick as the layout material to make sure the bond will work when brick is started at the finish grade line. Dry bonding is done by laying out the units on the base, dry (without mortar), and correctly spaced between the units for the mortar joints to determine the bond. See figure 8-1.

The brick or block is then pushed aside from the layout line, and mortar is spread on the base. The first course is laid to work full units. This may require a slight adjustment of

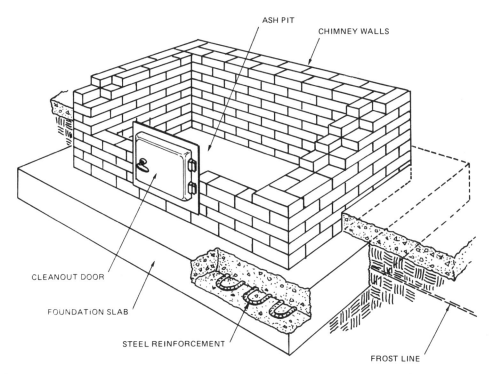

Fig. 8-2 Installation of cleanout door in the chimney base and ash pit

the size of the chimney, to make the bond work. Always make the chimney slightly larger, not smaller, if this is the case.

Type M portland cement mortar is recommended for building the chimney and fireplace (1 part portland cement, 1/4 part lime, 3 parts sand).

Lay the correct number of courses to the height of the cleanout door (usually 4 or 5 courses) for the ash pit. Install the cleanout door in the masonry work as shown in figure 8-2. The foundation of the chimney is then built to the first floor height and any furnace flues, thimbles, or the like are installed, as shown on the plan in figure 8-1.

INSTALLING THE ROUGH HEARTH

In the past, when the rough hearth was constructed, a brick arch was laid on a wooden form beneath the hearth. This arch was called a *trimmer arch*. The concrete or rough hearth brick was then placed over the arch and the arch carried the load or weight of the finished hearth.

In present day construction, brick trimmer arches are seldom used because they require more time to install. Instead, plywood or short lengths of lumber are installed under the hearth area and nailed to wood strips located on the headers that surround the hearth area in the floor. Steel reinforcing rods are then placed over the forms and tied together with thin wire, forming squares. Concrete is placed in the forms and vibrated so that it settles around and under the steel rods. The wood forms can be left in place after the concrete hardens or they may be removed from under the part of the hearth that projects out from the main chimney. As a rule, however, the builder usually leaves the forms in place. Figure 8-3, page 94, shows the method used today to construct a rough hearth for a fireplace.

94 ■ Section 2 Chimneys and Fireplaces

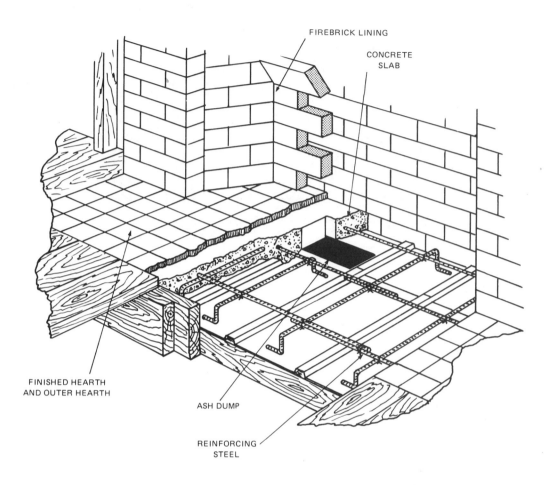

Fig. 8-3 Cutaway view of the installation of a rough hearth. Notice the reinforcing steel in the base of the hearth.

A form is built at the location in the inner hearth where the ash dump is installed. Use short pieces of wood or sheathing that can be knocked out easily after the concrete has hardened. This stops the concrete from coming through the ash dump hole.

Any flue liners coming from a furnace or basement fireplace should be set so they project through the hearth slab before the concrete is placed. The part of the hearth that projects out from the actual chimney is called the *outer hearth* (or forehearth). The inside hearth that the fire is built on is called the *inner hearth*. The rough hearth must be strong if it is to support the weight of the finished hearth without cracking. Figure 8-4 shows an overall section view of a typical fireplace and chimney.

THE FINISHED HEARTH AND ROUGH OPENING

The finished outer hearth can be constructed of quarry tile, brick, stone, or other similar materials. This hearth is built flush with the floor or raised off the floor. In many modern fireplaces the hearth is raised up from the floor in order to elevate the fire. The raised hearth also provides a bench seat around the fireplace. Either method is correct; the method chosen depends upon the design and the preference of the architect or homeowner.

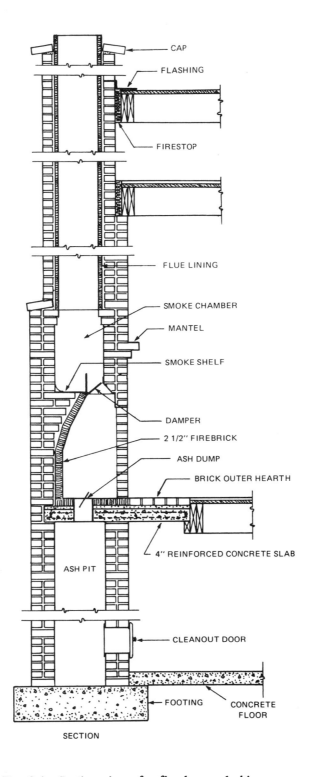

Fig. 8-4 Section view of a fireplace and chimney. Notice the location of the ash dump in the inner hearth.

96 ■ Section 2 Chimneys and Fireplaces

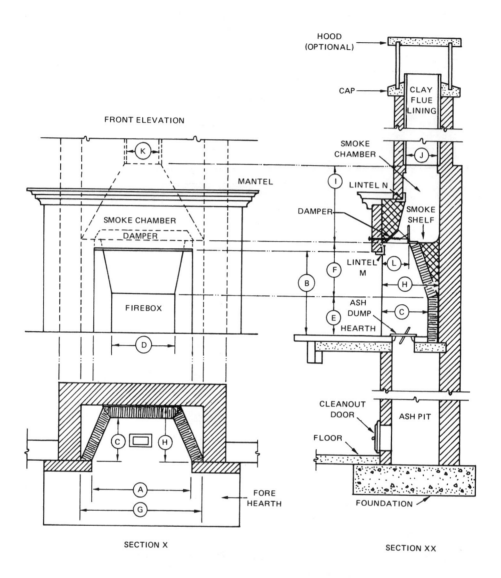

Finished Fireplace Opening						Rough Dimensions			Flue Lining Size		VESTAL MANUFACTURING COMPANY									
									Old Rectangular	New Modular	Damper Model			Damper Model		Damper Model		Ash Dump	Cleanout Door	Lintels	
A	B	C	D	E	F	G	H	I	J x K	J x K	Poker	Rotary	L	Chain	L	Steel	L			M	N
24	24	16	11	14	15	32	20	19	8½ x 13	8 x 12	24	124	10¼	824	9¼	624	9	49	88	36	42
30	29	16	17	14	20	38	20	24	8½ x 13	12 x 12	30	130	10¼	830	9¼	630	9	49	88	42	48
33	29	16	20	14	20	41	20	24	13 x 13	12 x 16	33	133	10¼	833	9¼	633	9	49	108	42	54
36	29	16	23	14	21	44	20	27	13 x 13	12 x 16	36	136	10¼	836	9¼	636	9	49	108	48	54
42	32	16	29	14	23	50	20	32	13 x 13	16 x 16	42	142	10¼	842	9¼	642	9	49	1210	54	60
48	32	18	34	14	23	56	22	37	13 x 18	16 x 16	48	148	10¼	848	9¼	648	9	49	1210	60	66
54	36	20	37	16	26	62	24	44	13 x 18	16 x 20	54	154	10¼					49	1210	66	72
60	38	22	42	16	28	68	26	45	18 x 18	16 x 20	60	160	11¾					49	1212	72	78
72	40	22	54	16	29	80	26	54	20 x 20	20 x 24	72	172	11¾					49	1212	84	90

Fig. 8-5 Illustration and table of fireplace construction principles. The illustration shows how various parts of the fireplace are built.

Before the hearth can be built, the rough opening of the fireplace must be determined. This is done to make sure that there will be enough room to install the finished hearth and firebox. (The rough opening should be built the same time as the chimney.)

Refer to the table in figure 8-5 (Standard Size Fireplace Dimensions in Inches). The finished size is given for different size fireplaces. When building a rough opening for a fireplace that is to be installed later, always allow 8 inches more than the finished width shown on the table. Allow 13 inches more than the opening shown on the table for the damper height from the finished hearth to where the damper will set. This provides enough room to build the finished firebox and set the damper.

For example, a standard fireplace will have finished dimensions of 36 inches x 29 inches. Adding 8 inches in width, the rough opening is 44 inches wide. The height of the rough opening should be 48 inches. An angle iron can be installed at this height to carry the unfinished masonry wall that will be built over the opening or a brick trimmer arch. The arch allows more working room for the mason when the damper is set. Either method can be used successfully. Figure 8-6 shows the rough opening for a standard fireplace.

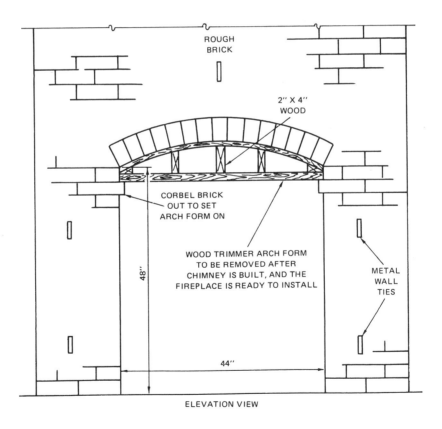

Fig. 8-6 Rough opening measurements for a 36-inch x 29-inch fireplace. A wood trimmer arch and plywood frame are used to lay the arch on. The form is knocked out later after the chimney is completed. The brick arch is better than a metal lintel.

ALLOWANCE IN ROUGH OPENING FOR RAISED HEARTH

If a raised hearth is to be built, the height of the rise must be added to the rough opening to allow for the difference. For example, if the raised hearth is to be 12 inches, then the rough opening must be 60 inches rather than 48 inches (48 + 12 = 60). This is very important to remember when building a raised hearth fireplace or the fireplace will not operate correctly. It is recommended not to build a raised hearth higher than 12 inches; if it is built higher, the fireplace appears out of proportion to the room.

To better understand how the various parts of the fireplace are built, see figure 8-5. Use the table for information as often as necessary throughout the rest of this unit.

LAYING THE HEARTH

The hearth is usually built of firebrick made of special clay that is highly resistant to heat. A standard firebrick measures 4 1/2 inches x 9 inches and is 2 1/2 inches thick. Firebricks are buff or off white in color and can be purchased at any building supply dealer who carries masonry materials. Firebricks in the firebox should be laid in either type M portland cement mortar or fireclay mortar, for best results.

The hearth is always laid out from the center of the fireplace. After determining the width of the opening to be used from the plans, mark off with a pencil the exact overall distance of the inner hearth. Refer to the illustration in figure 8-5.

Lay the outside course of firebrick. Make sure it is slightly longer than is needed, since the masonry jamb that is laid on top of it hides the rough edge. Use small (1/4 inch to 3/8 inch) head and bed joints and strike with a flat jointer, such as a slicker.

Lay the firebrick so they are half over each other. Continue laying the hearth until it is completed. Install the ash dump in the rough opening left when the rough hearth was built. This is usually slightly off the back wall of the firebox and in the center. The ash dump door is the same size as a firebrick and has a projecting flange that holds it in place. Figure 8-7 shows a completed hearth ready for the walls of the firebox to be laid out. Brush the hearth clean at completion and restrike all joints.

Fig. 8-7 A completed firebrick hearth ready for the walls of the firebox to be laid out

FIREPLACE OPENINGS

Correct proportioning of the opening greatly affects the operation of the fireplace. Height and width should be in proper proportion to each other. In general, the wider the opening the greater the depth and height of the firebox. The dimensions in figure 8-5 should be followed as a guide whenever laying out fireplace openings.

If the fireplace is designed mainly to burn hardwood logs, some builders find it desirable to increase the distance from the front to the back of the firebox by 3 to 6 inches. This allows the burning of larger logs without much affect on the operation of the firebox.

WALL THICKNESS

The fireplace walls should never be less than 8 inches thick. If made of stone, the walls should not be less than 12 inches thick. This is to prevent the inside walls of the fireplace from getting too hot and cracking. It is especially important for the walls of a fireplace and chimney built in the center of the home to be thick enough since the back of the fireplace wall often serves as the inside of a closet or projects into a hallway. If the wall only has a 4 inch thickness of brick, it can get very hot and possibly cause a fire. Metal ties should project out from the rough work so the finished fireplace can be tied to them when it is built.

LAYING OUT THE FIREBOX

The *firebox* is the part of the fireplace that contains the fire. It is built of firebrick and can either be laid in a stretcher position (as a brick is usually laid) or shiner position. A firebox is generally laid in the shiner position because it is more economical and still provides sufficient fire protection. The shiner position is accomplished by laying the firebrick on its longest surface (2 1/4 inch bed).

   NOTE: The size of the firebox can be determined, if a plan is not available, by referring to figure 8-5.

The first step in laying out the firebox is to find the center of the hearth. Hold a 2-foot framing square on that point and draw a straight line from that point to the back of the hearth. All measurements for the wall of the firebox are worked out from this line. Figure 8-8 shows how this is done. This example uses a standard 36 inch width opening.

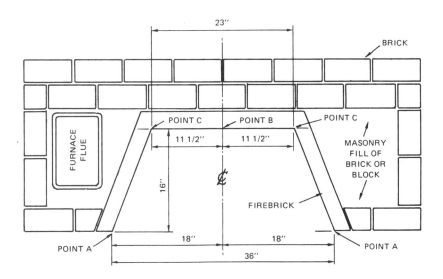

Fig. 8-8 Plan view of a standard 36-inch firebox for a fireplace. All dimensions are given for the layout of the firebox on the hearth. Notice that all measurements are taken from the center of the fireplace. The outside of the brickwork is not tooled, because it will be faced later with facebrick. This is the reason the firebrick extends to the face of the brick wall in the front.

Examining the drawing in figure 8-8, it can be seen that from each side of the centerline in the front, 18 inches is measured and marked with the point of a pencil. The total width, therefore, is 36 inches. See point A in figure 8-8.

Next, the depth of the firebox is marked off on each side of the centerline, from the front to the rear, a distance of 16 inches. See point B on the drawing. Use a pencil line to connect the back wall of the firebox. The 2-foot square can be used as a guide.

Now, measure out from each side of the rear centerline 11 1/2 inches. This totals 23 inches which (referring to the table in figure 8-5) is correct for a 36-inch fireplace opening. See point C in figure 8-8.

With the framing square laid flat on the hearth at point C to point A, draw a fine pencil line. This establishes the splayed (inclined or angled) side of the firebox. Repeat this on the opposite side and the walls of the firebox are laid out.

BUILDING THE WALLS OF THE FIREBOX

One course is laid in mortar, completely around the firebox, to establish the bond. This is then backed up with rough bricks as a filler, allowing approximately 1/2 inch for expansion.

The purpose of the angled sides of the firebox is to make it narrower than the front opening and to create a reflector shape which radiates heat out into the room. The slope of the back wall also helps the smoke and gases to curl upward and reflect the heat out from the firebox.

The angle of the side walls should not exceed 3 or 4 inches per foot of depth. It is a good idea to sprinkle some sand on the hearth after the firebox walls have been laid and tooled to prevent mortar from sticking as the work progresses.

The firebox walls are then built up 3 courses (approximately 14 inches). At this point the back wall is started in either a slope or roll that brings the firebrick up under the entire rear flange of the damper. It is very important to lay the head and bed joints as tight as possible because the fire will burn out large joints in a period of time. Refer back to figure 8-4 to see how the back of the firebox is rolled in for the damper.

Edges of firebricks that are cut to work bond against other firebricks should either be cut with the masonry saw or the hammer and chisel.

> **Eye protection should be worn when cutting any masonry unit.**

The firebricks are then rubbed with another piece of firebrick until the edges are smooth and straight. This is important because poorly fitted bricks allow fire to enter and may crack the firebox. Properly fitted firebricks also indicate good workmanship.

As the unexposed back of the firebox wall is built, it is a good practice to back parge it with mortar to help hold the bricks more securely in place. This also helps keep the wall from burning out too early.

The slope of the back wall, if built correctly, should not need any propping to prevent it from falling in while it is being built. Each course is built all the way around the firebox before the next course is started. This helps to keep all of the work tied together. Any wall ties projecting from the previously built chimney wall should be built into the wall as it is laid up.

Strike the mortar joints with a convex jointer. This type of joint is recommended because it gives a good seal and a pleasant appearance.

THE SMOKE SHELF

The smoke shelf is built by rolling the back wall of firebrick lining forward until it meets the rear flange of the fireplace damper. The depth of the smoke shelf varies depending on the size of the flue lining and the depth of the firebox.

Use a good, hard fired building brick. Lay it in back of the firebox wall in mortar as the back wall is built. As the height of the wall increases, the brick backing is laid in such a way that it becomes a *clipped header* (a header that has a small portion cut off of it). At the top of the shelf it becomes a full header brick. This is actually corbeling the backing behind the firebox wall until the top of the smoke shelf is reached. A small space should be left between the back of the firebrick wall and the backing material to allow room for expansion. The area at the top of the firebox is the hottest spot. This is where the hot gases and smoke concentrate when passing through the damper and up the chimney flue.

NOTE: Regardless of the type of masonry unit used to fill in back of the firebox and form the smoke shelf, it is very important that it be level with the last course of firebrick laid, where the damper rests.

The smoke shelf is one of the more important parts of the fireplace. If it is not installed properly, the fireplace smokes. The smoke shelf turns the wind coming down the flue back up the chimney, carrying the smoke and gases from the fireplace. If the smoke shelf is too high or too low, the smoke eddies around the damper area, resulting in a fireplace that functions incorrectly.

Figure 8-9 shows a cutaway view of the finished firebox with corbeling, smoke shelf, and damper in position on the top of the firebox. Any holes or voids on top of the smoke shelf should be

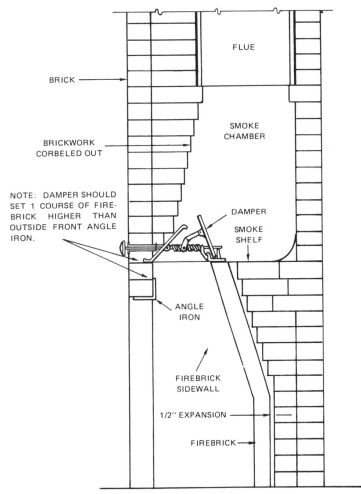

Fig. 8-9 Section view of the finished firebox showing corbeling, smoke shelf, and damper in position

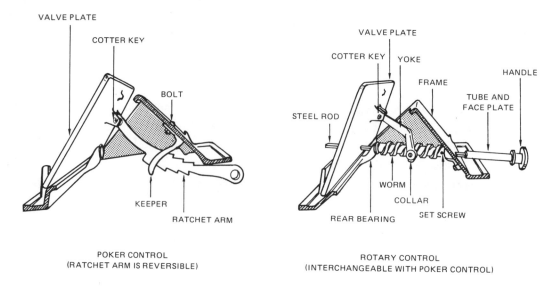

Fig. 8-10 Two types of dampers

filled with mortar. Troweling a coat of mortar over the entire area of the smoke shelf makes it smooth and waterproof. Be sure the top of the smoke shelf is level with the bottom of the damper.

FIREPLACE THROAT AND DAMPER

The throat is the part of the fireplace just above the opening. It connects the firebox with the smoke chamber. Correct shaping of the throat is one of the most critical parts of fireplace design and must be done correctly. A properly designed damper relieves the mason of the task of designing the throat.

Dampers for use in the fireplace have been designed and put on the market by various manufacturing companies. There are two main types of dampers: the poker and rotary control. Figure 8-10 shows the two types.

The poker type is operated by inserting a poker in the hole under the damper, in the ratchet arm, and pushing upwards. This action opens the damper to the desired width.

The rotary control damper is more popular. It is operated by turning a projecting handle which is built into the face of the finished fireplace. This causes the same opening operation as the poker type but a person does not have to work in the hot area of the firebox to open the damper.

The damper allows the draft to be shut off during nonuse periods, conserving other sources of heat or air conditioning. The fire burns best, and the maximum amount of heat enters the room, when the damper is opened just enough so the fire does not smoke but burns freely.

Since the damper is exposed to direct heat from the fire it must be made of a material that withstands a very hot fire over a long period of time. A cast iron damper is recommended. Care should be taken when installing a cast iron damper; if it drops it can break.

INSTALLING THE DAMPER

As a rule, the mason must assemble the damper before it is installed. Be sure that it is put together according to the manufacturer's instructions. All moving parts should work freely before the damper is installed in the fireplace. Once the damper is walled in, it is very difficult to make any corrections.

The worm which operates the lid of the damper (on a rotary damper) sometimes has rough metal edges left from the casting process. It is a good idea to file any imperfections on the worm until they are smooth. This permits smooth working operation of the damper. The poker type has a ratchet arm instead of a worm.

The setscrew holds the rotary worm securely against the handle and prevents it from pulling out the front of the fireplace. It should not be tightened until the damper is in position and the facing of the fireplace is completed. This is usually the last operation before the fireplace is finished.

To set the damper in the fireplace, first center the handle of the damper with the center of the fireplace. The handle should be out far enough so the face of the finished fireplace front wall comes in contact with the metal faceplate on the handle.

Set the damper in place in a mortar bed and tap down lightly to settle firmly. Do not tap too hard because the cast iron edge can break. The damper should be at least one course of firebrick higher than the angle iron lintel which covers the opening on the front of the fireplace. Including the mortar joints this distance is approximately 5 1/2 inches.

With the pointing trowel, smooth out the mortar around the edges of the damper at the point where it sets on the firebox wall. The edge flange of the damper is made so it fits back on the firebox neatly if the firebox is built correctly. Finally, point up the mortar joints on the underside of the damper, fully closing any holes. As a last check, try the damper to see if it opens and closes smoothly before laying any more masonry work.

BUILDING THE SMOKE CHAMBER

The space from the smoke shelf and damper up to the bottom of the flue lining is called the *smoke chamber*. Before any bricks are laid to form the smoke chamber walls, the damper must have a minimum of 1/4-inch clearance between both ends to allow for expansion.

A simple method of providing for this expansion space is to place fire-resistant materials (such as asbestos or fiberglass) at the end of the damper. This prevents the mortar from sealing in solid against the ends of the damper and provides for expansion. If this is not done, fireplaces crack in steps at the point where the lintel covers over the opening on the front of the fireplace.

FORMING THE SMOKE CHAMBER

Good, hard brick should be used to build the smoke chamber, as it will be fireproof. The back wall of the smoke chamber is built straight and parged with a coat of mortar. The distance from the bottom of the smoke shelf to the flue lining is stated for different size fireplaces on the table in figure 8-5. For a standard 36 inch wide fireplace, the distance is 27 inches. After slushing in (with mortar) all flatwork at the damper height, the brickwork is corbeled in from the front which rests on the edge of the damper and both sides. It is corbeled until it narrows and meets the area where the flue sets. Care should be taken that the corbeling does not interfere with the opening of the damper.

This brickwork is laid as full headers which bond and transmit the weight better without falling over on the damper. As the work progresses, it should be tied into the main chimney with metal wall ties.

The slope of the sides and front should be approximately 30 degrees. They should be corbeled uniformly to the center of the chimney. This prevents the draft from drawing from one side or the other and allows the fire to burn more evenly in the firebox.

The corbeling out of the brick courses should not exceed 3/4 inch per course. After all of the corbeling is finished and the flue lining is ready to be set, the underside of the corbeling should be smoothly parged with mortar. This is done so the smoke has an even flow with nothing to obstruct its passage to the flue.

The following is a good method for determining the correct slope to corbel the brickwork for the smoke chamber. Before laying any of the bricks in the smoke chamber area, locate the center of the chimney on the back inside wall. Using a rule, measure up from the

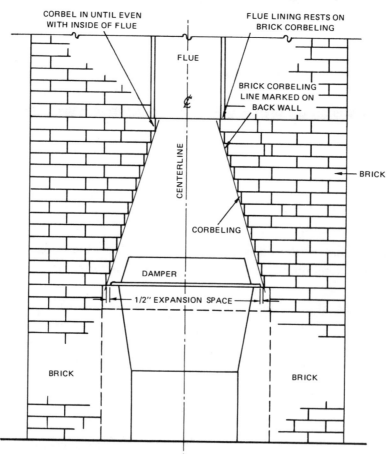

Fig. 8-11 Method of forming the smoke chamber walls. The brick is corbeled to a line drawn on the back wall from the damper to the area where the flue lining sets. Leave 1/2 inch expansion space at each end of the damper.

top of the smoke shelf to the point where the flue lining is to set. Then, mark the width of the flue on the wall. Next, draw a line from that point down to where the corbeling will start on the smoke shelf level. This can be marked clearly on the parged wall. The corbeling is laid out to match the line. This makes the task of dividing the corbeling much easier than random measuring or guessing the incline. Figure 8-11 shows how this is done.

> NOTE: In cases where the chimney has been built first, the wall that rests on the damper is built up and sealed off under the trimmer arch. The wall must be wedged and filled in tightly with brick and mortar. This seals off the smoke chamber.

Fig. 8-12 Completed firebox and rough work of the chimney. Notice the way the brickwork is sealed off under the trimmer arch. The fireplace is ready for facing work.

Figure 8-12 shows the completed firebox and rough work of the chimney around the fireplace area walled up under the trimmer supporting arch. This is an excellent method of finishing off the interior of the fireplace.

DETERMINING THE SIZE OF FLUE LININGS

The size of the flue lining needed for a particular fireplace is determined by using the opening of the firebox as a guide factor. The total size of the opening is found by multiplying the finished open width by the finished height. The result is the total square inches of flue lining.

The rule for a conventional fireplace (one side open only) is that the cross-sectional area of the flue lining is at least 1/12 the total area of the fireplace opening. For multiple-opening fireplaces (with two or more open sides) the cross-sectional area of each open side must be figured and added together. The cross-sectional area of the flue lining should be at least 1/10 the total area of all open sides for this type of fireplace.

The height and width of the opening has a definite function in the operation of a fireplace. If there is a question as to the size flue lining required for a fireplace, or if the determined size is on the borderline, always use the next larger size.

NONMODULAR AND MODULAR SIZE FLUE LININGS

Careful consideration must be taken when determining flue lining sizes because there are two types of dimensioning in use. In the past, flue linings were all of a nonmodular size (not based on the 4-inch modular grid). Today, modular size flue linings are also available from building supply dealers. The modular linings coordinate with dimensions of other masonry units more efficiently.

The table in figure 8-5 gives the old rectangular flue lining size which is still used in many parts of the country. The new modular size is also given. Do not substitute one style flue for the other without first checking the size of the cross-sectional area inside the flue lining. If working from a set of plans, follow the size flue indicated on the plan.

SETTING THE FLUE LINING OVER THE SMOKE CHAMBER

At the proper height, the flue lining should set in the center and against the back wall of the chimney, minus space for expansion. It should rest on the brick corbeling. The inside of the flue lining should be flush with the edge of the corbeling so it does not obstruct the draft.

If the flue lining must be set on an incline or angle, make it a gradual one, no more than 30 degrees from vertical, to prevent it from choking off. The rest of the chimney up to the roof is built as previously described.

BUILDING THE FRONT FACING OF THE FIREPLACE

Workmanship

The facing of the fireplace is very important since it reflects the ability of the mason. It is the part of the fireplace that can be seen, therefore, the work must be of the highest quality and neatness possible.

Dry Bonding

The first course of masonry work is laid off to a line across the front of the fireplace to ensure that it is perfectly straight. If brick is being used for the facing it should be dry

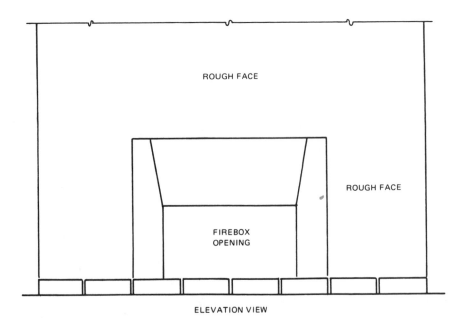

Fig. 8-13 Laying out the first course of facing brick dry to determine the bond when crossing over the opening. This must be done or a half of a brick can be in the center of the opening when built.

bonded across the fireplace from one end to the other. This is done to make sure it works whole bricks when it crosses over the lintel at the top of the opening. This is very important because a half brick can appear in the center of the lintel when crossed over if not bonded properly. If this happens the work must be torn down and rebuilt. Figure 8-13 shows dry bonding of the brickwork across the front of the fireplace.

As a rule, if stone is to be used for the facing, it is not necessary to dry bond. They run different sizes and can be made to work out without problems. Figure 8-14 shows the mason laying out the first course of stone on the fireplace. Notice the wall ties projecting from the rough face of the fireplace. They are used to tie the stonework together with the rough work.

Fig. 8-14 A mason lays out the stone on the face of the fireplace. This does not have to be bonded over the lintel because it is random stonework. It is lined up, however, across the fireplace with a line.

Setting The Lintel Over The Opening

The masonry piers are built up to the height where the lintel sets over the opening. This height can be determined by referring to figure 8-5 for the particular size fireplace being built.

Use a heavy angle iron lintel over the opening. The lintel should be at least 3/8 inch thick, 3 1/2 inches high, and 3 1/2 inches wide for the average fireplace. If the opening of the fireplace is 48 inches or more, a heavier angle iron must be used to prevent sagging. If the angle iron has been cut off with a torch, hammer the metal beads off of the edges of the iron before laying on the piers. This allows the bricks or stone to lay level on the angle iron.

The exterior angle iron should set one course of firebrick lower than where the damper sets. It bears a minimum of 6 inches on each side of the opening for proper support.

Mantels

The design of the mantel, if one is to be used, depends on personal preference. If it is a wooden mantel, the carpenter may supply and install it. In this case, the only thing the mason may have to do is to build wood nailing blocks into the masonry work at the appropriate height, so that the mantel can be nailed or fastened by the carpenter.

A brick or stone mantel is installed by the mason. The standard height from the finished floor to the top of the mantel is 54 inches. The fireplace can drop off at this point and be plastered or paneled, or continued of face brick the rest of the way to the ceiling. Figure 8-15, page 108, shows a finished brick fireplace with a raised hearth and a corbeled mantel.

108 ■ Section 2 Chimneys and Fireplaces

Fig. 8-15 Finished brick fireplace with raised hearth and brick mantel. The fireplace is not cleaned down yet.

Fig. 8-16 A stone fireplace with a raised hearth built in a basement that is to be developed into a recreation room

Raised Hearth

The raised hearth, figure 8-16, should be square with the main fireplace wall and the height of the inside hearth. It can be used as a bench seat around the fireplace.

Strike the mortar joints on top of the raised hearth with a flat tool. The slicker striking tool is recommended. Raised hearths are very popular for several reasons: they give a nice appearance because the firebox is raised to a higher level, heat is radiated at a higher level, and the hearth can be used as a seat.

Washing Down

The final task of building the fireplace is the cleaning of the masonry work. Wash down the fireplace with a solution of 1 part hydrochloric acid (muriatic) to 9 parts clean water. Use a nonmetallic container for the solution.

The fireplace should always be washed down and cleaned before the finished floor of the room is laid because the solution can damage the floor. Washing down is important since it is the finishing touch in building a fireplace.

A CHIMNEY WITH MORE THAN ONE FIREPLACE

Many homes have chimneys with two fireplaces and a furnace flue. In this case, the flues must be slanted on angles to divide them up equally in the chimney and to allow for the second firebox to be built. As mentioned previously, do not slant or lay the flue lining on more than a 30-degree angle or it can choke off. It is also very important to remember than only one flue can be used for each heat source.

Figure 8-17 shows how the flue linings are slanted in a chimney that has two fireplaces. This is done to provide space for the second fireplace and to divide them equally in the section above the second fireplace.

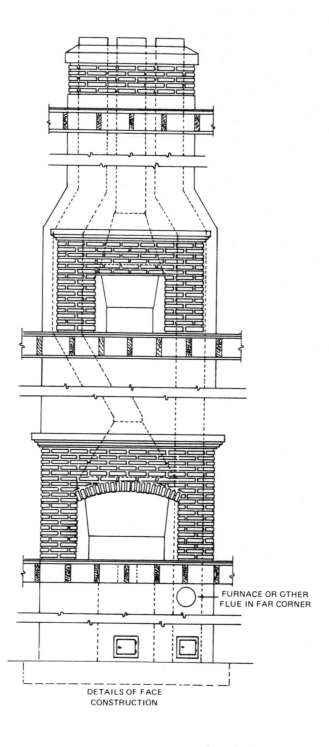

Fig. 8-17 Elevation of a typical chimney that has 2 fireplaces and a furnace flue. The slanted direction of the flue lining is indicated by the broken lines.

Fig. 8-18 Chimney that has been reduced in size by rolling the brickwork

Fig. 8-19 Chimney that has been reduced in size by cutting the chimney on a rake and paving the rake area with brick

Since the flue lining is drawn in, the size of the chimney above the second fireplace can be reduced to save materials. This can be done in several different ways. The two most popular methods are (1) to roll the brickwork in and lay a curved rowlock cap on it, or (2) to cut the chimney back on a rake and add a brick paving to run the water off. The rake can be on one side or both sides depending on the architect's preference or existing windows in the wall. A form of wood can be used for the rake guide or roll pattern and attached to the chimney until it is completed by the mason. Figures 8-18 and 8-19 show a reduced chimney built using the two methods described. This is done when the chimney is being built and the fireplace is installed later.

ESSENTIALS OF A FIREPLACE

If the information and techniques given in this unit are followed carefully, the resulting fireplace will burn properly and safely, and radiate heat into the room. Remember the following rules as a guide when building a fireplace.

- The fireplace should fit the size of the room.
- Each fireplace or source of heat must have its own individual flue to ensure good draft.
- Use type M portland cement mortar for below grade construction and type N for above grade.
- Lay out the size of the fireplace opening from the table in figure 8-5.

- Build the firebox with angled (splayed) sides and a sloping back.
- Provide a large enough smoke shelf, level with the bottom of the damper.
- Make sure that the damper operates smoothly before installing it in the fireplace.
- Position the damper so that it is directly under the center where the flue will be set.
- Slope the sides and front of the smoke chamber evenly. Do not leave pockets or obstructions to interrupt the smooth passage of smoke and gases.
- Select the flue linings on the basis of having 1/12 of the area of the fireplace opening.
- Remember that the larger the opening, the higher it must be.
- Use a good heavy angle iron over the finished opening of the fireplace.
- Use good work practices.

Before building the first fire, let the fireplace cure at least 7 to 10 days. Start the fire slowly and keep it back to the rear of the firebox. The air in the flue must heat up before the fireplace can draw smoothly. Never start a fireplace off with a big roaring fire because it can crack the exterior face of the fireplace or damage the firebox. Build the fire up gradually.

> Be sure that the damper in the fireplace is open before attempting to build a fire in the fireplace.

ACHIEVEMENT REVIEW
Part A

Select the best answer from the choices offered to complete each statement. List your choice by letter identification.

1. One of the major problems in building a fireplace that projects into the room is
    a. supporting the hearth.
    b. the damper rod may not be long enough.
    c. the firebox is more difficult to design and build.
    d. there is not enough working room for the mason.

2. The best all-around mortar to use for building a fireplace and chimney is type
    a. O.
    b. K.
    c. N.
    d. M.

3. If building a chimney on a brick veneer house, check to see if the chimney misses the windows and lines up with the hole cut in the roof. To do this, use a
    a. transit level.
    b. straightedge board.
    c. plumb bob.
    d. plumb rule to mark off the side of the building.

4. Raised hearths should not be built higher than
    a. 8 inches.
    b. 12 inches.
    c. 14 inches.
    d. 16 inches.

5. A standard fireplace that measures 36 inches wide by 29 inches high should have a rough opening to allow laying of the firebrick and damper of
   a. 44 inches by 48 inches.
   b. 48 inches by 52 inches.
   c. 52 inches by 60 inches.
   d. 54 inches by 62 inches.

6. The mortar joints on the top of the hearth should be tooled with a
   a. V jointer.
   b. convex jointer.
   c. slicker jointer.
   d. rake-out jointer.

7. The finished fireplace should be laid out from
   a. the outside wall.
   b. the edge of the chimney.
   c. the center of the chimney.
   d. the inside of the outside wall.

8. The sides of the firebox are built on an angle
   a. to conserve firebrick.
   b. for design and beauty.
   c. to reflect the heat outside.
   d. to enable the smoke and gases to draw up the flue.

9. The roll or slant in the back of the firebox should be started on top of the
   a. second course.
   b. third course.
   c. fourth course.
   d. fifth course.

10. The part of the fireplace that deflects the air coming down the chimney back up the other side of the flue is called the
    a. damper.
    b. firebox.
    c. smoke shelf.
    d. hearth.

11. Another name for the damper in a fireplace is the
    a. smoke chamber.
    b. smoke shelf.
    c. firebox.
    d. throat.

12. The area of the fireplace from the point where the damper sets up to where the flue lining is located is called the
    a. smoke chamber.
    b. hearth.
    c. smoke shelf.
    d. firebox.

13. Applying the formula discussed in the unit stating that the cross-sectional area of the flue lining is at least 1/12 the total area of the fireplace opening, which of the following modular size flues is needed for a fireplace with an opening of 36 inches x 29 inches?
    a. 8 inches x 12 inches
    b. 12 inches x 12 inches
    c. 12 inches x 16 inches
    d. 16 inches x 16 inches

14. Using the table of standard size fireplace dimensions in inches given in the unit, which of the following measurements is correct for the height from the damper to where the flue sets in a fireplace with a finished opening of 36 inches x 29 inches?
    a. 20 inches
    b. 24 inches
    c. 27 inches
    d. 30 inches

15. The standard height of the top of a mantel from the finished floor is
    a. 48 inches.
    b. 54 inches.
    c. 60 inches.
    d. 62 inches.

Part B

Identify the parts of the chimney and fireplace that are lettered in figure 8-20.

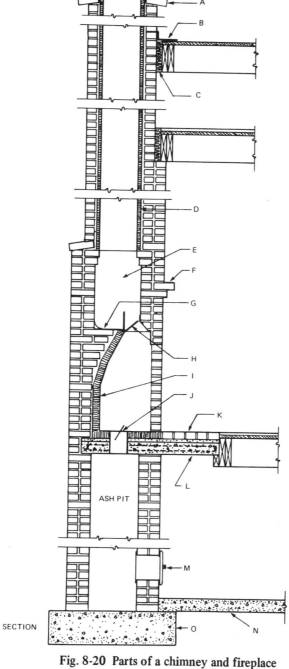

Fig. 8-20 Parts of a chimney and fireplace

PROJECT 7: BUILDING A BRICK FIREPLACE AND CHIMNEY

OBJECTIVE

- The student will lay out and build a brick fireplace and the section of the chimney directly over the fireplace area. This project is typical of a fireplace open on one side. It is the type found in most homes.

NOTE: Two students should work as a team on this project.

EQUIPMENT, TOOLS, AND SUPPLIES

mortar mixing tools and box
basic set of mason's tools
chalk box
2-foot framing square
mortar pan or board
mortar stand
supply of mortar materials
water
brick to be estimated from the plan by student: supply of good face brick for front of fireplace; supply of common brick for unexposed work inside the fireplace and chimney area
80 standard firebricks

ash dump door
one 48-inch angle iron (3 1/2 inches x 3 1/2 inches)
one 36-inch rotary control damper (poker type can be used if rotary is not available)
small amount of insulation to be used at end of damper for expansion
one 13-inch x 13-inch nonmodular flue lining (modular flue can be used if nonmodular is not available)
supply of bricks or 4-inch x 4-inch quarry tile for outer hearth. Estimate from the plans, figures 8-21 through 8-23.

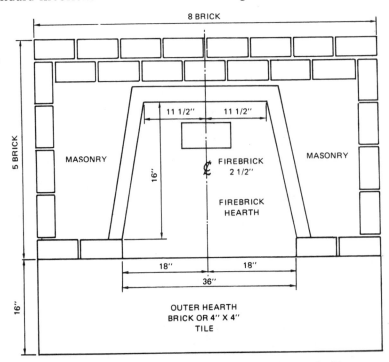

Fig. 8-21 Project 7: Building a brick fireplace and chimney (plan view)

Unit 8 Design and Construction of a Fireplace ■ 115

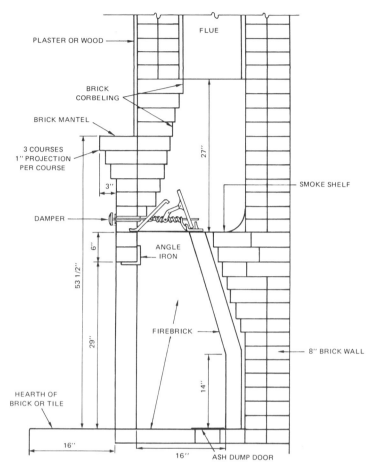

Fig. 8-22 Project 7: Building a brick fireplace and chimney (section view). Notice: Once the height for the flue is set, no purpose is served by building higher. This is why a broken line is shown.

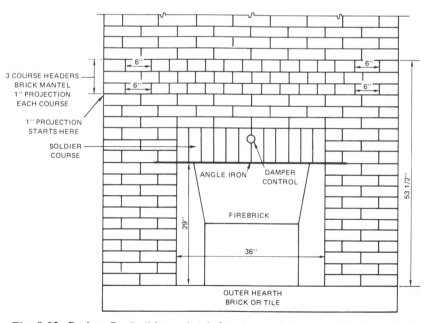

Fig. 8-23 Project 7: Building a brick fireplace and chimney (elevation view)

SUGGESTIONS

- When laying out the first course for the fireplace, be sure to use 2 or 3 different bricks of various lengths if there is going to be a variation in the bricks being used for the fireplace.
- Lime and sand mortar can be used instead of real mortar because it can be reused.
- Form all mortar joints fully since a fireplace is subject to intense heat.
- Keep all of the mortar joints of the firebrick small because heat leaks through a large mortar joint.
- Wear safety glasses whenever cutting any of the masonry units especially the firebricks since they are hard and brittle.
- Work all measurements from a centerline.
- After cutting any firebrick, rub the cut edge until it fits neatly against the brick previously laid.
- Do not mix more mortar than can be used in one hour.
- Strike the mortar joints as soon as they set enough so as not to cause burn marks from the steel striking tool.
- Practice good workmanship because the fireplace is the focal point of a room and should be of the best quality.
- It is not necessary to fill in on the sides of the firebox for a training project since all of this material must be reclaimed.

PROCEDURE

Refer to figures 8-21 through 8-23.

1. Mix the mortar to the stiffness desired.
2. Lay out the first course dry completely around the project according to the plans.
3. Lay the first course in mortar as shown.
4. Lay up the outside walls 3 courses high leaving room for the opening. Lay all standard courses of brick to No. 6 on the modular rule.
5. Working from the center of the fireplace, lay out the hearth to the correct size and height shown on the plans.
6. Install the ash dump in the hearth as shown.
7. Strike the hearth with a flat jointer (slicker).
8. Working from the centerline of the hearth, lay out the firebox walls as shown. Use the framing square to mark off the lines; be sure lines are true.
9. Lay out the first course of the firebox as shown on the plans.
10. Tool the mortar joints on the side and back walls of the firebox with a concave jointer.

11. Sprinkle a light covering of sand on the bottom of the firebox to prevent mortar from sticking.

12. Build the outside walls of the fireplace 6 more courses of bricks high and fill in on the inside with masonry.

13. Lay up the firebrick in the firebox until it is 3 courses high.

14. Continue building the firebox slanting or rolling the back to the throat size of the damper being used. See the plans.

15. At the completion of the firebox, parge the unexposed back wall with mortar.

16. Back up brick behind the firebox walls, being sure to leave 1/2 inch for expansion.

17. Level off the smoke shelf at the back of the fireplace with mortar where the damper will set. Parge a slight cove of mortar at the point where the back of the smoke shelf meets the back wall. See figure 8-22.

18. Lay up the outside masonry jambs in the front of the fireplace until they are 29 inches high. This is 11 courses of standard brick. See the plans.

19. Set the angle iron on the piers dividing the bearing up on each side of the opening.

20. Assemble and set the damper in mortar on the top of the firebox so that the flange rests securely on the firebox walls. Lay some insulation for expansion at each end of the damper.

21. Build up the outside of the fireplace walls to the point where the corbeling starts for the smoke chamber. At the same time, build in the handle of the rotary control damper flush with the outside wall. Do not mortar in solid the turning handle rod.

22. Build up the back and side walls of the fireplace and chimney racking away from the front of the fireplace to scaffold height.

23. Use a pencil to mark off on the parged back wall the correct slope of the corbeling for the smoke chamber. (This procedure was described in the unit.)

24. Lay the corbeling course as shown on the plans until it reaches 27 inches above the damper. This is where the flue lining sets on the corbeling. Plaster the projecting edges of the corbeling until smooth, as shown on the plan.

25. Continue building the front of the fireplace up to the bottom of the mantel height as shown.

26. Corbel out the mantel course as shown in figure 8-22, being careful to note how each course starts as related to the bond. Project each course out 1 inch. (Be careful not to let the projected headers droop down to the front as this will be very noticeable.)

27. Finish the front of the fireplace, tooling the mortar joints with either a convex or V jointer. Brush all work when it is sufficiently dried so that it does not smear.

28. Set the flue liner in place on the corbeling, and brick up around it as shown on the plans.

29. The smoke chamber should now be completely sealed off up to the damper. There is no point in building the chimney and fireplace any higher. (This depends on the preference of the individual instructor.)

30. Clean the loose sand from the inside hearth, and tighten the setscrew in the damper rod so it cannot pull out. Remove any loose mortar from around the damper or on the smoke shelf by reaching up and inside the damper.

31. Lay the outside hearth to the measurement shown on the plan, either of bricks laid flat or hearth tile. (Check with the instructor on this.)

32. Recheck the front of the fireplace for any holes, brush off all areas, and have the instructor evaluate the project.

Unit 9 MULTIPLE-OPENING AND HEAT-CIRCULATING FIREPLACES

OBJECTIVES

After studying this unit the student will be able to
- list the different types of multiple-opening fireplaces and state why a special damper improves performance.
- explain why the metal heat-circulating fireplace was developed and list some of its advantages over the conventional type.
- describe briefly the steps in constructing a metal heat-circulating fireplace.

The conventional fireplace is open on only one side. It is the most popular type of fireplace because it is the simplest to build. When fireplaces are open on more than one side they are called *multiple-opening fireplaces,* figure 9-1.

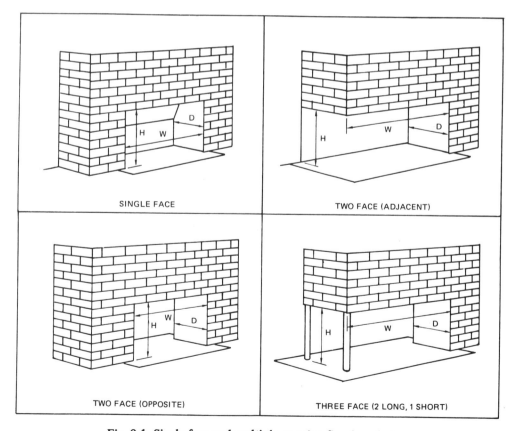

Fig. 9-1 Single face and multiple-opening fireplace designs

119

Although the multiple-opening fireplace (also called multiple face) may seem to be a modern design, it is actually quite old in origin. For example, the corner fireplace which has two adjacent sides open has been in use for several hundred years in the Scandinavian countries.

Another example is a fireplace in which the two opposite faces are open. It is used sometimes as a room divider or when a fireplace is desired on both sides of two rooms, using only one chimney.

Fireplaces with three or four sides open are usually designed by the architect to be the focal point of the home. When they are located in the center of the room with three or four sides open, they are known as *island fireplaces.* Although they are pleasing to look at, multiple-opening fireplaces present more problems concerning draft and operation than a conventional fireplace opened on one side.

Adequate draft must be obtained by using oversized flues and controlled face opening sizes. Cross drafts can go through the fireplace when front and rear doors in the home are opened at the same time. This can cause the smoke to blow out of the fireplace into the room. Glass fire screens on one or more of the face openings help to prevent this.

DAMPERS FOR MULTIPLE-OPENING FIREPLACES

Regular dampers have always been used for multiple-opening fireplaces. Sometimes two dampers are laid back to back with a masonry partition wall between them. This is the case when a fireplace opening fronts in two different rooms but uses the same chimney. Each fireplace requires a separate flue.

Special dampers are now on the market for use in multiple-opening fireplaces that function more efficiently than the standard throat damper. A regular damper has a long, narrow throat designed to make the fire burn evenly along the back wall of the fireplace with only one side open. Fireplaces with more than one open side require a damper shaped to promote even burning of the fire anywhere on the hearth. A high funnel shape and larger throat area help solve this problem.

The conventional fireplace discussed in unit 8 is closed on three sides and has a fixed area of opening. Fireplaces with more than one side open, therefore, have a larger area of opening, and the size of the flue and the area of the damper throat must also be bigger. A fireplace with more than one side open is affected more by any crosscurrents of air in the room. The multiple-opening fireplace must have a bigger flue and stronger drafts to overcome this problem. An example of a special damper for muliple-opening fireplaces is shown in figure 9-2.

ADVANTAGES OF THE MULTIPLE-OPENING DAMPER

The most important advantages of the multiple-opening damper is that it permits a choice of flue locations as

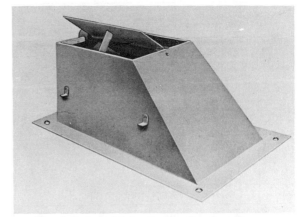

Fig. 9-2 A multiple-opening fireplace damper

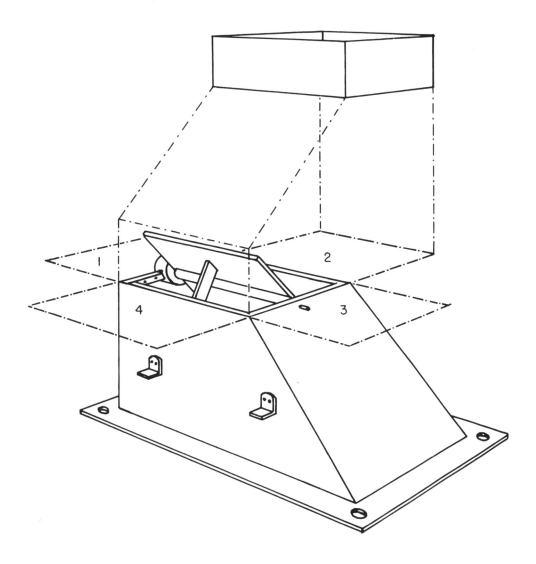

Fig. 9-3 An advantage of the multiple-opening damper is that it allows a choice of 4 locations for the flue lining.

shown in figure 9-3. When building a multiple-opening fireplace, often the flue cannot be located directly above the fireplace, so the flue lining must be slanted or angled. This obstructs the draft if it is not done correctly. Figure 9-3 shows how the smoke shelf can be built on any one of four sides when using the special multiple-opening damper. Reversing the damper allows four positions depending on the best location for the flue. It is important that all masonry be kept at least one-half inch away from all metal to allow for expansion and contraction.

Glass wool is supplied by the manufacturer with each damper. The glass wool insulation is placed in 1-inch thick pads along the four corners. This prevents the masonry work from coming in contact with the metal. It also keeps the proper expansion space. A thin mixture of mortar brushed on the steel helps to hold the glass wool in place until walled in by the mason.

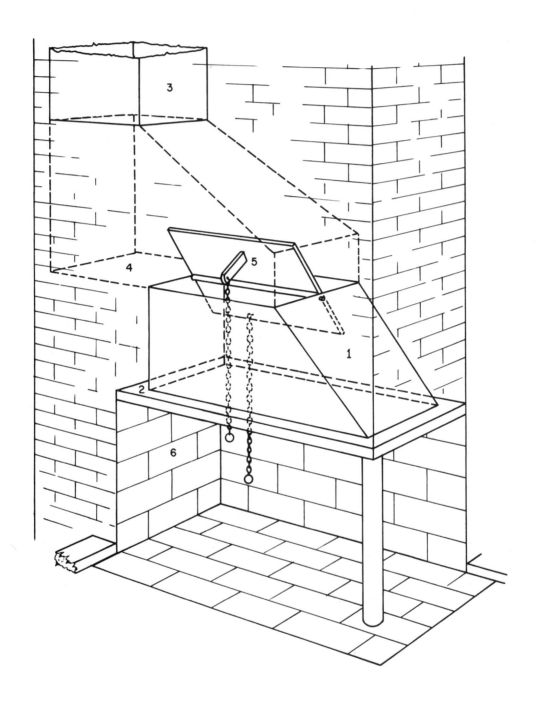

Fig. 9-4 Construction advantages of a multiple-opening fireplace. Numbers correspond to advantages listed on page 123.

Other construction advantages are shown in figure 9-4, and are explained below. (The numbers correspond to those on the figure.)

1. The damper is designed with a smooth, high funnel throat shape to offer as little obstruction to draft as possible.
2. It is designed to permit rapid laying of masonry work. All four sides have a strong lintel that can carry the load. Corner posts can be obtained to carry the outside corner of the damper when the fireplace is open on two sides.
3. The damper is designed to permit the use of different size flue tile depending on the area opening of the fireplace.
4. It allows for easy construction of the smoke shelf. If a conventional damper (as described in unit 8) is used in fireplaces with more than one opening, it presents a problem in forming the smoke shelf.
5. The removable valve plate on the damper has adjustable tension to permit draft adjustments. Once the adjustments are made it is rarely necessary to change.
6. A durable chain controls the damper opening. The multiple-opening damper is also good for fireplaces with only one side open, but with an opening that is usually high or deep.

DETERMINING THE FLUE SIZE FOR MULTIPLE-OPENING FIREPLACES

Fireplaces with more than one face or opening are not treated the same as those having only one face when determining the correct flue size. Flue size is obtained by adding together the sums of the square-inch areas of each open face of the fireplace. The cross-sectional area of the flue lining should be at least 1/10 the total area of all open sides for this type of fireplace.

Remember that any fireplace with more than one open side is much more difficult to build than a conventional one-side open fireplace. This is due to draft problems.

Multiple-opening fireplace construction differs mainly in the damper arrangement. The special multiple-opening damper provides more flexibility in constructing fireplaces that have more than one opening. Rules of construction for single-face fireplaces, with the exception of the special damper, also apply to the multiple-opening fireplaces.

DEVELOPMENT OF METAL HEAT-CIRCULATING FIREPLACES

Metal heat-circulating fireplaces have been in use for about 40 years. They were developed to answer the need for a fireplace that was more efficient in heat output. It was felt that the advancement of furnaces and heating plants would outdate the regular fireplace and possibly replace it in the home.

The problems of smoking, loss of heat up the chimney, and a limited amount of radiation from the firebrick caused designers to look for a better method of building a fireplace. The metal heat-circulating fireplace was developed as a solution to the problem.

The unit is designed to eliminate smoking and greatly reduce the loss of heat up the chimney. Two of the main causes of fireplaces not working properly are poor workmanship and not following the correct principles of construction. The chimney must be high enough for a good draft and the flue must be the correct size for the fireplace opening. Key parts such as the firebox, damper, smoke shelf, and smoke chamber also have to be built correctly or the fireplace does not function properly. The metal heat-circulating fireplace unit is scientifically correct in its design and construction. When delivered to the job it relieves the mason of building many of the key parts of the fireplace. Once the hearth is laid and the steel form set on the hearth, all the mason has to do is follow the instructions with the unit and continue to build the chimney. A mason is more likely to build a successful fireplace by using the metal unit.

CIRCULATION OF HEAT

Circulation is the most important principle in a metal formed fireplace unit. A metal formed circulation heating unit gives off more heat than a firebrick wall due to higher reflection. The walls of the metal unit are hollow with enough space between them for the passage of cool air. This cool air is taken from the floor level through a vent built into the masonry work, and drawn into this hollow space. The heating chamber extends all of the way around the sides and the back of the unit. Fire in the firebox area heats the steel walls. The air drawn in is heated and quickly rises. As the air expands and rises it is forced out of the heat chamber through vents built into the face of the fireplace. These vents can be built below or above the mantel height or ducted into any adjoining room to convey the heat where it is needed.

THICKNESS OF SIDES OF HEAT CHAMBER

In order for the sides and back of the metal formed unit to withstand the intense heat, it is designed much the same as a furnace wall. The sides are usually 3/16-inch steel plate. Outer walls that are next to the masonry work are not as thick because they do not have to withstand as much heat. The outer walls merely help to form the passageway for the flow of air and also serve as a guide for the exterior masonry to follow. The lighter weight steel also makes the price lower per unit than if made completely of heavy steel.

ECONOMY

Considering fuel shortages and high prices, the heat-circulating fireplace unit is desirable because it provides an emergency source of heat and helps lower the overall price of heating a home. That feature alone makes this type of fireplace attractive to the homeowner. In the spring and fall months when it is not necessary to fully heat the home, this type of unit can be used to keep the home comfortable.

It also offers long-range savings to the homeowner over a period of years. The output of heat is greater than from a standard fireplace, due to the movement of air through the heat chambers. A fan can be installed inside the venting area near the heat chamber to draw the air through the vent faster. When the sole source of heat is the fireplace (as in a basement fireplace, or summer lodge), a heat-circulating unit is much more efficient.

The metal formed unit can also be constructed faster, and requires less materials. The parts which require the most time for the mason to build are prebuilt at the factory. This faster construction results in a savings in labor for the builder or homeowner.

Unit 9 Multiple-Opening and Heat-Circulating Fireplaces ■ 125

The design and outward appearance is not limited since the unit is enclosed within the masonry work. Vents are finished off in such a way that they do not look offensive nor do they mar the finished job. Because of the strength of the steel and the fact that no welded joints are used across the back or corners, there is little possibility of cracking.

CONSTRUCTION DETAILS

Planning

The size unit to be used is determined by the size of the room, amount of heat desired, and location of the unit in the room. The size selected then determines the size of the chimney, foundation, and related details of construction. It is necessary to decide on the size of the unit before building any part of the fireplace or chimney, since it must fit properly into the chimney. As a rule, on new construction the architect specifies the size of the unit. Heat-circulating fireplace forms, figure 9-5, are available from building supply dealers.

Flues

The same rule applies to heat-circulating fireplaces that applies to regular fireplaces when determining the size flue lining to be used. Each unit must have a separate flue for proper draft. Any other flues built into the chimney for furnaces or heating plants are built the same way as for a regular fireplace and chimney; the only difference is that they must not interfere with the cold and hot air venting of the metal formed unit.

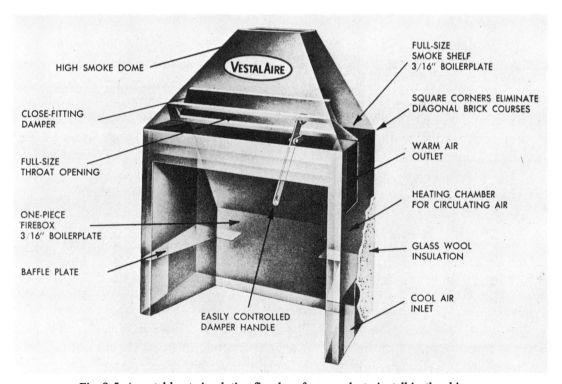

Fig. 9-5 A metal heat-circulating fireplace form ready to install in the chimney

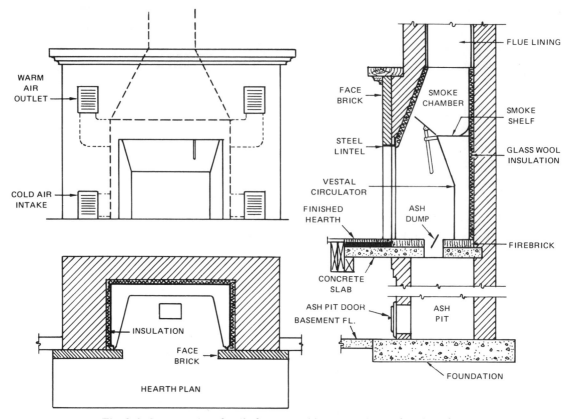

Fig. 9-6 Construction details for a metal heat-circulating fireplace form

GRILLES AND DUCTS

The location of the grilles for cold and hot air are determined by the location of the rough openings in the metal form. Air passages should be finished smoothly to help speed air and to prevent the loss of heat. The ducts can be made of masonry formed as a chamber to where the air enters the metal form. Ducts can also be metal formed if there is a special arrangement. For best results, sharp turns and corners should be avoided and the duct arrangement should be kept as simple as possible. See figure 9-6 for the construction detail of a grille and duct system.

INSTALLATION

Hearth

The hearth for a metal unit must be laid level in all directions. The fireplace form is centered on the hearth, even with the edge of the front course of firebrick. Mortar should not be used to level the unit because it cracks out from the heat.

Insulation

Insulation, usually glass wool, should be applied to the metal back and sides with a thin coating of mortar brushed on the unit. The cold and hot air ducts should not be covered with insulation. As the masonry work is built, it must not touch the metal at any place because of the expansion of the metal. Insulation is supplied with the unit.

Laying Out The Opening

Lay the first course in the front of the fireplace, making the front opening 1 inch to 2 inches less than the actual width of the steel opening. This allows the edges of the metal jamb to be hidden from view. Do not fill in any mortar at the point where the metal jamb and masonry work meet, since it can crack out.

Cold And Hot Air Ducts

The cold air ducts must be built in such a manner that no air passes directly from the cold air to the hot air ducts. The plans supplied with the unit show exactly how this is done. Air must circulate freely throughout the metal walls and out the hot air duct.

Electric fans that are made by the manufacturer of the unit can be installed in the cold air duct attached to the grille covers. They move the air more efficiently through the unit. All wiring or electrical connections must be built into the masonry away from the heat. Never locate the fan in the hot air duct because it will burn up.

Setting The Lintel Over The Firebox

The lintel should be set low enough to conceal the metal apron on the unit. This is considered when giving the heights in the installation instructions for the unit. Pack insulation behind and at the ends of the lintel to allow for expansion. The lintel is not a part of the unit so it will have to be supplied by the contractor.

Setting The Flue Lining

The corbeling of the brickwork in the smoke chamber is done in the same manner as for a conventional fireplace. The most important point to remember, however, is that the flue lining must set on the brick corbeling and not on the metal unit. This is very important, otherwise the weight of the flue will rest on the metal fireplace form.

The connection between the flue lining and metal form is made tight by packing insulation around the flue and the top of the metal form. The chimney is then finished in the same way as for a regular fireplace.

IMPORTANCE OF FOLLOWING THE MANUFACTURER'S INSTRUCTIONS

Since several different manufacturers make metal fireplace units, the installation may vary. A complete set of plans accompanies all metal fireplace units. Follow the plans exactly for best results.

The finished fireplace looks like a conventional fireplace, except for the grilles and the metal firebox. Figure 9-7 shows the completed metal heat-circulating fireplace.

Fig. 9-7 Finished metal heat-circulating fireplace in a home. Notice the grilles in the front of the fireplace for cold and hot air ducts.

ACHIEVEMENT REVIEW

Select the best answer from the choices offered to complete each statement. List your choice by letter identification.

1. The major problem with a fireplace that is open on two opposite sides and has fronts in two different rooms is that
   a. it is very difficult to build.
   b. crosscurrents of air cause it to smoke.
   c. the cost of building the fireplace is very high.
   d. both sides of the fireplace cannot be used at the same time.

2. The most important advantage of the multiple-opening damper is that it
   a. is made of heavier metal and lasts longer.
   b. is faster to install than a damper in a regular fireplace.
   c. permits a choice of flue locations in the chimney.
   d. is insulated to prevent cracking of the masonry work on the front of the fireplace.

3. The insulation can be attached to the damper during installation by brushing the metal with
   a. a light coating of tar compound.
   b. epoxy cement glue.
   c. a thin coat of mortar.
   d. contact cement adhesive.

4. When determining the correct size flue lining to use in a multiple-opening fireplace, the net inside sectional area of the flue lining and the total area of all open sides added together should be in the proportion of at least
   a. one to ten.
   b. one to twelve.
   c. one to fourteen.
   d. one to sixteen.

5. The most important advantage of the metal heat form fireplace is that
   a. the metal form outlasts the masonry fireplace.
   b. it is less expensive to buy and install in the masonry work.
   c. it is more efficient because the heat is circulated through the unit.
   d. it is easier for the mason to build because many of the key parts are prebuilt at the factory.

6. Metal was selected for the construction of the firebox mainly because it
   a. becomes hotter and reflects the heat better.
   b. is fireproof and lasts longer.
   c. is lighter in weight and is easier to handle when installed by the mason.
   d. can be built larger than a regular firebox and contains a larger fire more safely.

7. The metal firebox is completely insulated with
   a. asbestos.
   b. an air space.
   c. firebrick.
   d. glass wool.

8. The flue lining that is set above the smoke chamber over the metal fireplace form should rest on
   a. the top of the metal form.
   b. brick corbeling.
   c. a concrete shelf.
   d. a steel lintel attached to the metal form.

9. For best results, the cold air intake should be located at
   a. the middle of the unit.
   b. the bottom of the unit.
   c. the top of the unit.
   d. about 3/4 of the way up the face of the unit.

10. If other flues are built into the area of the fireplace near the metal heat-circulating unit, they should be separated by at least
    a. 4 inches of masonry.
    b. 8 inches of masonry.
    c. 12 inches of masonry.
    d. 14 inches of masonry.

SUMMARY – SECTION 2

- Early fireplaces were not very efficient because there was no damper to control the draft.
- Terra-cotta flue lining, highly improved portland cement base mortars, and firebrick have made the modern fireplace and chimney more durable and fireproof.
- Successful fireplaces operate on the principle of air coming down one side of the flue lining, striking the smoke shelf, and being deflected back up the other side of the flue lining. During this process air must also be sucked in from the room to create this draft.
- Chimney tops must be 2 to 3 feet above the top of the ridge of the roof to eliminate downdrafts.
- If there is more than one flue lining in a chimney, they should be separated by at least a four-inch brick wall.
- There must be a separate flue lining for each heat source, such as a fireplace and furnace.
- A space of at least 2 inches should be allowed between all woodwork and the chimney, for fire protection.
- The best method to use when building a fireplace and chimney is to build the chimney first, forming only the rough opening of the fireplace. The firebox and facing are built later.
- The base of the chimney should be dry bonded with brick so it works whole brick above finished grade.
- Type M portland cement mortar is recommended for building chimneys and fireplaces below grade, and type N above grade.
- Hearths can either be flush with the finished floor, or raised to provide a higher level of radiation and a bench seat near the fire.

- Fireplaces and chimneys should always be laid out from the center of the chimney.
- The angled (splayed) sides of the firebox and the slanted or rolled back help to deflect the heat into the room and make the smoke and gases rise up more smoothly into the smoke chamber.
- The smoke shelf should be level with the bottom of the damper.
- For the fireplace to radiate the most heat into the room, the damper should be regulated so that it does not smoke, but burns freely.
- The damper should be closed when the fire is out and the fireplace is not in use, to conserve heat in the house.
- The walls of the smoke chamber should be built on an incline from the damper to the point where the flue sets. This is done by corbeling the brickwork and plastering the underside smooth.
- When selecting a flue lining for a fireplace open on one side, the inside cross-sectional area of the flue lining must be at least 1/12 the total area of the finished firebox opening.
- Flue linings should always be set on brick corbeling flush with the inside of the flue.
- The mason must be especially neat when building the firebox and front of the fireplace since it is the focal point of a room and reflects the quality of the mason's work.
- If the fireplace requires cleaning, it should always be done before the finished floors are laid.
- Always be sure to open the damper before starting a fire in the fireplace.
- Multiple-opening fireplaces require a special damper because they are open on more than one side.
- One of the most important advantages of the multiple-opening damper is that it permits a choice of flue locations.
- Flue sizes for multiple-opening fireplaces are determined by adding the square inches of all openings. They should be in a proportion of not less than one to ten.
- Metal heat-circulating fireplaces were developed because a more efficient heat-radiating fireplace was needed.
- The metal heat-circulating fireplace operates on the principle of cold air entering the hollow metal wall of the fireplace form at the bottom. The air is heated and forced up and out through vents at the top of the fireplace. Fans can be installed in the ductwork to improve the flow of air.
- The prebuilt metal fireplace form is manufactured scientifically correct.
- The use of metal heat-circulating fireplaces has increased. This is due to the energy crisis and the ability of metal fireplaces to radiate more heat than a conventional fireplace, burning the same amount of fuel.

SUMMARY ACHIEVEMENT REVIEW – SECTION 2

Select the best answer from the choices offered to complete each statement. Refer back to the material presented in Section 2, if necessary.

1. One of the first Americans to write about fireplace construction was
   a. Thomas Jefferson.
   b. George Washington.
   c. Benjamin Franklin.
   d. James Monroe.

2. The popularity of the fireplace is increasing today due to
   a. a greater interest in the design and architectural effect offered in the main room of the home.
   b. the concern for energy conservation.
   c. the structural support it offers to the rest of the home.
   d. the cheery atmosphere it adds to the home.

3. The movement of draft in a chimney depends on the
   a. type of fuel being burned.
   b. size of the fire built in the firebox.
   c. height of the chimney and the temperature difference.
   d. size of the damper being used.

4. The best type of masonry material to use for areas that are subjected to intense heat from the fire is
   a. concrete block.
   b. stone.
   c. tile.
   d. brick.

5. The portland cement mortar that should be used in the construction of a fireplace and chimney is type
   a. M.
   b. N.
   c. O.
   d. K.

6. The purpose of building a firebox with angled sides is to
   a. add design and character to the fireplace.
   b. save firebrick.
   c. radiate the heat outwards better.
   d. make it easier to clean and remove ashes from the firebox.

7. The walls of a fireplace, if built of brick, should never be less than
   a. 4 inches.
   b. 8 inches.
   c. 12 inches.
   d. 14 inches.

8. To prevent the firebox from cracking from expansion and contraction, leave a space between the firebrick wall and the backing materials of
   a. 1/4 inch.
   b. 1/2 inch.
   c. 1 inch.
   d. 2 inches.

9. If the smoke shelf is to work correctly, it must be laid
   a. 1 inch lower than the bottom of the damper.
   b. level with the bottom of the damper.
   c. 1 inch above the bottom of the damper.
   d. 2 inches above the bottom of the damper.

10. The slope of the sides of the smoke chamber corbeling should be no more than
    a. 30 degrees.   c. 60 degrees.
    b. 45 degrees.   d. 90 degrees.

11. The ratio of the cross-sectional area of the flue lining for a fireplace to the area of the fireplace opening should be
    a. 1/6.   c. 1/3.
    b. 1/12.   d. 1/2.

12. The lintel that supports the facing materials of the front of the fireplace should be installed
    a. 1 course of firebrick below where the damper sets.
    b. 1 course of regular brick below where the damper sets.
    c. level with the damper.
    d. 1 course of firebrick above where the damper sets.

13. Fireplaces that have more than one opening are called
    a. island fireplaces.   c. multiple-opening fireplaces.
    b. conventional fireplaces.   d. heat-circulating fireplaces.

14. The main advantage of a special damper in a fireplace with more than one opening is that
    a. a bigger fire can be made in the firebox.
    b. the fire can be built anywhere on the hearth.
    c. it allows heavier construction and does not burn out as quickly.
    d. the design of the damper eliminates the building of a smoke shelf.

15. One important benefit of the metal heat-circulating fireplace for the homeowner is that
    a. the metal unit is stronger.
    b. it is more economical to have installed.
    c. the circulation of heat is more efficient.
    d. the metal unit is easier to keep clean and requires less upkeep.

16. The metal heat-circulating fireplace is more efficient in heat radiation because
    a. it is constructed scientifically correct at the factory.
    b. air space in the metal walls causes the heat to move through the unit and into the room.
    c. metal radiates heat better than bricks.
    d. it has a smoother finish and there is no place for the heat to be absorbed, such as a firebrick.

Section 3
Building Outdoor Structures

Unit 10 PATIOS AND BRICK PAVING

OBJECTIVES

After studying this unit the student will be able to

- describe the requirements for building a patio wall.
- list the various types of bricks and pattern bonds used in paving and flooring.
- describe how brick paving and floors are laid with mortar and without mortar.

TRENDS TOWARDS OUTDOOR LIVING

The use of patios is, of course, not new. The ancient Romans, Egyptians, and Italians had outside living rooms, dining rooms, and verandas. These structures were built to blend into the surroundings of the home. Flower beds, fountains, and curving walks highlighted the gardens and patios.

Modern technology has now given people even more leisure time than before. The tendency to spend more time outdoors has led to the increased demand for patios. This, in turn, has created a much larger market for the building of outdoor masonry structures. In addition to providing a relaxing atmosphere, an outdoor patio adds needed living space to the home at a minimum of cost.

For many years, home improvement work was ignored by the mason. Work such as patios was left up to the imagination and effort of the individual homeowner. Economic recessions have created a scarcity of masonry work at different times, and for this reason, the student mason should realize the importance of being able to do all of the various phases of masonry work. Many small home improvement contractors make a very good living from this type of work.

Building outdoor structures such as patios may be considered a more creative type of work than just following a set of plans for a building. The mason must be concerned with design and originality in the planning of the job. Not all patios come with a carefully laid out plan for construction. Homeowners are looking for suggestions from the mason to supplement their own ideas. The ability to build patios, walks, and paving can fill the gap for the mason when work is scarce. It is an extension of the basic skills learned in the masonry trade.

134 ■ Section 3 Building Outdoor Structures

MATERIALS, CONSTRUCTION PRACTICES, AND WORKMANSHIP

Masonry Units For Patios

Many different types of masonry materials can be used for patio construction. Brick is the most popular material because it requires little maintenance and is easy to construct. No painting is needed and the colors blend in well with the home and surroundings.

Before selecting any brick, three main factors should be considered: grade, size, and pattern. Other key factors are the amount and types of traffic it must bear (patio floor or paving), frost-and-thaw cycles, site drainage, and the design and appearance of the unit once it is laid. If bricks are used in a floor, they should be solid so that weather conditions cannot penetrate inside the unit. The brick should meet the requirements for grade SW (severe weathering) as defined in ASTM Standard Specifications for Facing Brick C216.

Salvaged, used bricks are not recommended for outdoor structures. The old, lighter pink bricks that are found in many homes are generally soft and are not to be reused in outdoor construction. When removed from these old homes and rebuilt into exterior structures they quickly start flaking and breaking into pieces in the face of the wall. This deterioration of old, soft bricks makes it necessary to tear out the brickwork at a great expense and install new bricks. This is the biggest problem when using old bricks to build new structures.

Although bricks are the most popular material, concrete masonry units can also be used for the construction of patios and floors. The slabs used in the floors may be obtained in different colors. Decorative concrete blocks can serve as a wall and are very beautiful in design. Some of these are made with holes arranged in different designs and are called *screen blocks,* figure 10-1. They are so-named because they screen out the surrounding area, and provide some privacy from adjoining properties. The openings in the blocks also allow cool breezes to pass through.

Stone can also be used, however, this type of construction is more expensive due to the amount of time required for cutting the material. The use of stone is discussed in Section 7.

Fig. 10-1 This concrete masonry screen block wall encloses the patio and provides privacy.

Foundations

Since patio walls do not generally support any weight other than their own, the footings and foundations are often incorrectly installed. The footing should be installed on undisturbed earth, not fill. As is the rule when building any outside wall, the footing should extend below the frost line. Concrete makes the best footing. It does not have to be reinforced with steel rods unless there is a special problem. The footings must be twice as wide as the wall that will be laid on top of them.

Workmanship

All head, bed, and collar joints should be filled solid with mortar. Changing weather conditions can cause rapid deterioration of the joints if any holes are left in them. Work should be level, plumb, and kept as free from mortar stains as possible.

Recommended Mortar

The recommended mortar for exterior walls that come in contact with the earth is type M. Type M mortar is composed of 1 part portland cement, 1/4 part hydrated lime, and 3 parts sand by volume. Type S or N mortar can be used above grade.

Capping The Wall

Some type of cap or coping, figure 10-2, must be laid on top of the walls to prevent water from entering from the top. Several different materials can be used for coping: stone, tile, concrete, or brick. Regardless of the type of coping used, it is a good practice to project the coping material out a distance of approximately 1/2 inch on both sides of the wall to form a water drip. Slope the coping slightly to prevent water from laying on top of the wall. Flashing under the cap is used depending on the weather conditions in the area, and is usually the decision of the builder.

Designing The Patio To Fit The Surroundings

The patio should fit into the area surrounding the home. If the house is in a wooded area, masonry work can be rustic with flowers planted in masonry beds that match the type

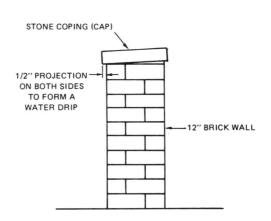

Fig. 10-2 Brick patio wall with a stone coping on top. The stone should extend out over each side of the wall 1/2 inch to form a water drip.

of plants in the area. Easy access from the home to a patio is important. Food is usually prepared in the home and brought out onto the patio. As a rule, the patio is near the kitchen area.

The design of the walls and paving can be as varied as the imaginations of the mason and homeowner. Regardless of the design selected, good workmanship is very important, since all of the work is on the outside and subject to changing weather and moisture.

BUILDING PATIO WALLS

The standard procedure in building a patio is to construct the walls first and then lay the paving work. Using this approach, the slope of the floors can be marked on the walls as a guide with a chalk line. This makes the work easier. The wall also serves as a brace or form to fit the edges of the floor paving against.

Types Of Walls

A patio wall, figure 10-3, can be straight, curved, or offset according to the builder's or homeowner's preference. The design can be very decorative or simply a straight running bond depending upon the effect desired. The wall should be at a height that allows a person to sit in a chair on the patio and be able to look out over the wall.

Drains In The Walls

Some type of drain should be built in the wall just above the finished grade to permit the drainage of any water that has built up behind the wall. Depending on the amount of water present or estimated, weep holes spaced every three bricks apart may be enough. If a large amount of water is present, drain tile (figure 10-4) may have to be installed in the wall. The type of drains and the amount needed depend on the problems of the particular job.

Fig. 10-3 A view of a brick patio wall and exposed aggregate floor. Notice how the planters and fireplace add to the beauty and design of the patio.

Joining The Patio Walls To The House

Some provisions must be made for finishing the walls where they meet the main wall of the house. Two common methods are shown in figure 10-5. In the first method, the patio wall is tied into the house wall by cutting out bricks in the existing wall and bonding the new work into it. When cutting out the brick, a mason's plugging chisel is used to cut out the mortar joints so the brickwork above and below is not chipped or broken. The major disadvantage of using this method is if the footing of the patio settles, the tie-in bricks will crack or break. Then the tie is worthless as well as unsightly.

The second method is used more often. Holes are drilled into the wall and two angle irons are attached, back to back, with expansion bolts. Then the brickwork is laid against the angle irons. The mortar joint against the wall is raked out to a depth of 1/2 inch and caulked with an elastic filler. An advantage of using the angle iron is that it allows the patio wall to expand and contract or settle without adversely affecting the wall it is tied into.

Fig. 10-4 Drain tile installed in a patio wall to drain water that has accumulated on the patio floor.

Other Features Which Add To Patio Design And Beauty

Optional projects such as barbecues, planters, or fountains can be built to add beauty and enjoyment to the patio. (Barbecues are discussed later, in the unit on outdoor fireplaces.) Planters can be constructed in a wide variety of designs. They protect decorative plants from foot traffic and contain the flowers in the patio area for enjoyment. Planters also make the patio blend more naturally into the outdoor setting.

Some type of drainage should be installed in the planter box to drain off excess water. Drainage can be accomplished with weep holes or drain tile, the same as in the patio wall.

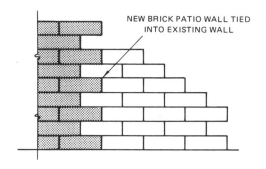

METHOD 1. TYING IN NEW PATIO WALL TO AN EXISTING WALL. ELEVATION VIEW.

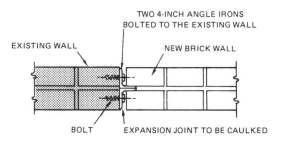

METHOD 2. BONDING EXISTING WALL AND NEW BRICKWORK USING ANGLE IRONS BOLTED TO THE OLD WALL. PLAN VIEW.

Fig. 10-5 Two methods of joining an existing wall with a new patio wall. Method 2 is preferred by builders.

Fig. 10-6 Stone waterfall and fountain built into a patio wall. Water is circulated from the pool with a pump.

It is also a good idea to parge the inside of the planter walls with mortar to lessen the chance of efflorescence. *(Efflorescence* is the leaching of soluble salts contained in the brick, staining through to the outside face of the brick wall. It appears as a whitish deposit on the surface. Moisture causes this reaction.)

A fountain or waterfall adds a cooling and relaxing touch to a patio. Stone and brick can be used to present a natural appearance. Figure 10-6 shows a stone waterfall built into the corner of a patio. The brickwork is completed first and then the concrete pool bottom is placed. The stonework is then laid and a plastic or rubber hose is installed inside the center of the stonework from the bottom to the top of the fountain, to recirculate the water. It is powered by a small, submersible pump which continuously circulates the water. The mortar joints in the stonework are rubbed out with a rounded broom handle and then brushed. This type of project tests the imagination and creativity of the mason.

BRICK FLOORING AND PAVING

For a long time, brick has been popular for paving because of the colors, patterns, and bonds available. Also, it presents a durable surface and ease of construction.

Brick is used in walks (figure 10-7), terraces, patios, roadways, and interior floors. Whereas wood rots out in time, masonry units such as bricks are almost indestructible and require little maintenance.

Sizes And Shapes Of Brick Paving Units

There are many sizes and shapes of bricks for paving and floors. Shapes range from rectangular or square to hexagonal. The most frequently used sizes and shapes are shown

Fig. 10-7 Brick walks at the University of Virginia. Various arrangements of bonds are included in these walks.

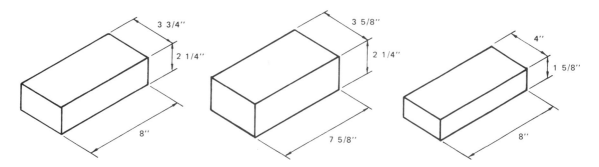

Fig. 10-8 Popular brick pavers used in patio floors

in figure 10-8. Pavers are made in thicknesses ranging from 1/2 inch to 2 1/2 inches. Widths range from 3 3/8 inches to 4 inches. Lengths range from 7 1/2 inches to 11 3/4 inches. There are also square shapes measuring 4, 12, or 16 inches on a side and hexagonal units made in 6-, 8-, and 12-inch sizes.

Patterns And Bonds Used In Paving

Colors, patterns, textures, shapes, and arrangement of the unit are all factors that can be varied to produce many designs. Figure 10-9, page 140, shows some of the most popular pattern bonds for paving.

Pattern bonds can be laid with or without mortar joints. Brick pavers, with length dimensions exactly twice that of their width (such as 4 inch x 8 inch or 3 3/4 inch x 7 1/2 inch), should be laid without mortar for pattern bonds. Examples of these are basket weave, herringbone, and running and stack bond mixed.

The patterns shown in figure 10-9 are developed with brick units having an exposed face of either nominal or actual sizes of 4 inches x 8 inches. *(Nominal* means a dimension

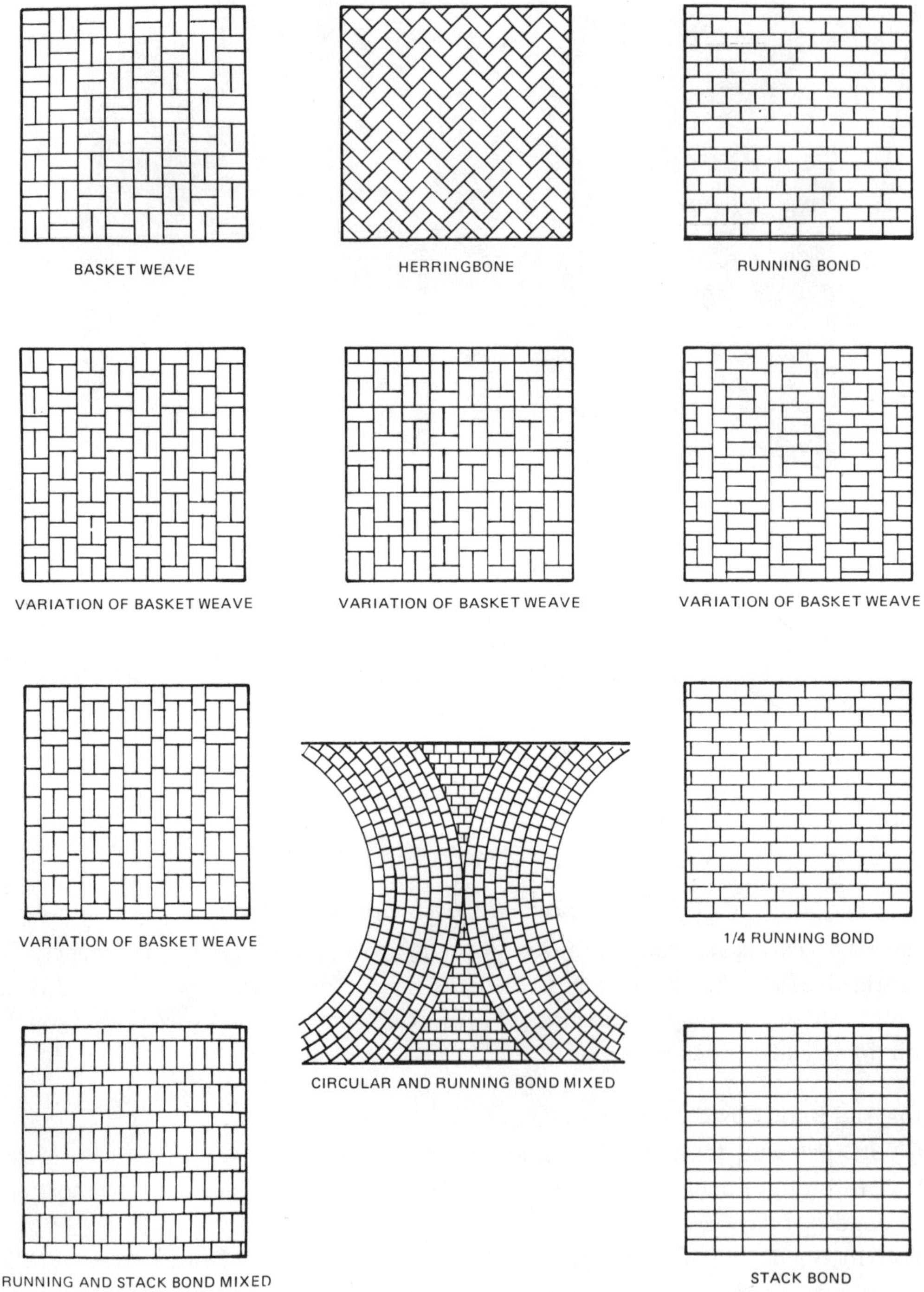

Fig. 10-9 Popular pattern bonds used for patio floors and paving in general

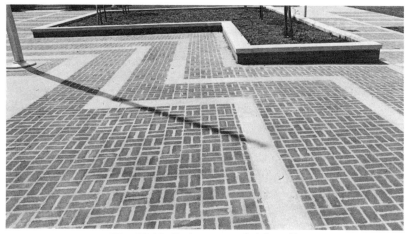

Fig. 10-10 Basket weave brick walk

greater than a specified masonry dimension because of an amount allowed for the thickness of a mortar joint not exceeding 1/2 inch.) The basket weave pattern is shown in a walk in figure 10-10.

The patterns that resist movement best in mortarless paving are the herringbone bond, figure 10-11, and the running bond. Walls, curbs, or planters surrounding the paved area help to restrain the movement so any of the designs shown can be used with proper planning.

Laying a border of bricks at the edges also helps prevent the shifting of mortarless paving or flooring. The edging can be a soldier course of bricks set in concrete or mortar, or some type of design that is in a different position than the paving bond. However, the edging should be laid before the paving bricks are installed so it stays in place. The edging also serves as a guide for elevation and slopes needed on the paving to drain surface water.

Two Methods Of Installing Paving Or Flooring

There are two ways of installing brick paving or flooring; laying with mortar (mortared) or without mortar (mortarless). Mortared brick paving is the traditional method and has been the most popular over the years. Mortarless paving, however, is becoming increasingly popular. Both methods have advantages and disadvantages, but there are two main reasons for the increased demand for mortarless paving: it offers ease of construction and there are no mortar joints to crack or break up. Mortarless bricks can also be laid faster and with less cost.

The selection of either type depends on the individual job requirements and the preference of the builder or homeowner. Masons must be able to install both types since both types will be encountered on the job.

Fig. 10-11 Herringbone paving in a courtyard

Construction Of Mortared Paving

Brick paving laid in mortar should be placed on a concrete slab base. The stability and noncracking of the mortar joints are dependent upon building on a good base that does not shift due to changing frost-and-thaw cycles. Good workmanship and full mortar joints are, of course, necessary for a successful job.

Drainage

Because exterior (outside) brick paving is subject to changing weather and moisture conditions, good drainage must be provided. Brick walks should be sloped, either to one side or to both sides. They are sloped to both sides by making a crown (high spot) in approximately the center of the walk. For drainage, patio floors are sloped away from the building, retaining walls, or other places that can trap water, or a gutter arrangement can be provided. The slope should be 1/8 inch to 1/4 inch per foot for most paving or floors. Larger areas may require a greater slope.

If there is a severe moisture problem, laying a layer of gravel under the concrete base before placing the concrete helps to drain trapped water away. The concrete base should also be installed with a slope. This makes it much easier to lay the bricks in mortar on a slope. Leave the concrete base surface rough and untroweled for a better bond with the mortar. Tamping with a garden rake provides a good working surface.

Edging

After the concrete slab is placed, the brick edging is laid. It bonds better if laid in a soldier or rowlock course and bonded to the edge with concrete or mortar. Figure 10-12 shows how a soldier course is laid in a typical paving job.

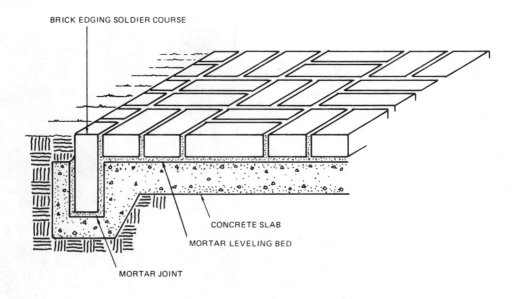

Fig. 10-12 Brick soldier course laid as an edging for the patio floor

Mortar

Type M mortar is recommended for outdoor paving, since the work is in contact with the earth. To keep out water, all joints must be well filled and tooled so there are no holes.

Laying The Bricks On The Concrete Base

There are three basic methods of installing brick paving with mortar joints. The first is the conventional way, spreading mortar on the base and applying the mortar to the unit using the trowel. Butter the brick with mortar, being sure to form a solid head and cross joint. Lay the bricks in the mortar bed, leveling so there is a slope to drain the water from the exposed face.

The most accurate method of leveling bricks in the mortar bed is by using a long wood straightedge, figure 10-13. Level from one side of the edging to the other,

Fig. 10-13 Leveling the bricks into position by using a wood straightedge with a level laid on top

tapping the bricks down or bedding up until they are all touching, with no high or low spots. A rubber crutch tip on the handle of the trowel helps prevent the trowel handle from splitting when tapping on the bricks. A hammer handle is also used in settling the bricks in position, as shown in figure 10-13.

Spread only enough mortar for the bed joints so that the bricks can be settled in position firmly without too much tapping. Tapping too hard on the bricks can result in cracking the units.

On a large or long floor, a nylon line can be stretched tightly before the brick is laid, to serve as a guide for the work. The line may sag somewhat, so the bricks must be laid slightly above the line, or a trig can be set up in the center of a long floor to keep the line stable. Tool the exposed mortar joints with a flat striking tool, such as a slicker.

> When cutting any units to fit, wear eye protection. A pair of rubber knee pads are helpful when working in a kneeling position.

The second method involves laying each brick on a mortar leveling bed with 3/8 inch to 1/2 inch of space left between each unit for head joints. This is followed by placing mortar grout between the spaces. The grout proportions of portland cement and sand are the same as for mortar, except that the hydrated lime can be left out. Care should be taken not to stain the surface of the grouted bricks because they will be difficult to clean.

The third method involves using a dry mixture of portland cement and sand as a base, in the same proportions as for grout. The bricks are laid on a damp cushion of this mixture.

Then the units are filled in between with the same mixture. After cleaning any excess mixture from the paving surface with a burlap bag, dampen the surface with a fine mist of water until the joints are thoroughly moist. The paving should be kept in a damp condition for a period of two or three days so it will cure.

Expansion Joints

Some provision must be made for the movement of paving or flooring when laid in mortar. The expansion joint for large areas should be located parallel to curbs and edgings, at 90 degrees or right-angle turns, and around any type of interruption or break.

Fillers for expansion joints, figure 10-14, must be compressible and made from materials that do not rot. Generally, pavement joint fillers are either made solid or preformed of materials such as Butyl rubber, neoprene, or other elastic compounds.

Mortarless Paving

The key to paving with mortarless bricks is the proper preparation and compacting of the subgrade and base before starting to lay any bricks. As when paving with mortar, ample drainage and slope must be provided to drain water if outside.

The edging for mortarless bricks should be laid in mortar, the same as when laying an edge for brick paving in mortar. The reason is that mortarless bricks tend to shift underfoot and the edging keeps the main floor from moving.

Base Materials

Except where there is a special subsurface drainage problem, regular sand or stone dust (stone screenings) can be used as a base. A leveling cushion bed of 1 inch to 2 inches of sand is spread out on the tamped earth and screeded off (leveled) with a wood straightedge, much like concrete is screeded. Fill any low spots as the screeding is done.

A layer of roofing felt or plastic sheets is then laid over the sand. There are several advantages of using the felt or plastic over the sand. First, it makes it easier to lay the bricks level on the base. Second, it tends to keep the subsurface groundwater from being drawn or sucked into the brick units, thus reducing the possibility of staining the bricks. Third, a layer of felt or plastic helps to prevent grass and weeds from growing up between

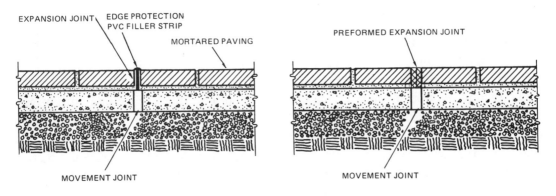

Fig. 10-14 Two types of joint fillers used to control expansion in patio floors

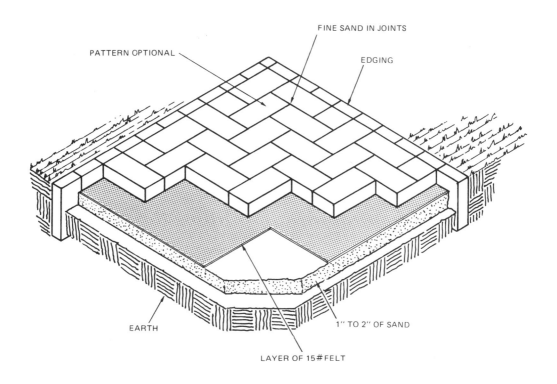

Fig. 10-15 Laying brick paving on a sand and felt base. This is called mortarless paving.

the joints. It is worth the extra time and money to install a layer of either of these materials when paving without mortar. Figure 10-15 shows brick paving with a sand and felt base.

In areas where there is much groundwater present or poor drainage, some drain tile and gravel should be provided under the base to carry the water away. Use gravel ranging in size from 3/4 inch to 2 1/2 inches as a cushion. After the gravel is placed, a leveling bed of stone screenings or pea gravel about 1 inch thick is placed on top. The drain tile is installed at the point in the base where the worst drainage problem is located. Drain tile is covered with a strip of felt roofing paper to keep out dirt. It is then covered over with the gravel. Drain tile must lead to a drain area away from the patio to carry the water away. If it does not, the drain tile becomes a trap for the water.

The brick paving is then laid on the stone screenings. It is also a good idea to provide a felt roofing base over the stone screening to reduce the growth of weeds. Figure 10-16, page 146, shows brick paving installed over a gravel base with drains and stone screening.

After the paving units have been laid in place and tightly fitted together, sand is swept into the joints. Another mixture that stays in place longer is dry sand and portland cement. Use a mixture of 1 part portland cement and 3 to 8 parts mason sand. After sweeping the mixture in place, use a fine water spray from a hose with a nozzle attached to dampen and set the mix. Exposure to the weather cures the joints in a short period of time. For best results, keep it damp for two or three days.

Only mortarless paving should be placed over a gravel or sand base. For paving laid in mortar, the base should be rigid on a slab or portland cement mixture base.

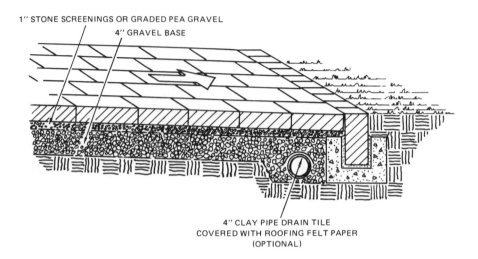

Fig. 10-16 Brick paving installed over a gravel base with drains and stone screenings

Interior Brick Flooring

Inside the home, brick flooring is being used more now than in the past. Brick flooring used to be laid only in the entranceway, recreation rooms, and sometimes in enclosed porches. The low maintenance cost makes interior brick flooring more attractive to the builder and homeowner.

Selection Of Material

A smooth, hard brick presents a good appearance and does not absorb moisture from spills or mopping up. The natural color of brick is also attractive. Brick in an interior floor does not have to withstand freezing and thawing as does brick outside. The decision of which brick to select depends somewhat on the size needed so the appropriate bond can be worked out. When selecting the brick color, remember that the job is permanent and cannot be changed later.

Two Methods Of Installing The Interior Floor

As with exterior paving, brick floors can also be laid in two ways. Brick floors can be installed over plywood subflooring and wood floor joists just as easily as over a concrete slab. However, special consideration should be given to the size of the floor joists. Brick pavers are heavier than regular flooring, so a stronger joist arrangement must be used. To accomplish this, joists can be placed closer together and 2-inch solid bridging should be used to further strengthen the floor. Joists can also be doubled for a stronger job.

The first method of installation is mortarless. Two layers of 15-pound felt are laid over the subfloor. Next, a 1/2-inch cushion bed of sand and portland cement is laid on the floor. The bricks are laid into position and tapped firmly in place. All bricks should be laid together as tightly as possible. In this way, the type of bond can be used that allows a tight joint. Two such bonds are the herringbone and running bond.

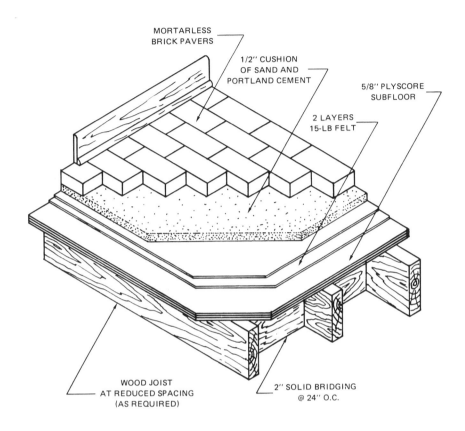

Fig. 10-17 Interior mortarless brick paving laid in sand over wood flooring

After all bricks are laid and leveled, fine sand can be swept in any cracks or holes. Portland cement and sand can be substituted here too, if desired. Figure 10-17 shows mortarless brick paving over wood floor joists.

The second method of laying an interior brick flooring is in mortar. The subfloor is installed flush to reduce the overall floor thickness. The wood floor is built as strong as with interior mortarless paving. Then a layer of plastic (polyethylene) is laid over the floor. A mortar bed is spread on the plastic in the thickness desired. The bricks are jointed with mortar and laid in the mortar bed. Leveling is done to the appropriate height. However, it is not necessary to slope the brickwork unless there is a drain in the floor as with outside paving.

The face mortar joints are tooled with a flat jointer the same as when striking the exterior paving. Figure 10-18, page 148, shows brick paving in mortar over wood floor joists.

A concrete slab can be used inside the home for a family or recreation room, especially if the room is on the ground level. The brick flooring is installed with a 1/2-inch cushion

148 ■ Section 3 Building Outdoor Structures

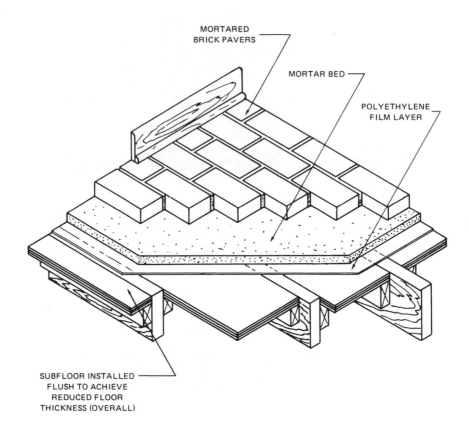

Fig. 10-18 Interior brick paving in mortar over a wood floor

base of portland cement and sand over the concrete. Care should be taken when laying units in mortar to keep them as clean as possible. As the mortar stiffens, brushing off and rubbing with a burlap bag removes most of the mortar from the surface of the bricks.

Cleaning Brick Floors

Interior floors should not be cleaned with acid solutions. If care is taken during the laying of the bricks to brush and wipe them clean, little or no cleaning should be needed. Brushing with clear water removes most of the stains, but do not use too much water for interior floors. Mortarless brick paving and flooring should only be rinsed off with clean water from a hose. Waxing after the floor is dry brings out the full color of the bricks.

Reinforced Brick Floors

Reinforced brick paving can be used to span an opening, or over a fill that may settle unevenly. Reinforcing the masonry can eliminate the need for a separate reinforced concrete slab or other rigid base. Metal reinforcement rods are used together with brick and mortar to make a strong floor. A reinforced floor is used only in special cases and is not the usual way of installing brick flooring. It is, however, practical over short spans.

Estimating Materials For Brick Paving

The estimating of materials for paving or flooring can be done by using the tables in figure 10-19. Table 1 is used for brick paver units for mortarless paving. Table 2 is used for brick paver units laid in mortar. Table 3 gives material quantities for 1/2-inch mortar setting beds.

It should be noted that not all size pavers are suitable for interior brick flooring. Always check with a local brick manufacturer or brick dealer before making a final selection.

Estimating Table 1[a, b]
Brick Paver Units
for Mortarless Paving

Face Dimensions (actual size in inches) $w \times l$		Paver Face Area (in sq. in.)	Paver Units (per sq. ft.)
4	8	32.0	4.5
3 3/4	8	30.0	4.8
3 5/8	7 5/8	27.6	5.2
3 7/8	8 1/4	32.0	4.5
3 7/8	7 3/4	30.0	4.8
3 3/4	7 1/2	28.2	5.1
3 3/4	7 3/4	29.1	5.0
3 5/8	11 5/8	42.1	3.4
3 5/8	8	29.0	5.0
3 5/8	11 3/4	42.6	3.4
3 9/16	8	28.5	5.1
3 1/2	7 3/4	27.1	5.3
3 1/2	7 1/2	26.3	5.5
3 3/8	7 1/2	25.3	5.7
4	4	16.0	9.0
6	6	36.0	4.0
7 5/8	7 5/8	58.1	2.5
7 3/4	7 3/4	60.1	2.4
8	8	64.0	2.3
8	16	128.0	1.1
12	12	144.0	1.0
16	16	256.0	0.6
6	6 Hexagon	31.2	4.6
8	8 Hexagon	55.4	2.6
12	12 Hexagon	124.7	1.2

[a] Table is based on BIA survey conducted in 1973. According to the survey approximately 38 sizes are currently manufactured.

[b] The above table does not include provisions for waste. Allow at least 5% for waste and breakage.

Estimating Table 2[a]
Brick Paver Units
for Mortared Paving

Brick Paver Units $w \times l \times t$	Paver Units per sq. ft.	Cubic Feet of Mortar Joints per 1000 Units	
		3/8-in. Joint	1/2-in. Joint
3 5/8 x 8 x 2 1/4[b]	4.3	5.86	–
3 5/8 x 7 5/8 x 2 1/4	4.5	5.68	–
3 3/4 x 8 x 2 1/4	4.0	–	8.0
3 5/8 x 7 5/8 x 1 1/4	4.5	3.15	–
3 3/4 x 8 x 1 1/8	4.0	–	4.0

[a] The table does not include provisions for waste. Allow at least 5% for brick and 10% to 25% for mortar.

[b] Running bond pattern only.

Estimating Table 3
Material Quantities for 1/2-in.
Mortar Setting Beds[a]

Mortar Type and Material	Cubic Feet[b] per 100 sq. ft.	Material Weight per cu. ft. in lb.
Type N (1:1:6)	4.17	
portland cement		15.67
hydrated lime		6.67
sand		80.00
Type S (1:1/2:4 1/2)	4.17	
portland cement		20.89
hydrated lime		4.44
sand		80.00
Type M (1:1/4:3)	4.17	
portland cement		31.33
hydrated lime		3.33
sand		80.00

[a] These quantities are only for setting bed. For mortar joint quantities see Table 2.

[b] The table does not include provisions for waste. Allow 10% to 25% for waste.

Fig. 10-19 Estimating tables for bricks and mortar for paving

ACHIEVEMENT REVIEW

Select the best answer from the choices offered to complete each statement. List your choice by letter identification.

1. The most frequently used bond for building brick patio walls is
   a. Flemish.
   b. running.
   c. stack.
   d. common.

2. The recommended mortar for exterior walls that come in contact with the earth is type
   a. M.
   b. S.
   c. N.
   d. O.

3. The whitish stain that sometimes appears on the outside of projects such as planters is due to the formation of
   a. acid.
   b. copper.
   c. salts.
   d. lime.

4. Bricks used for exterior walls and paving must meet grade requirements for
   a. SW.
   b. MW.
   c. HD.
   d. LW.

5. When laying paving without mortar, it is better to have as thin a head joint as possible. Of the following bonds, the best selection for a tight joint is the
   a. Flemish bond.
   b. stack bond.
   c. basket weave bond.
   d. herringbone bond.

6. The border edging brick on a patio should be laid in
   a. sand.
   b. stone dust.
   c. mortar.
   d. packed earth.

7. Brick paving or floors laid in mortar are strongest if laid on a base of
   a. sand.
   b. stone dust.
   c. concrete.
   d. a dry mix of portland cement and sand.

8. The correct drainage slope for an average patio floor is
   a. 1/8 inch to 1/4 inch.
   b. 1/4 inch to 3/8 inch.
   c. 3/8 inch to 1/3 inch.
   d. 1/2 inch to 5/8 inch.

9. The recommended mortar for exterior brick paving is type
   a. N.
   b. M.
   c. K.
   d. O.

10. The best tooled joint finish to use on brick paving or flooring is
    a. grapevine.
    b. concave.
    c. convex.
    d. flat.

11. The best way to keep all bricks on a patio floor at a straight height is to use a

    a. transit level.   c. wood straightedge.
    b. line level.   d. plumb rule.

12. It is recommended to remove excess mortar from brick paving before it completely hardens by using

    a. an acid wash.   c. a burlap bag.
    b. a soft cloth.   d. water from a hose.

13. When installing the sand base for mortarless paving, the correct thickness to spread it is

    a. 1/2 inch to 1 inch.   c. 2 inches to 3 inches.
    b. 1 inch to 2 inches.   d. 3 inches to 4 inches.

14. The main reason for installing roofing felt under the bricks in exterior paving is that

    a. the bricks are less likely to shift.
    b. it is faster to lay bricks on the felt.
    c. the sand joint does not wash out as quickly from rain.
    d. it prevents plant growth from coming through the joints.

15. After mortarless paving has been laid in place, the joints resist moisture and washing out longer if

    a. lime is added to the mix.   c. concrete is grouted in the joints.
    b. portland cement is added to   d. calcium chloride is added to the
       the mix.   mix.

PROJECT 8: LAYING BRICK PAVING IN A RUNNING BOND

OBJECTIVE

- The student will lay brick paving in a running bond, laid in sand.
  NOTE: This project can be laid in mortar or on a sand bed, (mortarless); it is the instructor's choice. The project is, however, written for laying the paving in sand without mortar.

EQUIPMENT, TOOLS, AND SUPPLIES

one wood box or form made from 2-inch x 6-inch lumber. The inside dimension should accommodate 5 bricks in length, laid tightly together. The box should be square. (No size is given due to variations in brick sizes.)

supply of paving bricks (solid bricks with no holes); estimate quantity from the plan in figure 10-20

supply of sand

mason's hammer

brick set chisel
plumb rule (48 inches long)
wood straightedge, approximately 60 inches long

mason's rule
pencil
brush

SUGGESTIONS

- Select a smooth brick since it is easier to work with.
- Dry bond the bricks out on the floor before constructing the wood box to make sure the box is the correct size.
- Wear safety eye gear when cutting any bricks.
- Rubber knee pads can be worn for more comfort (if available).
- Settle the bricks firmly in place with the hammer handle and straightedge.
- Rub all cut edges of brick cuts for a neater fit.
- Do not beat on the plumb rule; use a wood straightedge.

PROCEDURE

1. Construct the wood box to accommodate the bricks being used, making sure to brace the inside bottom of all corners, square.
2. Spread a layer of sand to the correct depth in the box.
3. Lay four bricks on each corner to the height shown on the plan in figure 10-20. (First dry bond to work full brick. The bricks on the corners are laid to different heights in order that the student can get practice in laying bricks with a slope to drain water away.)
4. Lay the first course on the highest end from edge to edge as tightly together as possible. Tap the bricks together with the handle of the hammer.
5. Cut two halves (bats) for the second course so the brick breaks half over the other. This is known as a *running bond*.
6. Continue laying each new course as shown on the plan, making the necessary cuts.
7. Level each course with the opposite end of the box to make sure that the paving is flush and has the correct slope. There should be no low or high spots in the paving.
8. Lay the paving until all of the bricks are laid in place.
9. Check the project on a diagonal with the straightedge to make sure all bricks are touching the level or straightedge.
10. Sprinkle a coat of sand on top of the project and sweep in any cracks in the head joints with a brush.
11. Have the instructor evaluate the project.

Unit 10 Patios and Brick Paving ■ 153

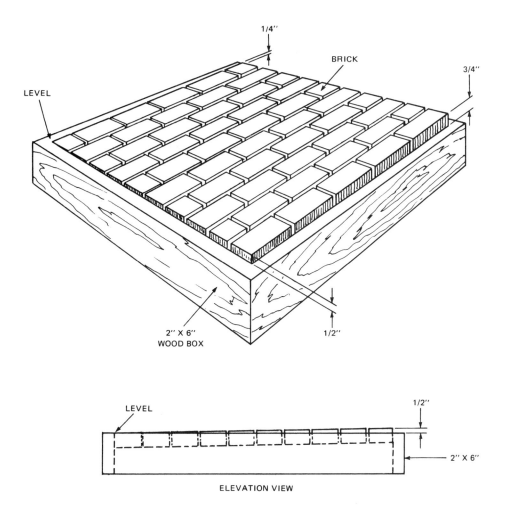

Fig. 10-20 Project 8: Laying brick paving in a running bond

Unit 11 PORCHES AND STEPS

OBJECTIVES

After studying this unit the student will be able to

- list the names of the parts of a porch and steps.
- describe the relationship between the riser and tread and why they are important in building brick steps.
- explain how brick steps are laid out and constructed.

Porches and steps are an important part of many buildings. The basic function of the steps is to safely get traffic from one level to another level. Although steps can be built of various materials, brick steps are the most versatile because of their size and ease of maintenance.

The mason is expected to be able to lay out and construct masonry steps. The theory and principles of porch and step construction are basically the same regardless of the type of masonry unit being laid.

Workmanship must be good and mortar joints must be well filled in order for the porch and steps to withstand changing weather conditions. A knowledge of basic math is needed to correctly proportion the steps so that each one is the same height and width. Poorly proportioned steps may cause a person to misjudge and stumble, resulting in an accident. This unit discusses in detail the building of brick steps.

THE PARTS OF THE PORCH AND STEPS

To prevent confusion when building a brick porch and steps, it is helpful to know the correct terms for the various parts. Figure 11-1 gives the parts of a porch and steps. The top of the porch floor is called the *platform* and is constructed before the steps are built. If the walls on each side of the porch are higher than the platform and serve as a rail, they are called *cheeks*. The porch and steps combined are also sometimes called a *stoop*. The part of the steps which a person steps on is called the *tread*. The *riser* is the upright part between two stair treads, and the vertical height of the step is called the *rise*.

FOOTINGS AND FOUNDATIONS

A porch and steps must rest on a firm foundation. The footing should be placed on undisturbed ground so the porch and steps do not settle. This is one of the most important points to remember in the construction of porches and steps. The footing must also be deep enough in the ground to be below the frost line.

Because the porch is so close to the building, many builders build the foundation of the porch up from the same level as the foundation footing to grade line. Pieces of concrete

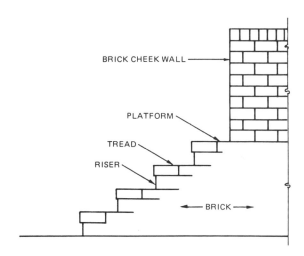

Fig. 11-1 Parts of a porch and steps.

blocks left over from the basement can be used if they are bonded in the mortar joints. The face brickwork should be started at least three courses below the top of the ground to allow for the natural settlement of the earth surrounding the steps and porch.

An alternate method of constructing the foundation of the porch is to place a footing approximately 2 feet below the finish grade level on solid ground. It must be out from the foundation to a point where the outside edge of the steps will start. As the building foundation is built up to the height where the steps would start, a concrete block lintel, figure 11-2, should be walled into the foundation and bridged out to rest on the concrete footing. This carries the weight of the porch safely over the excavated area around the basement wall. The excavation around the foundation is necessary to allow room for the mason to apply the parging and waterproofing.

The alternate method is more economical because the porch wall does not have to be built all the way from the basement footing. This eliminates the need for extra materials and labor.

TYPES OF MATERIALS TO USE

It is recommended to use grade SW brick since it will be in contact with the earth. Type M or N mortar is recommended for the porch and steps. If brick headers are not put in the walls of the porch to bond the brick facing and the block backing, metal wall ties should be used between the facing bricks and the concrete block backing.

Fig. 11-2 Concrete block lintel spans the area out from the foundation and rests on a footing on solid ground.

Stone can also be used to build porches and steps, figure 11-3. Stonework is, however, more costly due to the greater amount of time needed for construction. As a rule, concrete blocks by themselves are not preferred for constructing steps because of the size of the units.

BUILDING THE PLATFORM

The height of the finished platform (top of the porch) is determined from the top of the doorsill. The top of the doorsill is approximately 4 inches above the finished platform.

The porch should be built up to where the finished paving for the brick

Fig. 11-3 Stone porch and steps at the entrance to a house.

platform is to be laid. (The interior of the porch can be filled in with pieces of bricks and blocks left over from the construction of the building.) Then a border of bricks is laid in a rowlock position around the perimeter of the platform to hold the paving in place and match the rowlocks used on the steps. The paving is then laid level with the top of the border. A wood straightedge is used to make the paving level with the border.

The top of the platform should have a fall or slope from the point where the doorsill is located to the front and sides to drain off any water. The amount of fall depends on the size of the platform. Make sure when laying the platform that it is sloped away from the sill to prevent water from collecting in front of the doorsill.

LAYING OUT AND BUILDING THE STEPS

There are two methods of installing a footing for the steps. A concrete subbase can be placed in the shape of the steps minus the thickness of the finished brickwork. Then the bricks are laid as a veneer over the concrete. In the second method, a concrete slab is placed over the entire area where the steps will be built and the steps are built on the slab. Either of these methods is acceptable.

The concrete slab under the steps should be reinforced with reinforcing wire for extra strength. Do not trowel the concrete base smooth because the mortar bonds better if the surface is rough.

RISER AND TREAD RELATIONSHIP

The height of the riser and the depth of the tread are very important. Do not make the tread of brick steps less than 12 inches deep or they can be dangerous when covered with ice and snow. Pitch the step forward with a slope of about 1/4 inch per foot to drain off the water. The undersurface of the concrete footing for the steps should be level. The slope should be made in the bed joint under the tread bricks.

Risers should be in proportion to the treads to provide an easy step up and enough room to set the foot down without being cramped. Riser heights are usually between 6 and 8 inches for porch steps. Low risers and broad steps are preferred for garden walks. High risers and narrow treads should only be used when the horizontal distance is limited and nothing else can be done. In this case, install a handrail for safety.

The average height of a brick step is 7 1/2 inches. This consists of a brick stretcher with a rowlock plus mortar joints. The arrangement of the brick positions in the riser determines its height. For example, if a stretcher course with a header course is used, the height is 6 inches. Figure 11-4 shows the two different arrangements of brick steps and risers.

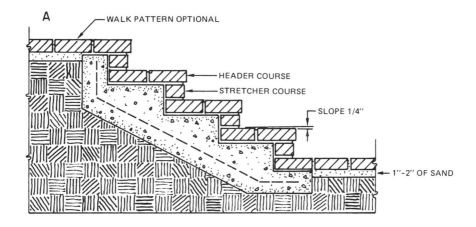

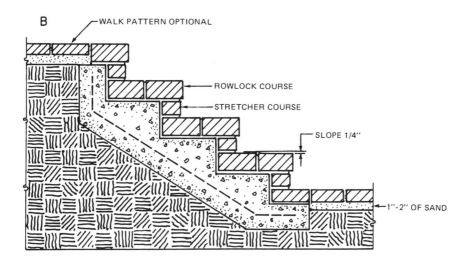

Fig. 11-4 Two different methods are shown for laying bricks on a brick step. Method A consists of a brick stretcher course and a header laid on top of it. Method B shows a stretcher course with a rowlock laid on top of it. Method B is the most used.

Fig. 11-5 The rowlock bricks are projected approximately 1/2 inch to form a water drip. This is a common practice when constructing steps.

It is a good practice to have the top front edge of the tread brick project 1/2 inch to allow the water to drip off. Figure 11-5 shows projected treads.

Generally, the length of the steps depends on the size of the sidewalk that leads to them. Steps are usually at least 3 feet in length. However, they can be the entire length of the porch. A longer step is preferred to a shorter step.

Brick steps also offer a practical solution when building a walk or path on a slope. In landscaping using masonry, a good rule is to make the width of the steps at least 4 feet, so that two people can use the steps at the same time.

The deeper the tread is, the lower the riser can be. This is what is meant by the relationship of tread to riser. It is important, however, that all risers be the same height. This includes the one used to step off onto the walk or platform. If they are not all the same height, a person using the steps could stumble and fall.

LAYING OUT THE STEPS

It is necessary to determine where the first step will start, and to make sure that all steps are even when finished. To do this, measure the total height from the top of the finished platform to the bottom of the surface of the proposed walk or landing. Since, as stated earlier, average brick steps are 7 1/2 inches in height, the figure 7 1/2 inches should be divided into the total inches of height. This gives the number of risers needed.

As a rule, the steps are built first as close as possible to fit the proposed finished grade. Then the walk is laid to the bottom of the first step. Figure 11-6 shows how the risers are laid out from the top of the platform.

Referring to figure 11-6 it can be seen there is a total distance of 30 inches from the top of the finished platform to the top of the walk. Dividing 7 1/2 inches into 30 inches gives

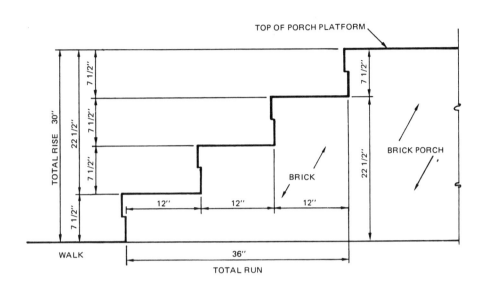

Fig. 11-6 This layout of brick steps shows how the dimensions of the risers and treads are calculated when constructing the steps. Notice that each step is the same rise and tread distance.

a total of 4, which indicates 4 risers are needed. This allows 3 brick steps to be built for a total of 22 1/2 inches. The remaining 7 1/2 inches is the riser which is formed from the top of the last step or tread to the top of the finished porch platform.

Refer, again, to figure 11-6 for laying out the treads. Since there are 3 steps, there are 3 treads. Each tread in this case is 12 inches deep. Since each tread is 12 inches deep and there are 3 treads, multiply these figures to get a total of 36 inches. Therefore, the distance from the porch wall to the outside of the first tread is 36 inches.

BUILDING THE STEPS

On the base, lay out the outline of the first step with a chalk line. Drive a stake into the ground near the outside corner of the first step. A level mark is put on the stake. This is done by leveling out from the top of the platform with a plumb rule. Then measure down with the mason's rule to where the height of the first tread should be and mark this point on the stake.

Laying The First Step

Dry bond the first course to make sure it works whole bricks, then lay in a full bed of mortar to the correct height. This is done by leveling to the mark on the stake.

The second course of bricks laid on a step is usually in a rowlock position. It can be projected out from the step approximately 1/2 inch to form a water drip, or laid flush. A typical step tread is 12 inches deep and slopes 1/4 inch from the rear to the front.

The best method of laying brick treads to keep them level and straight is to lay three bricks on each end of the step. Be sure to keep them level with each other. Then attach a

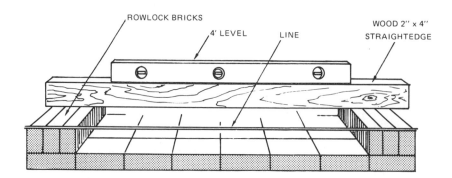

Fig. 11-7 Three bricks are laid on each end of the step. A line is attached to the front to keep the bricks level and in line. The slope of the step should be 1/4 inch to the foot. Notice the wood 2 x 4 and level used to level the back of the step.

line to the top front edge and lay the bricks to the line. The back edge of the rowlock can be kept level by laying a wood straightedge across and a level on top of it. Settle any high bricks down until level or raise any low bricks up until level. Figure 11-7 shows how this is done.

Spacing the bricks is done with a mason's rule to make sure the step works full bricks. At the point where the last brick *(closure brick)* is laid, two mortar joints must be allowed for, in addition to the thickness of the brick. This is important to remember, or the spacing will not be correct and the mortar joint will not be well filled. If head joints are too small they can leak. Good, well-filled joints must be used, figure 11-8.

Completing The Steps

The second step is laid the same way and should bond over the one below. The project plan at the end of this unit shows how this is done.

As the steps are being built, fill in any of the unexposed area with rough bricks or scraps of masonry units laid in mortar. In this way many of the waste pieces of masonry units left over from the job can be used. Remember, do not force the fill or mortar in the step because this can cause the step to bulge out.

End bricks laid on the steps should be of the solid type. Do not fill the holes in a brick with mortar and lay it on the end of the step. The mortar will eventually come out and this also gives a very poor appearance.

The mortar joints can be tooled (struck) either with a concave or flat slicker jointer to prevent water from penetrating. If the bricks have a

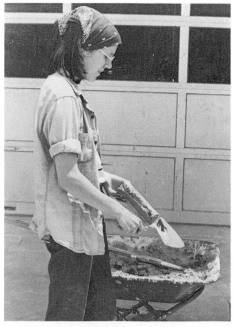

Fig. 11-8 Applying a well-filled mortar cross joint to a brick.

hard finish, they can be rubbed off with a piece of burlap bag when set to remove mortar stains. This lessens the amount of cleaning needed at the completion of the job.

Figure 11-9 illustrates two sets of brick steps as they appear when finished. In view A, the bricks are laid flat and in view B, the bricks are on edge.

Good workmanship is very important in the building of steps and porches. Also, construction of the risers and treads must be accurate if the job is to be done correctly.

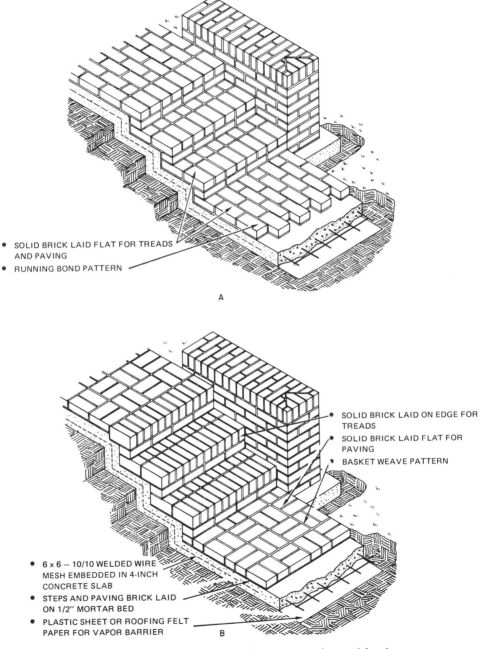

Fig. 11-9 Two brick steps. View A shows a stretcher and header combination and view B shows a stretcher and rowlock combination. The brick wall adjoining the steps is called the porch cheek.

ACHIEVEMENT REVIEW

Select the best answer from the choices offered to complete each statement. List your choice by letter identification.

1. The top surface of the porch is called the
   - a. tread.
   - b. riser.
   - c. platform.
   - d. run.

2. The part of the step a person steps on is called the
   - a. riser.
   - b. tread.
   - c. platform.
   - d. stoop.

3. The vertical part of the step between the treads is called the
   - a. riser.
   - b. tread.
   - c. run.
   - d. elevation.

4. The standard height for a brick step if a stretcher and a rowlock are used is
   - a. 5 1/2 inches.
   - b. 6 inches.
   - c. 7 1/2 inches.
   - d. 8 inches.

5. The depth of a brick step should be no less than
   - a. 8 inches.
   - b. 10 inches.
   - c. 12 inches.
   - d. 16 inches.

6. The correct projection for the top tread brick is
   - a. 1/4 inch.
   - b. 1/2 inch.
   - c. 3/4 inch.
   - d. 1 inch.

7. So that water can drain from the step, the slope should be
   - a. 1/8 inch to the foot.
   - b. 1/4 inch to the foot.
   - c. 1/2 inch to the foot.
   - d. 3/4 inch to the foot.

8. The width of brick steps for a porch should be determined by the
   - a. width of the front door.
   - b. length of the porch.
   - c. width of the sidewalk.
   - d. size of the masonry units being used.

PROJECT 9: BUILDING A BRICK PORCH AND STEPS

OBJECTIVE

- The student will lay out and build a brick porch and steps. Steps such as these are typical of the type built for most homes. The porch is built first and then the steps.

EQUIPMENT, TOOLS, AND SUPPLIES

mixing tools
mortar pan or board
supply of hydrated mason's lime
supply of sand

supply of water
bricks and concrete blocks to build porch and steps (to be estimated from the plan by the student); concrete blocks are used as fill for the steps
2-foot framing square
chalk box and line
pencil
mason's trowel
mason's hammer
brick set chisel
18-inch level
4-foot level (plumb rule)
ball of nylon line
line pins
mason's modular or course counter rule
convex striking iron (forms a concave joint)
flat slicker striking iron
brush
burlap bag

SUGGESTIONS

- Use full mortar joints.
- Wear eye protection when mixing mortar or cutting bricks.
- Use the 18-inch level when determining the slope of the steps because it is easier to handle.
- Select unchipped bricks for the steps since they will show on all sides.
- It is better to select a hard smooth brick for the project (if available).
- Carefully check the height of every riser and the width of each tread to make sure they are correct before building the next step.
- Do not spread mortar too far ahead when laying the step tread bricks because this part of the work is slow.
- Check the back of the tread bricks with the level frequently as the line is being run in.
- Strike the joints as soon as they are thumbprint hard rather than waiting until the entire step is built.
- Make the mortar stiff enough to allow step bricks to be pressed together without too much pressure.

PROCEDURE

1. Strike a chalk line on the base or floor the length and width of the porch. The porch is built before the steps are started.

2. Lay out the first course of bricks dry (without mortar) to the size of the porch. Make the porch work whole bricks without cutting; this may involve increasing the dimensions slightly. Use 3/8-inch mortar head joints for the dry bonding.

3. Lay the brickwork on the porch in a running bond as shown in the plan, figure 11-10, page 164. Use No. 6 on the modular rule as a guide.

164 ■ Section 3 Building Outdoor Structures

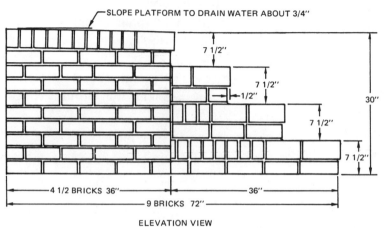
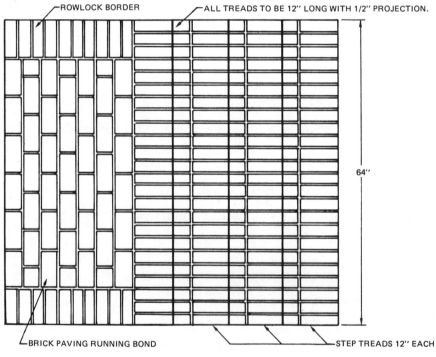
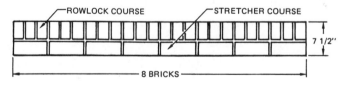

Fig. 11-10 Project 9: Building a brick porch and steps.

4. Make the last course on the porch edge a rowlock and project it 1/2 inch out from the edge of the porch.
5. Slope the top of the porch platform from the rear to the front 3/4 inch to drain off water.
6. Lay the porch top in a running paving bond up to and against the border rowlock bricks.
7. Strike the mortar joints on the walls of the porch with a convex jointer which will make a round or concave joint.
8. Strike the mortar joints on top of the porch with a flat joint using the slicker, and brush slightly.
9. Lay out the outline of the steps with a chalk box. See the plan, figure 11-10, for dimensions.
10. Establish a level point on a stake near the outside edge of the first tread. This is obtained by leveling out from the top of the porch platform and down to the height of the first tread. See the plan.
11. Lay out the first stretcher course of brick as shown on the plan.
12. Fill in behind the stretcher course with rough brick or pieces of bricks.
13. Lay 3 rowlock bricks level with each other on each end of the stretcher course. Project them out 1/2 inch over the course below and slope the bricks 1/4 inch per foot to drain off water. Use solid bricks on the ends of the tread.
14. Attach a line to the front edge of the bricks and fill in between. Use the mason's rule as the bricks are being laid to make sure the step works whole bricks.
15. Lay another rowlock brick behind the first one the same way to complete the tread. See the plan.
16. Behind the tread on the outsides of the step, lay a rowlock header back to the porch wall. Fill in the hollow portion behind the rowlocks.
17. Measuring back from the edge of the tread, mark off a 12-inch line with a pencil. Make it all the way across the step.
18. Lay the second step the same way as the first.
19. Lay the third step the same as the other two, however, it is necessary to cut and lay half brick (bat) rowlocks behind the first rowlock to form the 12-inch tread.
20. Point up any holes with mortar. Tool the mortar joints with the large convex jointer. (Tool the mortar joints when they are thumbprint hard.)
21. Brush the steps off lightly when they are dry enough that they do not smear. An old burlap bag can also be used to rub off the bricks after they have dried sufficiently. This removes mortar stains.
22. Have the instructor evaluate the project.

Unit 12 GARDEN WALL CONSTRUCTION

OBJECTIVES

After studying this unit the student will be able to

- list the important considerations to follow when building garden walls.
- describe the use of pilasters in garden wall construction.
- describe the principle and construction of a serpentine wall.

INTRODUCTION

Garden walls are built mainly to separate the grounds of an estate or border a property. They add charm as well as pleasant areas to expand the home proper. A garden wall can be very simple in construction or very elaborate. They can be built in a straight line, curved, or with a series of piers and pilasters, according to the taste of the architect and homeowner. Garden walls can be built of brick, stone, concrete masonry units, and so forth. This unit deals with the building of brick garden walls, because their construction utilizes some of the most decorative pattern bonds found in brickwork.

MATERIALS AND WORKMANSHIP

Garden walls are subjected to varying extremes in exposure as weather conditions change from season to season. Rain, snow, frost-and-thaw cycles, direct sun, and combinations of all of these conditions, test the durability of a garden wall. The quality of materials, design, and workmanship all determine whether the masonry wall will last.

BRICKS FOR GARDEN WALL CONSTRUCTION

Bricks for garden walls should meet the requirements for a grade SW (severe weathering) of the ASTM Standard Specifications for Facing Brick C216 (where exposed to view) or ASTM C62 (where not exposed such as below grade line).

Salvaged, used bricks are not recommended for garden wall construction. While the rustic appearance of a used brick is appealing, unless it is hard and in a good solid state it deteriorates rapidly when laid in an exterior garden wall, figure 12-1. The pink "salmon" bricks (which are softer because of their location in the old brick kilns) are an example of the type *not to use* in garden walls.

MORTAR

The recommended mortar for reinforced and nonreinforced brick garden wall construction is type S, consisting of 1 part portland cement, 1/2 part hydrated lime, and 4 1/2 parts sand by volume. This mortar conforms to ASTM Standard Specifications for Mortar for Unit Masonry, C270. It is a very durable high-bonding mortar.

WORKMANSHIP

Good work practices must be followed. All mortar joints are filled solid, including the collar joint which is between the thickness of a double brick wall. The striking of all mortar joints is also of prime importance, because any holes in the wall will allow the entrance of water.

The wall should be level and plumb unless designed differently for architectural effect. The work should be neat and kept clean of mortar stains. This helps eliminate the amount of cleaning to be done at the end of the job.

Regardless of the weather or season, the wall should be covered with a canvas or plastic sheet while being constructed. The more moisture that can be prevented from entering the wall, the less likely that efflorescence will take place. A board or plank laid on top of the wall at the end of the workday is an unsatisfactory covering, since a driving rain can penetrate under the board.

The thickness of the mortar head joints and the bed joints should be the same. Varying sizes of head joints indicate poor workmanship.

Fig. 12-1 Old, used bricks that have been relaid in a garden wall. Notice how the faces of the bricks in the center have flaked off and are deteriorating. The only way to correct this is to cut out the old brick and replace with new brick.

FOOTING AND FOUNDATIONS

Although garden walls do not support any load other than their own, the same rules that apply to any foundation apply to garden wall foundations. The footing must be below the frost line and the masonry materials used as the base or foundation should be sound and laid in well-filled mortar joints. Concrete blocks make an excellent base for a brick retaining wall. Concrete is usually used for the footing.

COPING

The appearance of the top of the wall may differ depending on the type of cap or coping laid on the wall. There must be some type of coping if water is to be prevented from entering the wall.

Some general recommendations to follow in laying the coping on a garden wall are
- Project the top course on both sides of the wall approximately 1/2 inch to provide a drip.
- Fill all mortar joints and strike them with a flat finish, using a slicker jointer or the flat blade of a trowel.

- Slope the coping slightly to prevent the water from laying on top of the wall.
- If the cap is brick and more than 8 inches in width, it should be a bonded cap.

STRAIGHT WALL TYPE

The simplest type of garden wall is a straight brick wall with no piers or pilasters. These are generally low in height; if high, they must be reinforced to withstand the lateral stress caused by the wind and weather changes. It is recommended that this type of wall be at least 8 inches thick.

The proper mortar joint finish can help to make the wall blend in with the surroundings and give a more pleasing appearance. For example, a

Fig. 12-2 Mason tooling mortar joints with a convex striking tool. The convex jointer creates a half-round finish in the mortar.

V joint looks good with a rough finished brick. If a light color colonial brick is used, as in colonial restoration work, a grapevine joint creates a pleasant effect. In contrast, a smooth brick wall looks good with a concave joint. Regardless of the mortar joint finish chosen, good workmanship must be practiced when tooling the joints, figure 12-2.

ADDING PILASTERS TO A GARDEN WALL

A *pilaster* is a masonry column that is built into a wall at selected intervals and projects out from the face of the wall on one or both sides. Pilasters can be added to a garden wall for design purposes and for extra strength, figure 12-3. The brick wall should be bonded into the pilaster by interlocking the units, as shown in figure 12-4. Pilasters also serve as a brace and add depth to the wall. They do, however, cost more than a wall without breaks. Gates and decorative lighting can be installed to add to the general appearance of the wall.

Fig. 12-3 A brick garden wall built on a curvature surrounding an estate. The brick pilasters are used to strengthen the wall.

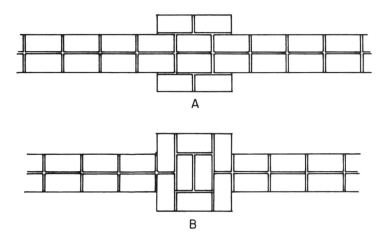

Fig. 12-4 Layout views of two sections of garden wall showing how the pilaster is bonded onto the wall (A), and tied on alternating courses of bricks (B).

A pilaster can be filled with mortar, concrete, grout, and steel rods to provide even greater strength, especially if the wall is high, long, or subjected to strong winds. Footings under pilasters should always be larger than for a regular wall to distribute the increased weight of the filled pilaster.

When building a garden wall that has pilasters, it is very important to bond the wall and pilaster together by tying the masonry units into each other. Refer to figure 12-4 to see how the pilaster can be interlocked with the masonry units of the wall to form a tie. Pilasters usually project out from the face and back of a garden wall 4 inches, but this distance may vary depending on the size of pilaster.

A *pier* is an unattached column of masonry that is built between openings. In garden wall construction, a series of piers are often constructed with a chain connected between them to serve as a type of fence.

Patterns And Designs Of Brick Garden Walls

There are many designs or patterns that can be built in garden panel walls. They are, of course, more expensive to construct than a straight wall due to the time required to lay the intricate patterns. The patterns appear more pleasing if separated by pilasters at varying intervals. The garden wall should be at least 8 inches wide unless it is specially designed on a curvature. The following three designs are typical of pattern bonds used in decorative brickwork.

Basket Weave. One of the most popular bonds is the *basket weave* in which three bricks are alternated in a soldier and stacked stretcher position. It is even more effective if the basket weave design is slightly recessed from the face of the wall. Figure 12-5, page 170, shows that the basket weave bond can be laid regular (view A) or on a diagonal (view B).

170 ■ Section 3 Building Outdoor Structures

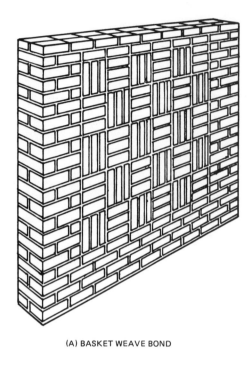

(A) BASKET WEAVE BOND

(B) DIAGONAL BASKET WEAVE BOND

(C) HERRINGBONE BOND

Fig. 12-5 Basket weave and herringbone bonds that are popular in garden wall construction.

Herringbone Bond. The *herringbone* bond is laid in a zigzag pattern on a 45-degree angle. Figure 12-5, view C, shows how this pattern looks when built into a brick wall.

Diamond Bond. The *diamond* bond, figure 12-6, is one of the more difficult patterns to lay. One brick out of position can cause the entire bond to be wrong, destroying the effect of the diamond pattern. Diamond designs are formed in the panel wall by following a diamond bond plan which shows the location of each brick forming the diamond design. The diamond pattern can be highlighted by using light or dark stretchers for the bricks that form the diamond.

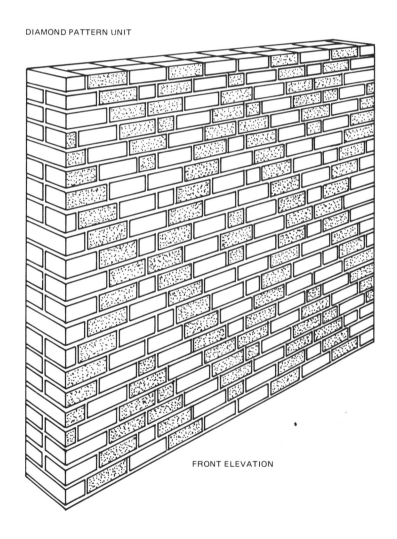

Fig. 12-6 A brick panel laid in a diamond pattern. Notice that the arrangement of the bricks forms a diamond design.

172 ■ Section 3 Building Outdoor Structures

Fig. 12-7 Serpentine brick wall at the University of Virginia built from designs by Thomas Jefferson.

SERPENTINE WALL

The *serpentine wall* derives its name from its shape, which resembles a serpent or snake. Early examples of serpentine walls in the United States are in Williamsburg, Virginia and at the University of Virginia in Charlottesville, figure 12-7.

This unique curving design provides lateral (sideways) strength to the brick wall so that it can usually be built 4 inches thick without any additional support. Since the wall depends on its shape for lateral strength, the degree of curvature is very important.

The following general rule is based upon the successful performance of many serpentine walls over the years. The radius of the curvature of a 4-inch wall should be no more than twice the height of the wall above finish grade, and the depth of the curvature should be no less than half the height. Figure 12-8 gives the details of construction of a typical serpentine garden wall. Although serpentine walls can be built of other masonry units and stone, bricks are more suited to the curving shape of the wall, due to their size.

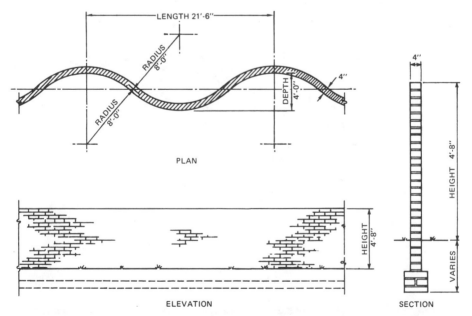

Fig. 12-8 Details of construction for a typical serpentine garden wall

ACHIEVEMENT REVIEW

Select the best answer from the choices offered to complete each statement. List your choice by letter identification.

1. The major problem with using salvaged, used bricks in garden wall construction is that the
   a. colors fade.
   b. bricks are all different sizes.
   c. bricks are softer than new bricks.
   d. supply of bricks is limited.

2. The recommended mortar for use in garden wall construction is type
   a. M.
   b. S.
   c. O.
   d. N.

3. The best joint finish to use on top of a garden wall is
   a. concave.
   b. flat.
   c. raked.
   d. grapevine.

4. The practice of not covering the top of a garden wall at the close of the workday can lead to
   a. efflorescence.
   b. brick deterioration.
   c. settlement of the wall.
   d. excessive shrinkage of the mortar joints.

5. If a raked joint is used on the face of a garden wall its depth should be no more than
   a. 1/4 inch.
   b. 3/8 inch.
   c. 1/2 inch.
   d. 5/8 inch.

6. A brick column that is built in a garden wall and projects on one or both sides is called a
   a. chase.
   b. pier.
   c. pilaster.
   d. post.

7. A popular brick bond used in garden wall construction in which three bricks are laid in a soldier position and alternated with three bricks laid in a stretcher position is called a
   a. stack bond.
   b. herringbone bond.
   c. diamond bond.
   d. basket weave bond.

8. The design of a serpentine wall is
   a. square.
   b. rectangular.
   c. curved.
   d. in the form of a triangle.

9. A serpentine wall is usually built of
   a. bricks.
   b. stones.
   c. concrete blocks.
   d. tile.

10. The design of a serpentine wall depends on its shape for
    a. compressive strength.
    b. lateral strength.
    c. tensile strength.
    d. reverse strength.

PROJECT 10: BASKET WEAVE PATTERN IN A PANEL WALL

OBJECTIVE

- The student will lay out and build a brick panel wall in a basket weave pattern. This project gives the student practice in working with a variety of brick positions that form a pattern.
  NOTE: Patterns such as these are used in garden walls, fireplaces, and building fronts.

EQUIPMENT, TOOLS, AND SUPPLIES

mixing tools
mortar pan or board
brick hammer
plumb rule (2 feet and 4 feet)
2-foot framing square
V sled runner jointer
modular rule
brush

brick set chisel
trowel
pencil
chalk box
mason's line
pair of line blocks
one 8-inch x 16-inch concrete block to place the level in when it is not in use

NOTE: The student will estimate the number of bricks used in the project from the plan in figure 12-9. Based on the number of bricks, the student also estimates the amount of mortar needed for the project.

SUGGESTIONS

- When laying out the project, extend the chalk line longer than the actual size to serve as a reference line if mortar blots out the layout line.
- Mix only the amount of mortar that will be used, to avoid waste or premature setting.
- The bricks that form the basket weave design should be handpicked for size.
- Brick coursing is represented by No. 6 on the modular rule.
- Make all cuts with the brick set chisel or masonry saw.
- For a more dramatic effect, stretcher bricks can be tooled with a V jointer and the basket weave bricks with a 1/4-inch raked joint. This decision is the choice of the instructor.
- Follow good safety practices when cutting any bricks and keep the work area free of chips or pieces of bricks.
- This project can be built by two students as a team due to the large amount of materials involved.

Unit 12 Garden Wall Construction ■ 175

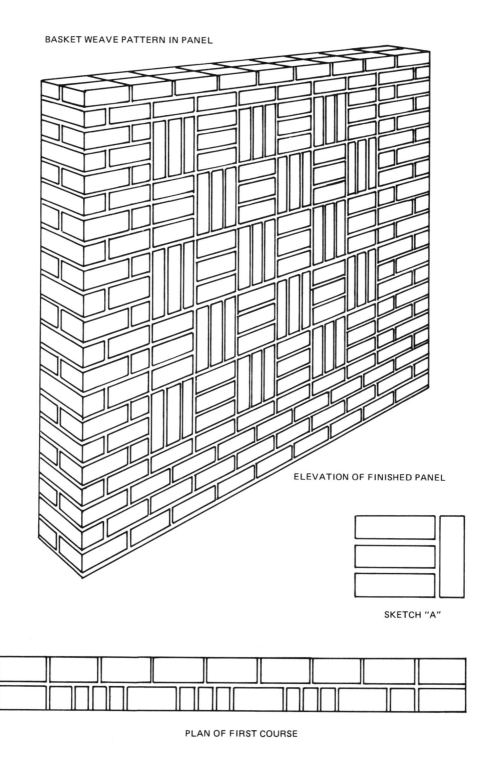

Fig. 12-9 Project 10: Basket weave pattern in a panel wall.

PROCEDURE

1. Mix the mortar in the work area. (If in a shop situation, use a mix of 1 part hydrated lime to 3 parts sand.)
2. Lay out the project on the floor using a chalk line and steel square.
3. Lay up the project where the panel work starts with the aid of a level and line.
4. Strike the work with a V jointer and a brush.
5. Lay up three courses of brick 12 inches long (1 1/2 bricks) on each side.
6. Attach the line and lay a course in the basket weave pattern as shown on the plan in figure 12-9.
7. Plumb all vertical joints to make sure they are straight.
8. Attach the line to the back of the project and fill in the remaining back wall in a running bond.
9. Continue to lay up courses as in steps 5, 6, 7, and 8.
10. Tool the mortar joints when they are thumbprint hard with either a V joint or raked joint.
11. When the top of the basket weave panel is reached, lay the 2 remaining courses of stretchers to the line across the panel as shown on the plan.
12. Check all work to be sure it is plumb, level, and free from holes. Finish striking and brushing.
13. Have the instructor evaluate the project.

PROJECT 11: DIAMOND BOND PATTERN IN A PANEL WALL

OBJECTIVE

- The student will lay out and build a brick panel wall in a diamond bond. This project gives the student practice in working with an intricate bond that utilizes stretchers and headers in a pattern.

  NOTE: Pattern bonds such as these are used in garden walls, fireplaces, and ornate building fronts.

EQUIPMENT, TOOLS, AND SUPPLIES

mixing tools
mortar pan or board
brick hammer
mason's trowel
plumb rule (2 feet and 4 feet)
2-foot framing square
grapevine jointer (suggested but
  left up to discretion of instructor)
modular rule

brick set chisel
chalk box
mason's line
2 line blocks
pencil
brush
one 8-inch x 8-inch x 16-inch block
  to place the level in when it is not
  in use

NOTE: The student will estimate the number of bricks by studying the plan in figure 12-10. Based on the number of bricks, the mortar can also be estimated.

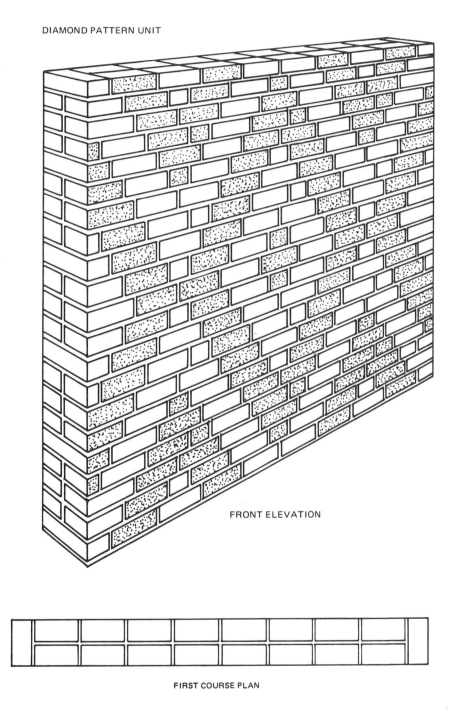

Fig. 12-10 Project 11: Diamond bond pattern in a panel wall.

SUGGESTIONS

- Lay out the first course dry to determine the bond.
- Lighter or darker bricks can be used to form the diamond pattern.
- Mix only enough mortar ahead that is expected to be used, to avoid waste.
- Make the cuts one course ahead.
- Strike up the mortar joints as soon as they are thumbprint hard.
- Space materials approximately 2 feet back from the wall line.
- Follow good safety practices when cutting any masonry units by wearing eye protection.
- Study the plan for each course before attempting to lay that course. One mistake can ruin the pattern of the bond.
- Check each course after it is laid to make sure it is correct according to the plan.
- It is recommended for two students to work as a team due to the size and complicated bond of the project.

PROCEDURE

1. Mix the mortar and load it on the board or pan. (Use mortar consisting of 1 part lime to 3 parts sand if working in a shop situation.)
2. Lay out the project on the floor with a chalk box and square.
3. Lay out the first course dry as shown on the plan in figure 12-10.
4. Lay the first course in mortar.
5. Lay up a 4-course lead at each end of the project, according to the plan.
6. Lay the bricks to the line on both sides of the wall. Cut the bricks to size for the diamond pattern face panel, as shown on plan.
7. Use the No. 6 scale on the modular rule for all coursing.
8. Tool the mortar joints as soon as they are thumbprint hard.
9. Check all vertical joints with the plumb rule to make sure the pattern is maintained.
10. Continue to build up short leads and fill in the line until the project is finished.
11. Strike the remaining joints and brush the project.
12. Recheck the bond for accuracy as shown on the plan.
13. Have the instructor evaluate the project.

Unit 13 RETAINING WALLS

OBJECTIVES

After studying this unit the student will be able to

- describe where retaining walls are used and list some of the major problems in their construction.
- define the two types of retaining walls and how reinforcement is installed to strengthen the wall.
- explain the construction procedures of a typical retaining wall.

PURPOSE OF A RETAINING WALL

The main purpose of a retaining wall is to hold back and contain the earth or fill materials. Low retaining walls (under 6 feet in height) are an example of this type of wall. This is the type of retaining wall the mason comes in contact with most often. Masonry retaining walls higher than 6 feet require special planning and reinforcement because of the pressure of the earth filled in behind the wall.

Retaining walls protect the slopes and banks around a property and can be the answer to terracing the land where there are grade problems or a steep slope. They can and should blend in with the landscaping plan. When a cut is made in a hillside, this often causes erosion, resulting in a ditch or loss of the topsoil. The retaining wall can prevent this erosion by holding the moisture in the soil and preventing the movement of the earth.

USES OF RETAINING WALLS

Some of the more common uses of retaining walls are to maintain the grade at property lines and to restrain earth at driveways (especially the type of driveway that leads to a garage in the basement). They are also used for containing and holding earth around trees where the grade has been raised, keeping slopes from eroding, and forming terraces or garden areas.

A retaining wall is built out in the open with no protection from the extreme conditions of heat and cold. It is subject to harsh weather such as rain and snow. In addition, it is expected to hold back a great amount of weight on only one side. Therefore, it has to be built much stronger than an ordinary wall. Good drainage must be provided behind and through the wall. Well-filled mortar joints and good workmanship are important. Retaining walls are often the most neglected part of the masonry construction. They are usually an afterthought to the design and construction of a building, and therefore are not built well.

WHY RETAINING WALLS FAIL

The retaining wall is the most likely part of the masonry to require repairs in a short period of time. There are several reasons for this. First, the wall may not be designed

180 ■ Section 3 Building Outdoor Structures

Fig. 13-1 A poorly built retaining wall. Notice that the wall is cracked. Anchor bolts with plates attached are used to help keep the wall in position.

Fig. 13-2 This retaining wall is leaning and is about to collapse due to lack of drains at the bottom of the wall.

strong enough for what it is expected to do. Remember that no two jobs are exactly alike. Each job requires individual planning and designing. Second, the workmanship is often poor and shortcuts are taken in the construction procedures. Third, if no drainage is provided, pressure builds up from surface water that settles in back of the wall.

Adding all of these factors together, the wall is doomed to failure before it is completed. Figures 13-1 and 13-2 are examples of poorly constructed retaining walls.

REINFORCING THE WALL

All retaining walls should be reinforced with steel and grout, depending on the pressure they must contain. The only exception to this is if the earth that is being retained behind the wall is stable. If the earth is stable, it does not exert pressure on the wall even if it is in a saturated condition. However, this condition seldom exists, and therefore all retaining walls should be reinforced to resist earth pressure.

There can be a small amount of reinforcing steel if the wall is short or if the earth is stable. The absence of reinforcing steel and grout is the most abused practice in the construction of a retaining wall. It is better to spend a little more money in the beginning and build a successful wall, than to have to tear down and rebuild the complete wall.

Large, complex retaining walls are designed by the engineer or architect. The plans that are supplied to the mason often do not contain construction plans for the wall. Its construction is left up to the judgment and experience of the mason working on the job.

The information in this unit helps the student mason lay out and construct a simple low retaining wall. This type of wall is found on most small jobs.

FACTORS TO CONSIDER WHEN BUILDING A RETAINING WALL

Fill

If all loose fill is being placed behind the wall, there is likely to be more pressure as the earth settles. Rainwater soaks through the loose dirt more quickly than earth that has been undisturbed. Drain tiles or weep holes are built into the wall just above the grade line on the face of the wall. These allow the water to drain away and relieve the pressure on the wall.

Rain Spouts

Rain spouts may empty in the area of the retaining wall. This allows much of the water that has been collected from the roof to run in behind the wall. There are solutions to this problem. In one method, the leader from the downpipe is piped under the ground to a point where it can be drained away from the house and away from the retaining wall. In another method, the leaders can be extended by adding longer lengths of pipe. Either method works but the underground pipe looks the best. Although this is not the work of the mason, the mason should suggest to the homeowner or builder to use one of these methods if the retaining wall is in a problem area. This is especially true when repairing or rebuilding a retaining wall that has failed so it will not fail again.

Size Of The Wall

Retaining walls should never be built less than 8 inches thick. All retaining walls should be reinforced with steel and concrete. The best all-around size for a masonry retaining wall is 10 inches thick. This size wall is built with a 4-inch wide masonry unit on each side of the wall, and a 2-inch reinforced center.

Metal Wire Joint Reinforcement

If steel rods are not being used, metal wire joint reinforcement should be installed in the mortar bed joints every 16 inches in height, figure 13-3, page 182. The addition of this wire is an inexpensive way to increase the resistance to cracking or shifting of the bed joints. Steel rods and grout must be used if there is much back pressure. Metal wire joint reinforcement and steel rods are available from building supply dealers.

FOOTING

The wall will not be any stronger than the footing it rests on. A good quality concrete mix of 5 bags of portland cement per cubic yard of concrete is recommended. This tests at 2,500 psi. The footing should be wide enough to support the wall and resist tipping over from the pressure exerted from the backfill. Figure 13-4, page 182, illustrates how a cantilever wall and footing help to resist back pressure from the fill.

182 ■ Section 3 Building Outdoor Structures

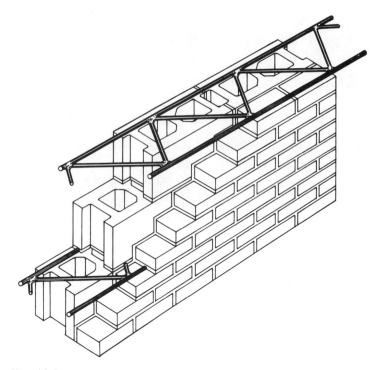

Fig. 13-3 Metal wire joint reinforcement laid every 16 inches in height strengthens a masonry retaining wall.

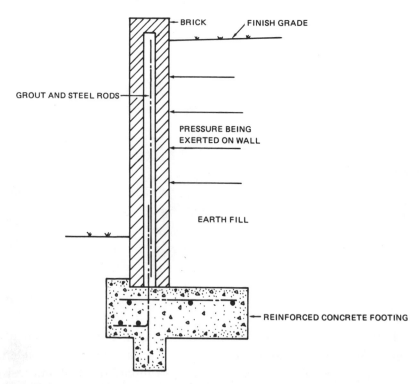

Fig. 13-4 A cantilevered wall and footing help to resist the back pressure from fill.

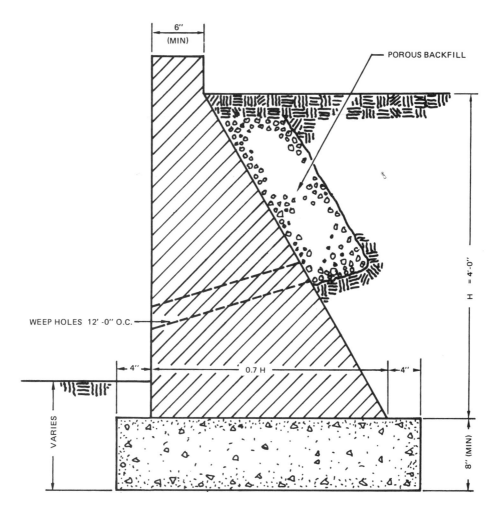

Fig. 13-5 A gravity retaining wall including weep holes. The weep holes allow the water to escape from the back of the wall.

TYPES OF RETAINING WALLS

There are two general classifications of retaining walls: gravity and cantilever. The gravity retaining wall, figure 13-5, is not reinforced with steel or grout. The weight and mass of the wall must resist the pressure of the earth being retained. The weight of the wall equals or exceeds the weight of earth being held.

A gravity wall is built plumb on the face side. The inside (earth side) is built wider at the base. As the wall is constructed, it is built on an incline, sloped or stepped inwards up to the point where the surface of the backfill is located. When completed the gravity wall serves as a wedge to retain the earth backfill.

A cantilever retaining wall has an extra wide footing that ties into the earth on the inside bottom of the wall to resist the pressure. The use of reinforcement and the wider footing make the cantilever very strong. There are several variations of cantilevered walls but figure 13-6, page 184, shows a typical reinforced masonry cantilever wall.

184 ■ Section 3 Building Outdoor Structures

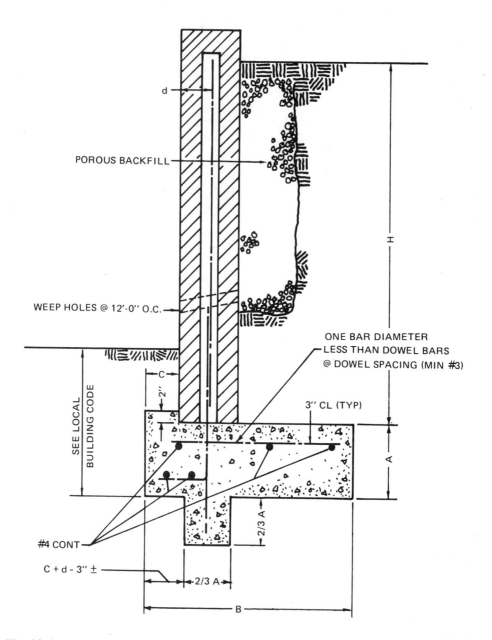

Fig. 13-6 A reinforced masonry cantilevered wall. Notice the wide spread of the footing so it will tie back into the earthen bank.

WALL SHAPES THAT RESIST PRESSURE

Battered

A design that is very popular for stone walls is to batter the wall back. *Batter* means to build the wall on an incline (sloping inward) from the bottom to the top of the wall. The theory is that the sloping in of the wall will help the wall resist pressure from the backfill. The rake or incline of the batter is determined by making a batter form board. The form is built plumb on the front and a board is nailed on the back to fit the slope being followed.

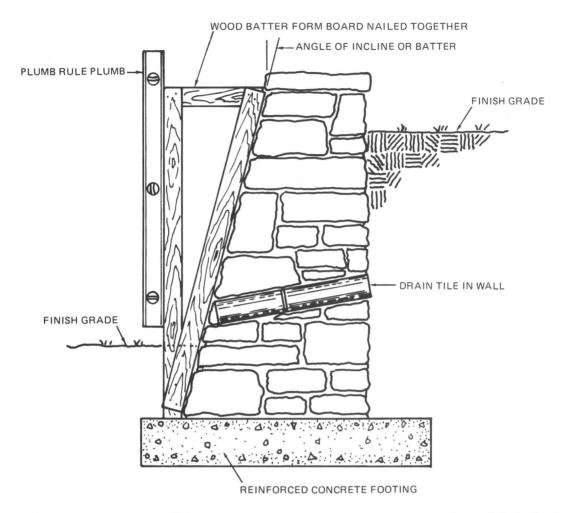

Fig. 13-7 Battered stone wall built by using a wood batter form board. The wall is inclined on the face to the top while the back is built approximately plumb.

The plumb rule is held against the front of the form and the masonry work is fitted to the angled back. Figure 13-7 shows how the batter form is used for building a retaining wall.

Stepping Back Into The Bank

Another method of building a retaining wall begins by building up the front wall plumb. While this is being done, intersecting wing walls are stepped at a 90-degree angle into the fill area. At the point where the intersecting walls are stepped back into the earth, a footing should be provided for them to set on, the same as when building any masonry wall. Reinforcement should also be installed in the wall. The stepped walls are tied into the main retaining wall every other course by interlocking the units in mortar rather than using a metal tie. Figure 13-8, page 186, shows a retaining wall that has a stepped masonry wall built into the earth bank behind the wall.

Adding Pilasters For Greater Strength

Masonry pilasters (projecting piers that are built into the wall) can be built at intervals along the back of the retaining wall to provide greater strength. This is usually done when

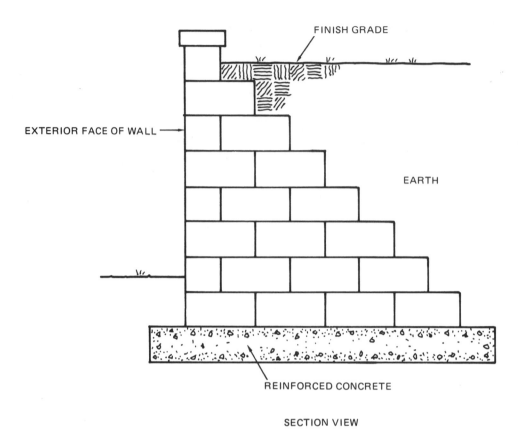

Fig. 13-8 Retaining wall that has a stepped masonry wall built back into the earthen bank to act as a brace. This is a type of gravity wall.

there is a great amount of pressure from the fill. The pilaster should be tied into the main wall every other course by means of a masonry tie. It is also recommended to reinforce the wall and pilasters with steel and grout for more strength. This is called a *counterfort* type of cantilever wall. Figure 13-9 shows a retaining wall that has pilasters built in back of it.

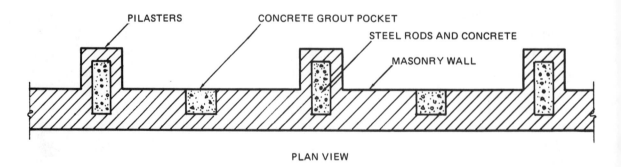

Fig. 13-9 Counterfort wall that has reinforced pilasters built in back of the wall for extra strength. In addition, there are group pockets of concrete spaced between the pilasters. These are for a retaining wall where there is much pressure on the back of the wall.

MATERIALS

Brick

Retaining walls can be built from different types of masonry units to match the landscape or existing structure. Brick is a very popular choice. Select a hard, kiln-fired brick, grade SW, Specifications ASTM C216 or C62, facing brick or building brick. Soft brick does not last in a retaining wall. Information is available on the classification of bricks from the dealer where they are purchased.

Since bricks are small, they are very adaptable to designs and arrangements of bond. The colors and textures also make them a good choice for a retaining wall. Usually, old or salvaged bricks are not good for retaining walls because they deteriorate (flake off and break up) rapidly once exposed to the weather.

Concrete Masonry Units

Concrete masonry units also make a very practical, inexpensive retaining wall. A concrete masonry wall can be constructed more rapidly than a brick wall due to the size of the units. Concrete masonry units for retaining walls should meet the requirements of Specifications For Hollow Load-Bearing Concrete Masonry Units: ASTM C90-69. Units having two or three hollow cells are used. (The unit with three cells is more popular because the larger cell space allows easier placement of vertical reinforcement rods and grout.) A concrete masonry retaining wall also requires drains installed through the wall to relieve the pressure.

Stone

The use of stone makes a very attractive retaining wall because it blends into the natural setting of the landscape, figure 13-10. Any relatively hard stone is suitable. The design and shape of the stone used depends on individual preference. A thick stone wall can either be built on the gravity wall principle or reinforced for extra strength. Drains should also be provided through the stone wall.

Mortar

Mortar should be type M consisting of 1 part portland cement to 1/4 part hydrated lime and between 2 3/4 and 3 parts masonry sand proportioned by volume. The water should be clean and the sand free from silt.

Steel Reinforcement Rods

The steel rods should be free from rust. They should also meet ASTM specifications for deformed bars and steel wire.

Grout

Grout is used for reinforced retaining walls, and can be either fine or coarse. Fine

Fig. 13-10 Rubble stone retaining wall and walk.

grout is used for smaller spaces and coarse grout is used for larger spaces. The following grout proportions by volume are stated in ASTM C476.

Fine grout consists of 1 part portland cement, 1/10 part type S hydrated lime, and 3 parts sand mixed with enough water to allow it to be poured into the wall. Grout should be mixed for at least 5 minutes and should be used within 2 1/2 hours after mixing.

Coarse grout is for large spaces. It consists of 1 part portland cement, 1/10 part hydrated lime, and 2 1/4 to 3 parts fine aggregate and 1 to 2 parts coarse aggregate, mixed with enough water to allow it to be poured into the wall.

Drain Tile

Four-inch clay or concrete clay tile should be used. Twelve inches is the recommended length for tiles for the walls.

CONSTRUCTING THE FOOTING FOR A LOW RETAINING WALL

Place the footing below the frost line on a firm undisturbed bed. If the soil is soft or silty clay, it is advisable to place 4 to 6 inches of well-compacted sand or gravel before placing the footing. Footings for masonry retaining walls should be reinforced both horizontally and vertically.

As the footing is being placed, insert short lengths of steel rods. Allow them to project up vertically at least 12 inches above the top of the footing. When the first course of masonry units is laid, the mortar is slushed in tightly around the rods. If vertical wall reinforcement rods are to be installed, the next rod should be overlapped with the stub rod. The purpose of the short stub rod is to prevent the wall from shifting forward on the footing. Do not trowel the top of the footing smooth because it bonds better to the mortar if left slightly rough.

The footing is one of the most important parts of the wall. Therefore, all precautions should be taken to make sure it is installed correctly. Failure of the footing means failure of the wall. Figure 13-11 shows the correct installation of stub rods in the footing.

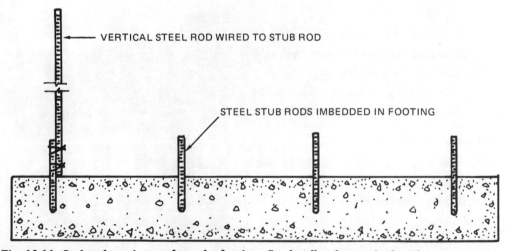

Fig. 13-11 Stub rods project up from the footing. Steel wall rods are wired to these stub rods to give greater strength and to keep the wall from shifting on the footing.

BUILDING THE WALL

Use a full mortar joint under the first block course to ensure a good bond. If concrete blocks are being laid, bed all of the cross webs on every course for greater strength. This is not done in a regular wall. In a wall reinforced with vertical steel rods and grout, the cross webs where the grout is being placed do not need to be bedded with mortar.

When a brick-faced retaining wall is being laid, the common bond should be used. There is a full header every sixth or seventh course to bond the wall together with the backing materials.

Install the reinforcement wire in the mortar bed joints every 16 inches and mortar in fully. Overlap each length of wire a minimum of 6 inches if more than one piece is being used.

INSTALLING THE DRAIN TILE

Drain tile, figure 13-12, is laid in the wall at grade level to drain water collected from the back area of the wall. Mortar the tile in place with a slope to the front so the water can run through. A copper or aluminum screen can also be installed over the opening of the tile from the earth side. This prevents the stone or earth from clogging up the tile. Large crushed stone can be used instead of the screen to lodge against the tile. It is important that the stone surround the pipe and not the soil.

WATERPROOFING THE WALL

The back of the retaining wall that is below grade should be waterproofed with 1/2 inch of parging mortar or asphalt. This helps prevent water from saturating the wall. It also prevents mineral salts from leaching through the wall and appearing as efflorescence or staining. Waterproofing also helps prevent moss or other fungus plant life from growing on the face of the wall.

EXPANSION JOINTS

Long retaining walls should be broken into panels 20 to 30 feet long by means of vertical expansion joints. Joints should be designed to resist shear and lateral forces while allowing a longitudinal movement.

CAP

A masonry cap is laid in mortar on top of the wall to prevent the water from standing or going down inside the wall. The following are all suitable caps for retaining walls: Bricks projected as a rowlock, preformed stone, or concrete. The cap should project out 1/2 inch on each side of the wall to provide a drip.

Fig. 13-12 Drain tile installed in a retaining wall surrounding a patio. Notice that the drain tile are slightly above the finish grade.

REINFORCING A CONCRETE RETAINING WALL

The retaining wall is reinforced to make it stronger than a conventional wall. The resulting wall combines the beauty of the masonry wall with the strength of a reinforced concrete wall.

The masonry units act as a form. The concrete grout and steel rods are installed inside the cells of the blocks. Care should be taken to maintain the hollow cell until the steel rods are inserted and grout is poured in the wall. Since mortar that has squeezed out from the joints keeps the grout from filling in around the steel rods, the drippings should be broken loose with a long strip of wood before they harden. Cleanout holes should be left at the bottom of the wall to remove the drippings and then patched with masonry before pouring the grout.

Some type of horizontal reinforcement should be laid in the wall lengthwise every 16 inches in height. Vertical rods are not installed until the masonry wall has been built and grouting has been done.

HIGH-LIFT GROUTING

When building walls higher than 6 feet and using grout to fill, it is called *high-lift grouting*. The wall should set at least 24 hours before any grouting takes place. Grout should be poured in layers of 4 feet, and one hour setting time should be allowed between successive layers. Compact the grout tightly by puddling or vibrating.

LOW-LIFT GROUTING

Both low-lift and high-lift grouting can be used with brick or block work. When grouting a low brick wall (usually under 6 feet in height), it is recommended to pour at 8- to 18-inch intervals. Puddle or rod to ensure complete filling of the wall. The minimum grout should be 3/4 inch around all vertical reinforcement.

Brick units, concrete masonry units, and stone can all be reinforced. The techniques vary but the goal of all methods is a stronger retaining wall. Figure 13-13 shows a typical reinforced concrete block wall. Refer to the table in figure 13-14 for the various sizes of walls.

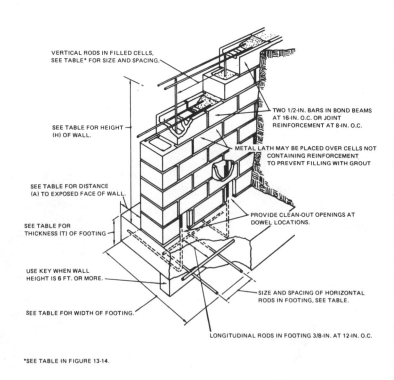

Fig. 13-13 Reinforced concrete masonry retaining wall.

REINFORCED CONCRETE MASONRY RETAINING WALLS					
8" WALLS					
Height of Wall (h)	Width of Footing	Thickness of Footing (t)	Distance to Face of Wall (a)	Size and Spacing of Vertical Rods in Wall	Size and Spacing of Horizontal Rods in Footing
3'-4"	2'-4"	9"	8"	3/8" @ 32"	3/8" @ 27"
4'-0"	2'-9"	9"	10"	1/2" @ 32"	3/8" @ 27"
4'-8"	3'-3"	10"	12"	5/8" @ 32"	3/8" @ 27"
5'-4"	3'-8"	10"	14"	1/2" @ 16"	1/2" @ 30"
6'-0"	4'-2"	12"	15"	3/4" @ 24"	1/2" @ 25"
12" WALLS					
6'-8"	4'-6"	12"	16"	3/4" @ 24"	1/2" @ 22"
7'-4"	4'-10"	12"	18"	7/8" @ 32"	5/8" @ 26"
8'-0"	5'-4"	12"	20"	7/8" @ 24"	5/8" @ 21"
8'-8"	5'-10"	14"	22"	7/8" @ 16"	3/4" @ 26"
9'-4"	6'-4"	14"	24"	1" @ 8"	3/4" @ 21"

Fig. 13-14 Table of various reinforced concrete masonry retaining walls

RETAINING WALLS AROUND TREES

The main function of a retaining wall is to hold back banks or earth on slopes. Sometimes, retaining walls are also needed around trees when the grade has been raised. Minor fills (6 inches or less in depth) do not harm the tree or rob it of air, water, or minerals needed for its growth.

One of the most popular types of retaining walls is the circular retaining wall. The round design of the wall makes it strong since the stress is put equally on all parts of the wall. The wall also adds design and beauty to the property. Figure 13-15 shows a circular retaining wall around a tree. These types of walls are also called *tree wells*.

Fig. 13-15 Circular brick retaining wall, more commonly called a tree well.

192 ■ **Section 3** Building Outdoor Structures

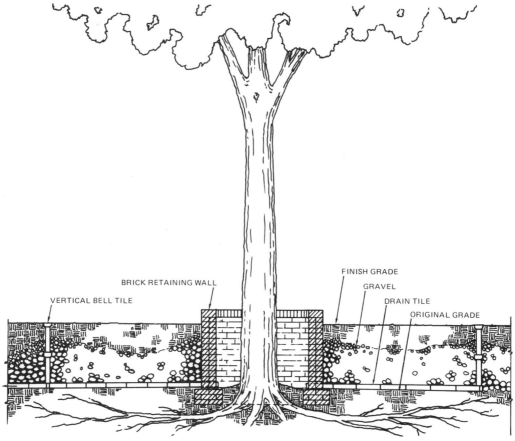

Fig. 13-16 This masonry tree well retaining wall is used when the tree is deep below the level of the existing grade.

WHEN FILL IS NEEDED AROUND A TREE

If it is necessary to fill in around a building or raise the grade, the tree's roots may be buried too deep in the ground to survive. To prevent the roots from drowning, a retaining wall is built around the base of the tree. If the roots are very deep, a layer of gravel and a system of drain tile may have to be installed over the roots. The drain tile should slope away from the tree and be installed through the retaining wall so that the water does not fill the wall and drown the tree. A problem such as this should be solved by a landscape architect. Figure 13-16 shows a masonry tree well with drain tile and stone on the outer edges. This is for a tree that is buried deep in the ground.

LOWERING THE GRADE

A tree also has to be protected from too much lowering of the grade. Lowering the grade can be harmful to the tree unless proper attention is given to cutting the roots, pruning the branches, stimulating root growth, and watering. The best way to protect the tree from this type of situation is to build a retaining wall that terraces the grade. Figure 13-17 shows one method of building a retaining wall that forms a terrace and maintains the original grade line.

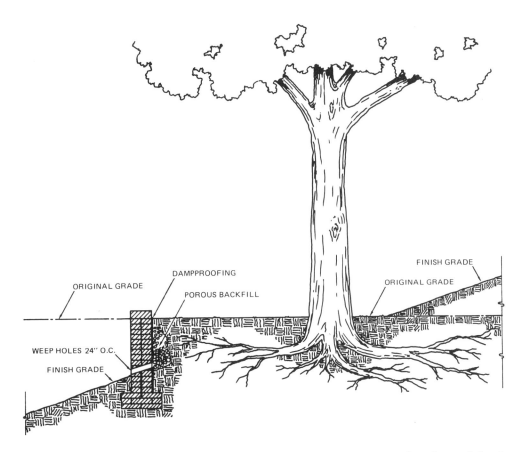

Fig. 13-17 A brick retaining wall is used to hold the earth around a tree when the grade has been lowered. This is done to protect the roots with enough earth.

Remember that the rules of good construction for building any retaining wall should be followed when constructing a tree retaining wall. It is important that footings be well built, below the frost line.

BACKFILLING

Backfilling with earth behind the retaining wall should be done with care. If reinforcement and grout has been used in the wall, do not backfill for at least 7 days to give the wall a chance to cure. Heavy equipment should not be operated next to the wall because the wall may push in or out. If there is a long period of rain, it is necessary to decide how long to wait before backfilling.

Stone should surround the drain tile where it is installed in the wall and up to the point where the rough grade is in back of the wall. This is done to ensure that the water drains down through the stone and out through the drain tile. The safest method of backfilling around a retaining wall (if the wall is not too large), is using a dirt shovel and tamping down by walking on the earth near the wall to settle it in place. While there are different types, designs, and kinds of materials that can be used for building retaining walls, safe working practices and good workmanship should always be followed during construction procedures.

ACHIEVEMENT REVIEW

Select the best answer from the choices offered to complete each statement. List your choice by letter identification.

1. Low retaining walls are classified as walls no higher than
   a. 4 feet.
   b. 6 feet.
   c. 8 feet.
   d. 10 feet.

2. The recommended thickness of a retaining wall is not less than
   a. 4 inches.
   b. 6 inches.
   c. 8 inches.
   d. 12 inches.

3. Masonry retaining walls should be reinforced with metal wire reinforcement in the bed joint every
   a. 8 inches in height.
   b. 16 inches in height.
   c. 20 inches in height.
   d. 24 inches in height.

4. To prevent the masonry units from shifting on the footing it is advisable to
   a. pour a large amount of grout in the top of the footing.
   b. install metal rods vertically in the footing.
   c. lay the first course on the soft concrete and let the units project partially above the footing.
   d. project steel rods below the bottom of the footing.

5. One variation of constructing a retaining wall is to batter the wall. Batter means to
   a. beat the wall with a length of lumber to make it irregular in shape.
   b. build the wall on an incline opposite to the thrust of the backfill.
   c. fill the wall with mortar and vibrate in solid.
   d. corbel the wall on both sides to increase the width for greater strength.

6. The most successful method of providing drainage through the retaining wall is to
   a. insert short lengths of a garden hose in the wall.
   b. lay drain tile behind the wall in stone.
   c. build drain tile in the wall on a slope.
   d. install weep holes every 3 units in length.

7. If mortar is used as waterproofing on the inside of the retaining wall, it is recommended to parge with a coat of mortar
   a. 1/4 inch thick.
   b. 1/2 inch thick.
   c. 3/4 inch thick.
   d. 1 inch thick.

8. Reinforced concrete masonry walls should not be filled with grout until the wall has set for a period of
   a. 8 hours.
   b. 16 hours.
   c. 24 hours.
   d. 36 hours.

9. One hour is allowed between each pouring of grout. The depth of each pour should not exceed
   a. 18 inches.   c. 4 feet.
   b. 2 feet.   d. 6 feet.
10. The main problem in constructing tree wells is to keep the tree from drowning. This can best be done by
   a. building ditches leading away from the tree.
   b. installing drain tiles.
   c. using French drains.
   d. building on a slope.

PROJECT 12: BUILDING A 12-INCH BRICK AND CONCRETE BLOCK RETAINING WALL

OBJECTIVE

- The student will lay out and construct a 4-inch brick facing wall backed up by an 8-inch concrete block wall. The project also includes the installation of 4-inch drain tiles in the wall. This type of retaining wall is typical of the ones built around the home to hold back banks.

EQUIPMENT, TOOLS, AND SUPPLIES

mixing tools
supply of clean water
mortar pan or board
set of basic masonry tools
wire cutters
three 4-inch drain tile, each 12 inches long

eight feet of masonry joint reinforcement wire
standard bricks, lime, and sand — student will estimate the amount needed for the project from the plans, figure 13-18

SUGGESTIONS

- Strike the chalk line a little longer than the actual size of the project. This provides a reference line to use as a guide as the project is built.
- Use 3/8-inch mortar head joints.
- Wear eye protection whenever mixing mortar or cutting masonry materials.
- Use full mortar joints between all of the masonry work.
- Keep any scraps of materials out from underfoot.
- When the plumb rule is not in use, place it upright in the cell of a block to keep it from breaking.
- Strike the mortar joints as soon as they are thumbprint hard.
- The project is designed so that the proposed finish grade line is level with the bottom of the rowlock cap on the inside of the wall. Therefore, the concrete block on the back of the wall is not tooled but is parged with mortar up to the bottom of the rowlock cap.

196 ■ Section 3 Building Outdoor Structures

NOTE: This project is not reinforced with rods and grout because it will be torn down and the materials reused in a shop situation.

PROCEDURE

1. Mix the mortar, 1 part lime to 3 parts sand, and load the mortar pan or board in the work area.
2. Strike the line for the brick wall on the floor.
3. Lay out the first course as shown on the plan, figure 13-18.
4. Continue to lay up the project until it is 3 courses high. Build a small lead on each end and attach the line to lay the wall in between.

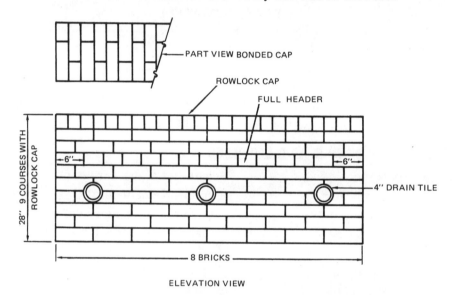

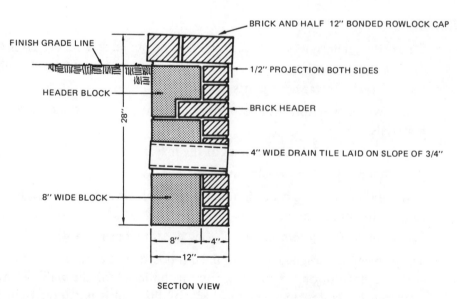

Fig. 13-18 Project 12: Building a 12-inch brick and concrete block retaining wall.

5. Tool the mortar joints when necessary.
6. Clean off all mortar from the inside of the brickwork. Strike a chalk line 12 inches back from the face of the wall for the block work.
7. Lay the 8-inch block backing wall to the chalk line and level with the brickwork.
8. Lay a length of wire joint reinforcement on the wall 12 inches wide. This ties the wall together and adds extra strength.
9. Install the drain tile as shown on the plan.
10. Fitting the bricks around and over the tile, build the wall 6 courses high.
11. Back up the concrete block work fitting around the drain tile.
12. Cut a 6-inch piece of brick on each end of the wall and lay the header course, bonding the bricks and blocks together.
13. Continue building the brickwork 2 more courses high. This is the height where the rowlock will be laid.
14. Back up the brickwork with a header block. If the header block is not available, a 4-inch block and 2 courses of bricks can be used instead.
15. Lay the rowlock cap on the top of the wall. Project it out on the front and back 1/2 inch to form a drip.
    NOTE: Because the wall is 12 inches thick, it requires a brick and a half to cover the top of the wall. Make the cuts with a brick set or on a masonry saw, if available. Be sure the cuts are all the same size since the difference will be very noticeable. Bond the cap so it staggers the joints.
16. Tool the mortar joints and brush the work.
17. Have the instructor evaluate the project.

Unit 14 OUTDOOR FIREPLACES

OBJECTIVES

After studying this unit the student will be able to

- list the important factors to consider when building an outdoor fireplace.
- describe how the foundation of an outdoor fireplace is constructed.
- explain how the firebox and grill are installed in an outdoor fireplace.

Outdoor fireplaces can be very simple or complex in construction depending on the owner's preference or needs. There are many different types and sizes. Outdoor fireplaces range in size and design from small and simple barbecues to large, elaborate fireplaces with grills, counters, fuel storage space, food work areas, ovens, and incinerators for the burning of paper and light materials. Before building an incinerator, be sure to check the local burning laws for restrictions on open burning.

Fireplaces can be built into one of the outside walls during construction of the house or they can be built separately. Often a barbecue or fireplace is built as a functional part of the patio or chimney, figure 14-1. The outdoor fireplace has expanded the market for mason's skills, a market that was not well developed years ago. The building of outdoor fireplaces and grills is a valuable source of work that is very satisfying and financially rewarding to the mason due to the custom design and high quality of workmanship that the owner will demand. It is also a very important part of the home improvement business (for example, patio construction). Work like this helps add to the mason's income so the mason does not have to depend only on new residential and commercial construction.

Fig. 14-1 A barbecue is built into the exterior chimney wall, utilizing the chimney of the house.

FACTORS TO CONSIDER WHEN BUILDING A BARBECUE OR OUTDOOR FIREPLACE

A *barbecue* is a raised masonry structure for the smoking, drying, or preparation of meat over an open fire. It is used for cooking all types of food, usually with charcoal.

Some important things to consider when building a barbecue are the cooking needs, appearance and design, good location so the wind current can carry the smoke away, accessibility to the kitchen of the home, cost, and the following of local building regulations. Some barbecues require a building permit depending on the size and cost. Building codes should be checked if there are any questions.

Location

The location is very important in order to get the fullest use of the barbecue or fireplace. A level, well drained, shaded area near the kitchen is ideal. A barbecue built in the open with no shade is not practical for use during hot summer weather. Plan for enough space to work around the cooking area and be sure the barbecue is not in the direct path of family traffic.

Know the prevailing wind pattern and position the barbecue or fireplace so that smoke and sparks will not blow on the picnic table or in the face of the cook. It is also a good idea not to build too close to trees or shrubs since the heat can affect them.

Materials

Outdoor fireplaces can be built from a large variety of masonry materials. Bricks (figure 14-2), concrete masonry units, and stone are the materials used most often.

Make sure the type of masonry unit selected is sound, sufficiently hard and if brick, a kiln-fired unit. Old, used bricks are not recommended for the construction of outdoor fireplaces. The old, pink used bricks, called salmon bricks, should not be used because they are soft and deteriorate very quickly. This makes it necessary to tear down the structure and rebuild it at a high cost.

Fig. 14-2 This brick barbecue fireplace is built in a circular design. Notice that there is no chimney; a chimney is not needed on a low barbecue.

200 ■ Section 3 Building Outdoor Structures

Fig. 14-3 This stone barbecue is built as a part of a stone wall around a patio. Notice the brick lining within the cooking area.

Concrete blocks can be used for outdoor fireplaces and barbecues and are fast to construct because of the size of the unit. They are less expensive than brick or stone due to lower labor costs. It is important, however, to line the immediate area of the firebox with firebrick when using concrete block for the structure.

Stone adapts to outdoor design more readily than other masonry materials because it is a natural part of the landscape. It can be laid in many patterns and has an attractive appearance when completed, figure 14-3.

Mortar

It is recommended to use type M mortar for the building of outdoor fireplaces and barbecues. This type of mortar withstands exposure to the elements better than other types of mortar.

Ovens

Steel ovens, figure 14-4, that fit into an outdoor fireplace are available from manufacturers of fireplace accessories. They are built of heavy-gauge steel with grates, grills, and door frames of long-lasting cast iron. Steel ovens are resistant to warping and burning out from the heat. Ample draft controls are located in the fire doors and ash pit doors. The oven extends the use of an outdoor fireplace as a cooking center. When the cookout season is over, the unit can be removed from the fireplace, figure 14-5, and stored until the next year.

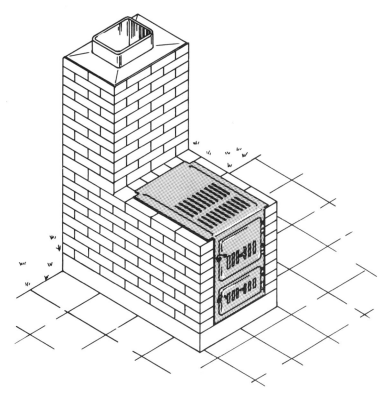

Fig. 14-4 A brick fireplace with a steel oven built in.

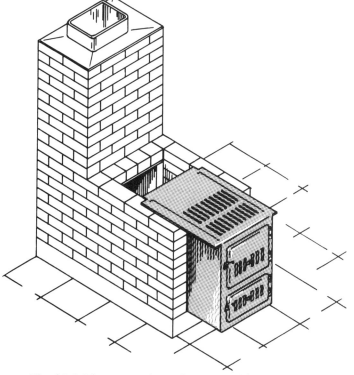

Fig. 14-5 The oven unit can be removed for winter storage.

202 ■ Section 3 Building Outdoor Structures

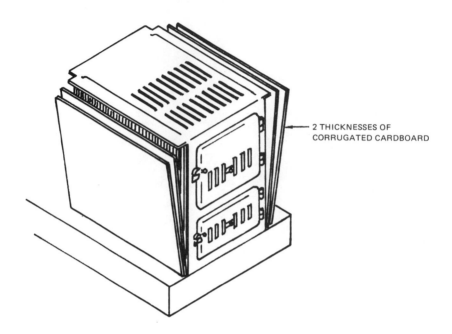

Fig. 14-6 The cardboard is installed around the oven to allow for expansion.

If the oven is to be removable, provisions should be made for this when the opening of the fireplace is constructed. The simplest method is to place two thicknesses of cardboard spacer, figure 14-6, between the unit and the masonry work as the fireplace is being built. The cardboard burns out with the first fire and permits the oven unit to slide out for storage. The space remaining also allows for contraction and expansion of the steel unit.

BUILDING THE FOUNDATION

Regardless of the size or design of the fireplace, a good footing and foundation are necessary. Because the size of an outdoor fireplace and barbecue is relatively small, often the proper attention is not given to installing a suitable base.

The rules for installing the foundation of an outdoor fireplace are the same as for any masonry structure. The concrete footing must be below the frost line where the fireplace is being constructed. Floating slabs that rest on the ground will crack in time due to freezing and thawing cycles of the earth. This happens even when the slabs are reinforced with steel. It is also a good idea to put reinforcing wire in a foundation for an outdoor fireplace because the frost affects the edges more than in a regular foundation.

The part of the fireplace that is below the finish grade can be built of concrete block, stone, rough brick, or concrete. The materials should be sound, but chipped units can be used for the foundation since they do not show.

If all concrete is placed for the foundation, use a 5 bag mix per cubic yard. Install steel reinforcement 1/4 to 1/2 inch in diameter laid in a checkerboard fashion on 6- to 12-inch centers. Tie the steel together with tie wire at each intersection. Reinforcing concrete

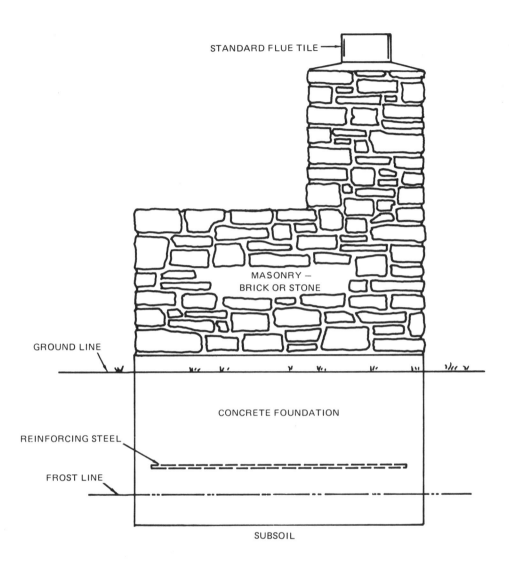

Fig. 14-7 Section view of an outdoor fireplace showing construction details.

wire is also acceptable instead of using steel rods. These materials can be obtained from any local building supply dealer. Figure 14-7 shows a section view of an outdoor fireplace giving details of construction.

LAYING OUT THE WALLS ABOVE GRADE

Lay out the outside walls dry (without mortar) to establish the bond. As discussed previously, if an oven is going to be installed, the hearth must be built first and the oven positioned before any outside walls are built. This is to allow for expansion of the oven. Figure 14-8, page 204, shows the construction plan for a brick outdoor fireplace.

Mortar joints should always be well filled to prevent the leakage of water inside the masonry work. Eye protection should also be worn when mixing mortar or cutting any of the masonry units.

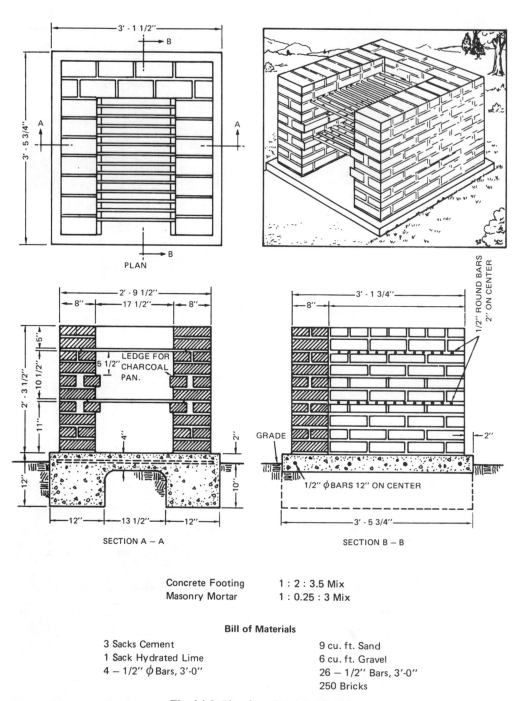

Fig. 14-8 Plan for a brick barbecue.

BUILDING THE WALLS

After laying the first course in mortar, build the corners three or four courses high. Then the wall should be filled in using a short line as a guide. Excess mortar is returned to the mortar pan and retempered to use again. Do not mix more mortar than can be used in a two-hour period. Mortar joints should be tooled as soon as they are thumbprint hard.

BUILDING THE FIREBOX

There is no one special method for building the firebox. The method depends on the materials being used and the design. Although the firebox does not have to be lined with firebricks, it lasts longer if firebricks are used. The size of the top opening is determined by the grill being used, the depth of the grill, the type of grate, and the fuel that will be used for cooking.

HEIGHT OF THE GRILL

The grill should be high enough so that it is easy to work from, but not so high that the chef must bend over it uncomfortably. The ideal height for the top of the grill is between 24 and 30 inches. Referring to figure 14-8, the height of the fireplace above the foundation is 2 feet, 3 1/2 inches (or 27 1/2 inches). The dimensions of the opening depend upon the area of the cooking surface that is desired by the owner.

FITTING THE GRILL INTO THE FIREPLACE

The grill size should be decided before the fireplace opening is built. An average size firebox opening is given on the plan in figure 14-8.

The metal grill can be ordered from a building supply dealer who carries such items. Usually, however, most outdoor fireplaces or barbecues use either racks made of steel rods or a heavy-gauge wire made at a local welding shop. Stainless steel racks are the best because they do not rust. If an oven is being used, the grill is a part of it.

> **Old refrigerator racks can be dangerous because they have a coating on the metal that is not safe to use when cooking.**

The grill can be installed in the fireplace in several ways. One method is to rake out the mortar bed joint at the selected height and slide in the grill from the front. Another technique is projecting the masonry work out as a shelf and laying the grill on the projections. The second method allows easy removal of the grill for cleaning and storage when it is not in use. Do not wall the rack permanently in position with mortar because it may rust (unless made of stainless steel) after a period of time. Refer back to figure 14-8 to see how the rack is laid on the masonry shelf. The important thing to remember about the grill is to select the size desired first, and then build the fireplace to fit.

SPACING OF THE GRILL

The distance the grill is spaced from the fire depends on the type of fuel. If charcoal is used, a distance of 9 to 11 inches is a good cooking height. If wood is being used, a good distance is between 20 and 22 inches.

Regardless of the type of fuel chosen, good judgment must also be used when deciding the distance from the grill to the fire. One important consideration is the intensity of the fire. This is the reason why adjustable heights for the shelves should be built into the fireplace. Referring again to figure 14-8, notice that there are different heights at which the grill can be located.

FRONT OF THE FIREBOX

The front of the firebox can be built in three different ways. In the first method, the firebox opening remains completely open in the front. This allows for a good draft, easy access to the fuel, and easy cleaning of the firebox after use. This is the most popular method of building the firebox.

In the second method, a metal draft control door is installed in the front of the fireplace. This allows the draft opening to be opened and closed as the situation requires. A good metal door to use is one with sliding louvers or slots so it can be easily opened or shut. The door, however, should be built hinged to the opening to allow easy removal of the ashes.

For the third method, masonry slots are built into the enclosed front of the firebox. This can be one or two bricks left out to provide a draft. After the fire is burning well, a brick can be inserted into the slot to control the draft. This works well as long as the food on the grill does not block the draft. The major problem with this method occurs when the firebox must be cleaned out and the ashes must be removed from the top or through an ash dump door at the bottom. This is the least preferred of the three methods described. As a rule, it is not advisable to put the ash pit under a hearth in an outdoor fireplace.

CHIMNEYS FOR AN OUTDOOR FIREPLACE

It is not necessary to build a chimney for a low barbecue-type fireplace unless it is enclosed on a patio that has a roof. Once the charcoal becomes hot there is only a little smoke. The wind gets rid of the small amount of fumes and odor.

For larger fireplaces that burn wood, it is advisable to build a chimney to carry the smoke away. An example of this is the incinerator type of outdoor fireplace shown in the plan in figure 14-9.

The chimney also increases the draft for the fire in a larger fireplace. The height of the chimney depends on the amount of smoke emitted and the size of the fireplace being built. As a rule, the height of a chimney for an outdoor fireplace is at least 7 feet so the smoke does not bother anyone. It is also a good practice to install a flue lining in the chimney so that the chimney does not burn out as quickly. A chimney does increase the cost of building the fireplace so it is necessary to decide if the chimney is really needed. A chimney is usually built on more elaborate fireplaces, as shown in figure 14-10, page 208.

ESTIMATING

Rule-of-thumb estimating is all that is necessary for a small project such as this. The following are some figures to remember:
- 125 standard bricks can be laid with one bag of masonry cement.
- There are approximately 7 bricks to the square foot. This allows for waste.
- A thousand bricks can be laid with one ton of sand.
- If type M cement mortar is used, follow this formula:
  1 part portland cement to 1/4 part lime to 3 parts sand.
  Type M cement mortar can be used for brick, block, or stone.
- Concrete for a foundation is estimated by cubic yards.

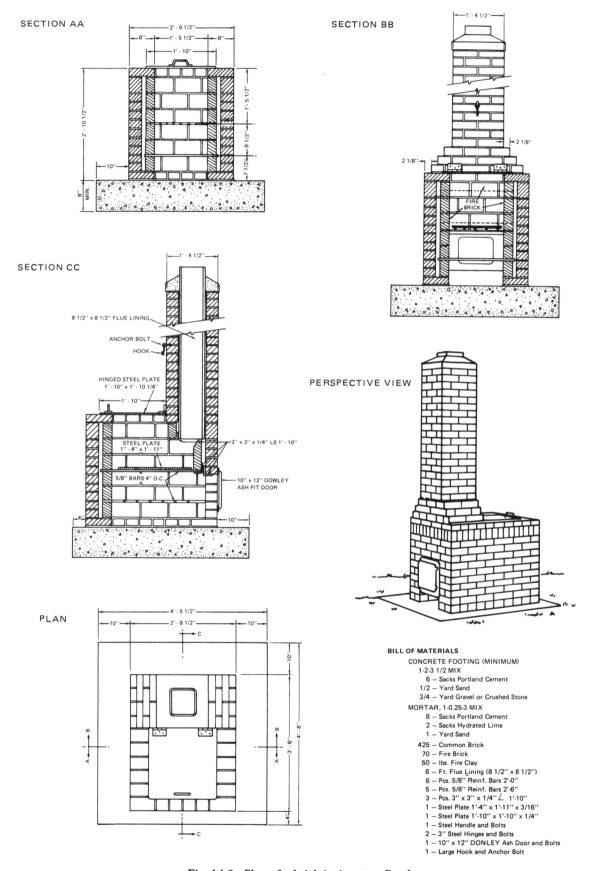

Fig. 14-9. Plan of a brick incinerator fireplace.

208 ■ Section 3 Building Outdoor Structures

Fig. 14-10 A chimney is usually built on elaborate outdoor fireplaces, such as the one shown, which also contains an oven.

CURING AND CLEANING THE MASONRY WORK

After the outdoor fireplace is completed, allow seven days before building a fire to let the fireplace cure properly. The fireplace or barbecue can be cleaned with a solution of 1 part hydrochloric (muriatic) acid to 9 parts clean water or a proprietary (a manufactured prepared cleaner) cleaning compound. Both solutions can be obtained from a building supply dealer. Be sure to follow the instructions on the label for proper use of these compounds.

> Eye protection and rubber gloves should be worn when using any of these solutions.

Regardless of the type of fireplace or barbecue built, good workmanship and the correct basic masonry skills should be followed. Full mortar joints are very important when building any outside exposed masonry work. The building of outdoor fireplaces and barbecues is an important part of the masonry trade and should be learned by the student mason.

ACHIEVEMENT REVIEW

Select the best answer from the choices offered to complete each statement. List your choice by letter identification.

1. The term barbecue refers to
   a. an outdoor fireplace with a chimney.
   b. an incinerator for the burning of paper.
   c. a small fireplace, usually with no chimney.
   d. the metal oven that fits into the outdoor fireplace.

2. As a rule, the major problem with used bricks is that
   a. they are different in size.
   b. the colors vary too much.
   c. they are too hard and can crack from the fire.
   d. they are too soft and cannot withstand the weather.

3. The best mortar to use for building outdoor fireplaces and barbecues is
   a. type N.   c. type S.
   b. type M.   d. type O.

4. The most important reason for leaving a small space around a metal oven in an outdoor fireplace is to allow for
   a. ventilation.   c. cleaning behind the unit.
   b. expansion of the metal.   d. more air for combustion.

5. The best type of metal grill to install in a fireplace is made of
   a. copper.   c. stainless steel.
   b. galvanized metal.   d. metal rods.

6. For cooking over a charcoal fire, the grill should be
   a. 9 to 11 inches from the fire.   c. 13 to 15 inches from the fire.
   b. 11 to 13 inches from the fire.   d. 15 to 17 inches from the fire.

7. To carry the smoke away and prevent sparks from falling on the cook, it is recommended to build a chimney for an outdoor fireplace at least
   a. 4 feet high.   c. 7 feet high.
   b. 6 feet high.   d. 8 feet high.

8. When estimating bricks for an outdoor fireplace by rule of thumb, the standard number of bricks to the square foot is
   a. 4.   c. 10.
   b. 7.   d. 12.

PROJECT 13: BUILDING A BARBECUE

To practice building a barbecue type of fireplace or an incinerator with a chimney, use the plans given in figures 14-8 and 14-9 for a guide in construction.

A list of materials is given with each project. If this project is being done in a shop situation, use hydrated lime made of 1 part lime to 3 parts sand so that the mortar can be reused.

SUMMARY – SECTION 3

- The footing for a patio is as important as the footing for any other masonry structure. It is installed below the frost line.
- Bricks selected for patio walls and floors should be type SW.
- Type M mortar is recommended for masonry below grade. Type N mortar is recommended for masonry above grade.
- The patio wall should be built first and then the floor should be laid.
- The floor should be built with enough slope to drain water.
- Good workmanship and full mortar joints are mandatory.
- There are two major methods of installing paving work: with mortar or without mortar.
- Mortared paving should be laid on a concrete base.
- Mortarless paving is laid on either stone dust or sand. Asphalt roofing paper is laid directly under the paving to discourage plant growth.
- Large brick paving floors should have expansion joints.
- Edging bricks for both types of paving should be laid in mortar to prevent the floor from shifting at the edges.
- The most popular types of paving patterns are the running, herringbone, and basket weave bonds.
- Sand is swept in the joints at the completion of the work in mortarless paving.
- A flat tooled joint should be used in paving laid in mortar to prevent water from entering.
- Brick paving can be laid inside as well as outside.
- Reinforced paving is used over small excavations or unstable soil.
- Caps should be laid on a patio wall and projected about 1/2 inch to form a water drip.
- Steps consist of two basic parts: the tread where the foot is placed; and the riser, which is the vertical part from one tread to the next tread.
- There is a relationship of height to width that must be followed in step construction, otherwise the steps can be dangerous.
- Steps must be built on firm, undisturbed soil with a good footing.
- Type M mortar is recommended for steps and porches.
- The platform (top surface of the porch) should be built with a slope to drain water.
- The average rise (height) of a brick step is 7 1/2 inches. The tread is 12 inches.
- The top brick on a step should be projected 1/2 inch to form a water drip.
- Steps should be laid with the help of a line.

- A slope of 1/4 inch is recommended for brick step treads from the back to the front of the tread.
- Flat work such as step treads should be tooled with a flat joint.
- Well-filled mortar joints are critical in step construction.
- The porch and steps should be cleaned with a hydrochloric acid solution after they have cured.
- The mason must know basic arithmetic to lay out the steps correctly.
- Bricks selected for garden walls should be of the SW (severe weathering) type.
- Salvaged or old soft bricks should not be used in garden wall construction unless they are hard.
- Some type of coping should be installed on top of garden walls to prevent the entrance of moisture.
- Pilasters built into the garden wall provide extra strength in addition to a pleasing appearance.
- The basket weave bond is very popular for panel garden wall construction.
- The unique design of a serpentine wall (which is built on an alternating curve) provides lateral strength and may be built of a single-thickness masonry unit.
- Good workmanship and full mortar joints are absolutely necessary if the garden wall is to stand the test of time.
- The retaining wall which the mason comes in contact with most often is the low type, usually 6 feet or less in height.
- The purpose of a retaining wall is to hold back earth.
- All retaining walls should be reinforced to some extent, depending on the job requirements.
- A footing is installed below the frost line for all retaining walls.
- Short lengths of steel rods are projected vertically out of the footing to tie the masonry wall and to prevent shifting on the footing.
- The two general classifications of retaining walls are the gravity and cantilever types.
- Masonry retaining walls can be built of brick, concrete masonry units, or stone.
- Type M mortar is recommended for retaining walls.
- Steel rods are placed in the wall and grout is placed around them when a reinforced wall is required.
- Grout should not be placed in the wall until it has cured.
- Drain tiles are used through the retaining wall for drainage. The wall is filled in the back with crushed stone for good drainage and to relieve pressure from water.

- The side of the retaining wall that the earth is filled against should be waterproofed with either mortar or an asphalt tar compound.
- Backfilling retaining walls should not be done until the wall has cured.
- Special retaining walls (tree wells) are built around trees when the existing grade is raised.
- The building of outdoor fireplaces is a valuable source of work for the mason.
- The type of low outdoor fireplace used for cooking that has no chimney is usually classified as a barbecue.
- The location of an outdoor fireplace should be planned with respect to wind direction, drainage, and shade.
- Footings should be installed below the frost line for all outdoor fireplaces.
- Steel ovens and accessories are available for outdoor fireplaces.
- Lining the firebox of an outdoor fireplace with firebrick helps prevent the fireplace from burning out.
- The spacing of grills in a fireplace depends on the type of fuel being used and the type of food being cooked.
- If a chimney is built on a fireplace, it should be at least 7 feet in height.
- Fires should not be built in an outdoor fireplace until the mortar has cured, which usually takes at least seven days.
- Local burning laws should be consulted before building an outdoor incinerator.

SUMMARY ACHIEVEMENT REVIEW, SECTION 3

Part A

Complete each of the following statements or answer the question referring to the material presented in Section 3.

1. Old, used bricks are not recommended for use in patio walls because _____.

2. Bricks that are hard and recommended for use in patios are classified as _____.

3. Concrete block walls that are built around the patio to give privacy are called _____.

4. Type M mortar is recommended for use in the construction of patio walls. The formula for type M mortar is _____.

5. When laying the cap on a patio wall, it should be projected on each side a distance of _____.

6. When laying mortarless paving, a layer of roofing felt is laid on top of the sand before the bricks are laid. This is done for the purpose of _____.

7. Bricks that are laid in mortar for a patio floor should be leveled into place by using a _____.

8. Whether the patio is laid with mortar or mortarless, the edging should be laid in _____.

9. When there is an opening or when installing a brick paving over loose fill, the floor should be _____.

10. In mortarless paving, the joint is filled with _____.

11. A set of steps is built with 5 risers. The rise is 7 1/2 inches for each step. What is the total height of all the steps? _____

12. The depth of tread for each step is 12 inches. What is the total distance of the treads? _____

Part B

Select the best answer from the choices offered to complete each statement. List your choice by letter identification.

1. The material used for a reinforced retaining wall is called
   a. mortar.
   b. grout.
   c. cement.
   d. concrete.

2. The best all-around size for a retaining wall is
   a. 4 inches thick.
   b. 8 inches thick.
   c. 10 inches thick.
   d. 12 inches thick.

3. The type of retaining wall that depends on its bulk or mass to hold back the earth is called a
   a. cantilever wall.
   b. counterfort wall.
   c. gravity wall.
   d. reinforced wall.

4. A masonry retaining wall that is built on an angle opposite to the earth fill is called a
   a. gravity wall.
   b. battered wall.
   c. racked wall.
   d. cantilever wall.

5. Drain tiles that are used in a retaining wall should have a diameter of
   a. 4 inches.
   b. 6 inches.
   c. 8 inches.
   d. 10 inches.

6. The bond used in the construction of a brick retaining wall that utilizes a brick header tie every sixth or seventh course is called a
   a. Flemish bond.
   b. running bond.
   c. common bond.
   d. stack bond.

7. Concrete masonry retaining walls that will be filled with grout should set for a period of
   a. 8 hours.
   b. 12 hours.
   c. 16 hours.
   d. 24 hours.

8. When concrete blocks are used for an outdoor fireplace the firebox should be lined with
   a. firebrick.
   b. asbestos.
   c. metal lining.
   d. tile.

9. The concrete footing for an outdoor fireplace should be a mix of
   a. 4 bags to the cubic yard.
   b. 5 bags to the cubic yard.
   c. 6 bags to the cubic yard.
   d. 7 bags to the cubic yard.

10. When estimating the number of standard bricks to the square foot of a 4-inch wall, the figure used is
    a. 5.
    b. 6.
    c. 7.
    d. 8.

Section 4
Modular Coordination

Unit 15 UNDERSTANDING MODULAR COORDINATION

OBJECTIVES

After studying this unit the student will be able to
- define modular coordination.
- explain how nominal dimensions are used in the layout of a masonry wall.
- list the various mortar joint thicknesses used for modular masonry products.
- describe the advantages of a modular system from the mason's viewpoint.

DEFINING THE TERM MODULAR COORDINATION

Modular coordination is the method of applying the principles of mass production to reduce the cost of construction work. It has been known for many years that much money could be saved in the construction business through the use of this system.

Basically, modular coordination means two things. First, it means that the various parts that go into the construction of a building such as the masonry units, doors, windows, mechanical equipment, spacing of lumber to receive drywall, paneling, ceiling tile, etc., fit together with little or no cutting or altering. This is the same idea that is used in automobile assembly and many other manufactured articles. In the assembly line, the piston fits the cylinder, the door fits the opening in the frame, and so forth. This part of modular coordination is called the coordination of the building products. The spacing of a wood stud 16 inches on center in a partition wall in carpentry work is an example of this theory.

Secondly, modular coordination involves dimensioning building plans so that the dimensions (measurements) are exact multiples of the dimensions of standard modular products. For example, if the building is to be constructed of bricks whose standard measurement (center to center of mortar joints), is 8 inches, then the nominal overall dimensions of the building would be a multiple of 4 inches. This is known as correlating building dimensions with the standard sizes of building products. Drafting methods have been developed for this purpose through the use of dots, arrows, and grid lines. They are called *modular measure, modular dimensioning,* or *modular drafting.*

The main reason that the entire construction industry is not using a complete modular system is because this method has not been accepted by all manufacturers of building products, building designers, and contractors and builders who construct the buildings. Each of these groups has operated independently in the past. Only recently have some group leaders decided that they should all be working together to standardize the industry. Standardization can result in reduced building costs, increased production, more profit for the contractor, and more work for everyone.

A BRIEF HISTORY OF MODULAR COORDINATION

In 1938, a group of industrial leaders interested in modular construction met to discuss this idea. This group was called the American Standard Association. The meeting resulted in the organization of a project to standardize building products and building designs on a modular basis.

The project was formally set up in 1939 under the sponsorship of the American Institute of Architects and the Producers Council. Since that time, much progress has been made in the development of modular sizes of building products and building design.

Many building organizations now support modular coordination. Except for the nonmodular standard brick (3 3/4 inches x 2 1/4 inches x 8 inches) and some oversized brick (3 3/4 inches x 2 3/4 inches x 8 inches) many of the bricks produced and used in the United States are sized to fit the modular system. There are still many nonmodular bricks made; however, more than half of all bricks produced in the United States are modular in size. Manufacturers also use modular sizes for concrete masonry units.

Other products such as structural lumber, dry wall, and paneling are made in modular sizes. A similar Canadian standard has also been adopted. The forthcoming change to the metric system will help advance standardization by requiring all building products to be built to the same scale based on the meter.

BASIS FOR MODULAR COORDINATION

The base unit for modular coordination is a 4-inch square grid called the *module*. This is applied to the building in all three dimensions, height, length, and width. International standards have established 4 inches as the basic module for countries using foot-inch measure, and 10 centimeters (which is approximately equal to 4 inches) for countries using the metric system. This grid is used by architects and engineers when determining the dimensions of the building and by manufacturers when making their products. The cooperation between the two groups permits the planning and installation of the products together on the job without expensive cutting or alterations.

Products manufactured to modular sizes fit an opening when measured with a rule in multiples of 4 inches. Examples of this are openings of 8 inches, 12 inches, 24 inches, 32 inches, and so forth.

The most significant difference between the plans for a nonmodular building and the plans for a modular building is that all dimensions on the modular plan are in multiples of 4 inches. There are no fractions of an inch as commonly found in a nonmodular plan. The modular building is laid out on an imaginary grid composed of lines that are spaced 4 inches on center in all directions. Then, all parts of the structure are related to this grid. Figure 15-1 shows a plan layout of a brick wall with a door laid out on the 4-inch grid.

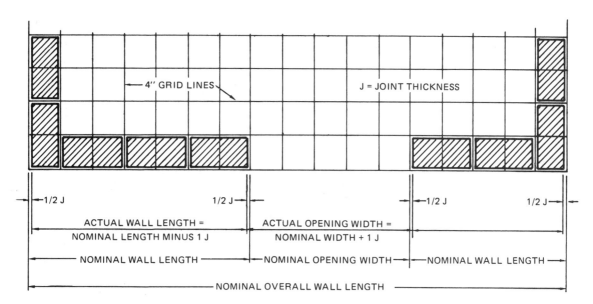

Fig. 15-1 Plan layout of an opening in a masonry wall laid out on the 4-inch grid. Notice how the bricks are laid out to fit in the grid.

The 4-inch actual grid lines are not usually shown on the working drawing. However, most correctly detailed modular plans show the grid lines on all large-scale details such as door and window sections. Here the actual relationship of the individual parts to the grid lines are shown. In a window detail, for example, the actual position of the masonry jamb (a *jamb* is the side of a window or door) is determined in relation to the grid dimensions shown on the plan. Figure 15-2, page 218, shows this grid relationship.

THE IMPORTANCE OF NOMINAL DIMENSIONS

In modular design, the *nominal dimension* of a masonry unit (such as a brick or block) is the actual size or manufactured dimension of the unit plus the thickness of the mortar joint to be used. That is, the size of the brick is designed so that when the size of the mortar joint is added to any of the brick dimensions (thickness, height, or length), the sum equals a multiple of the 4-inch grid. For example, a modular brick whose nominal length is 8 inches has a manufactured dimension of 7 5/8 inches if it is to be laid with a 3/8-inch mortar joint. Standard brick, with a nominal height of 2 2/3 inches, has a manufactured height of 2 1/4 inches if it is to be used with a 3/8-inch joint. Thus, 3 courses of standard bricks laid in 3/8-inch mortar joints will lay up to a total of 8 inches, or twice the basic 4-inch grid.

All of the masonry units listed on the mason's modular rule level off together every 16 inches in height. This allows the mason to bond the units together at that height using wall ties or a masonry header. Figure 15-3, page 219, shows 6 courses of standard bricks laid in 3/8-inch mortar joints, and 2 courses of concrete blocks laid in a 3/8-inch mortar joint. Notice that they are level at a height of 16 inches, which is a multiple of the 4-inch grid.

In the modular system, the nominal dimension of a masonry unit is always from center to center of the mortar joints. Therefore, the nominal dimension is the thickness of one

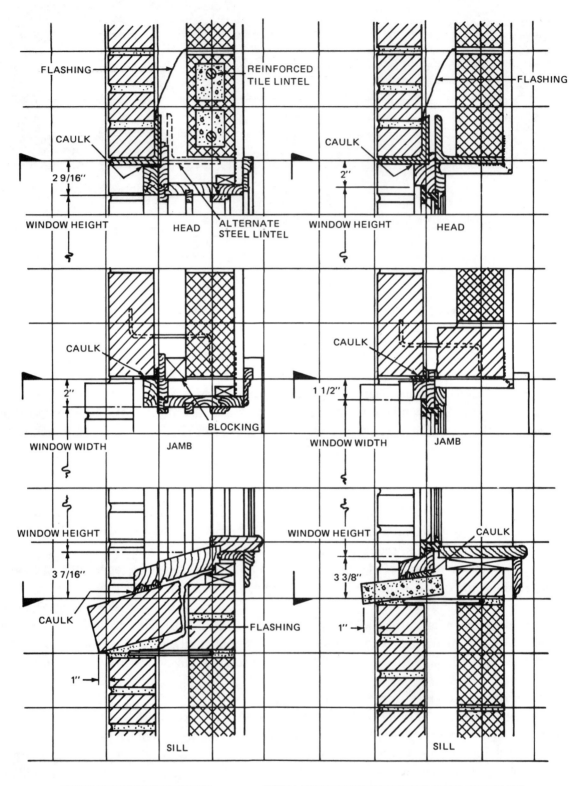

Fig. 15-2 These window details are drawn on the 4-inch grid. Notice how all materials are worked into the grid.

mortar joint greater than the specified dimension of the unit. This is true because the measurement is the sum of the unit plus two half mortar joints. Figure 15-4 shows how the nominal dimension is drawn to the center of a mortar joint.

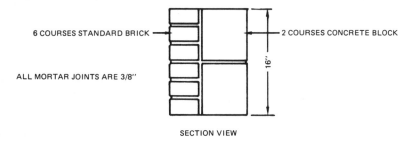

Fig. 15-3 Six courses of standard bricks laid with 3/8-inch mortar joints equal the height of two courses of regular concrete blocks with 3/8-inch mortar joints. Notice that both are level at the height of 16 inches, which is the standard height where the wall is bonded together.

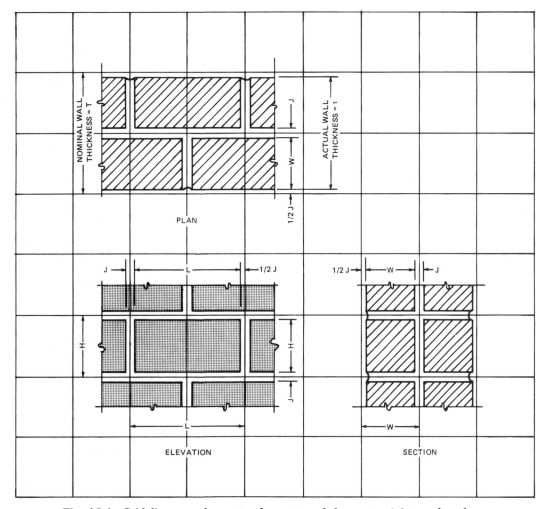

Fig. 15-4 Grid lines are drawn to the center of the mortar joints rather than to the end of the unit itself. This is an application of the modular system.

The actual size of a masonry unit is sometimes slightly different from the specified dimensions of the unit, as in the case of a fired brick that may expand or contract. However, there is a plus or minus tolerance for modular masonry units which the manufacturer must conform to, or the unit may be rejected on the job by the architect. A minor difference is expected in the size of bricks or fired units and the difference is made up in the mortar joints when the mason lays them in the wall.

As explained earlier, there is a nominal joint relationship between masonry units of different lengths. A nominal brick 8 inches long is manufactured 7 5/8 inches long to be laid with a 3/8-inch joint; a nominal 12-inch brick is made 11 5/8 inches long to be laid with a 3/8-inch joint; and a concrete block that is stated to be 16 inches long is really 15 5/8 inches long to accommodate a 3/8-inch mortar joint.

MORTAR JOINT THICKNESS

In the masonry industry, three mortar joint thicknesses are used as standards for various grades of masonry units: 1/2 inch for building (common) bricks and structural clay tile; 3/8 inch for face bricks and concrete blocks; 1/4 or 3/8 inch for unglazed facing tile; and 1/4 inch for glazed facing tile or glazed bricks. The glazed facing tile is number 3 on a modular rule, because 3 courses with the bed joints will lay 16 inches in height. Figure 15-5 shows various scales on the modular rule applied to masonry units when laid in a mortar bed joint.

UNDERSTANDING NOMINAL DIMENSIONS ON A MODULAR BUILDING PLAN

Nominal dimensions are used for masonry layout and spacing on a modular plan. This means that the actual width of a masonry opening is the nominal width plus the thickness of one mortar joint at each jamb. The actual distance between an external (outside) corner and an internal (inside) corner is the same as the nominal distance minus a mortar joint at the external corner plus a mortar joint at the internal corner.

When laying out modular masonry, the corner is started one mortar joint from the building line which is fixed by the nominal dimensions on the plan. From this point the layout procedure is the same as for a nonmodular unit. However, since the nominal length of all masonry units is a multiple of 4 inches, the layout is simplified and can be checked with a mason's rule or steel tape.

Vertical dimensions can be laid out on the story pole in the same way, since course heights are multiples of 4 inches or simple fractions of multiples of 4 inches. For example, assume that 3 courses of brick is 8 inches including the mortar joints, or 3 courses of glazed facing tile is 16 inches including the mortar joints. Since overall nominal dimensions of a building do not contain fractions of an inch, this makes it much easier to estimate quantities from the plans.

GRID LOCATIONS OF MASONRY WALLS ON A PLAN

Nominal masonry dimensions, as previously explained, are from the center of one mortar joint to the center of the next mortar joint. They are used for both vertical and horizontal layouts.

Horizontal layouts, including wall openings, involve only 2-inch and 4-inch multiple dimensions. For vertical dimensions, fractions of inches are avoided except for standard bricks which are 2 2/3 inches and glazed tiles which are 5 1/3 inches. Since thirds of an inch are not given on the ordinary foot rule, 5/16 inch is used for 1/3 inch and 11/16 is used for 2/3 inch. The inaccuracy of this figure is very small and insignificant, unless it is repeated

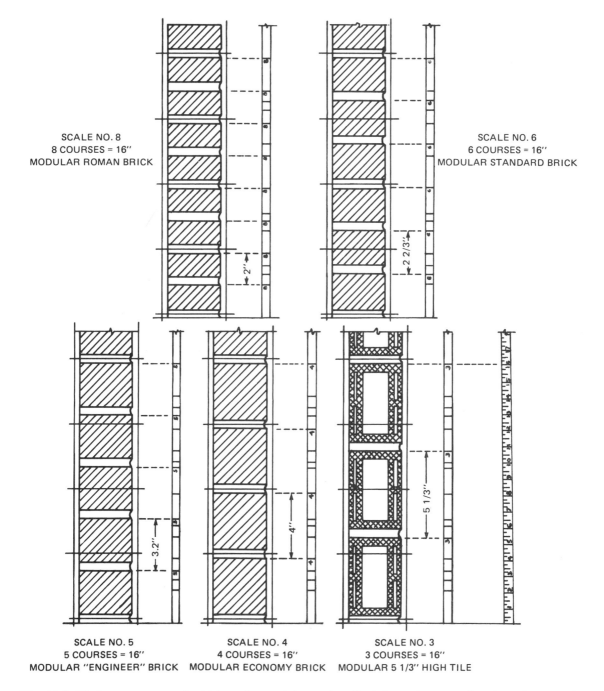

Fig. 15-5 Various masonry units laid on the modular grid. All of those shown are listed on the mason's modular rule.

many times. Figure 15-6 shows the grid location of mortar joints when constructing walls of different size modular units. The walls are centered between grid lines.

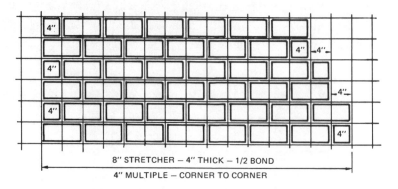

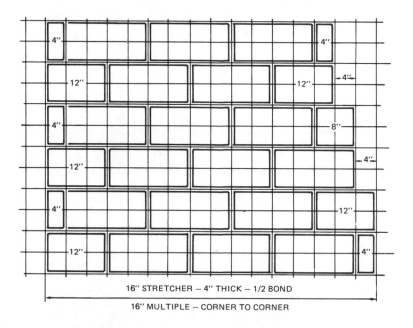

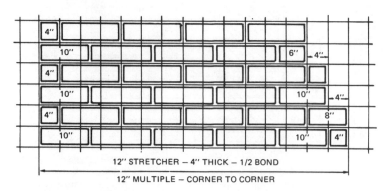

Fig. 15-6 Application of the 4-inch grid with different sizes of masonry units. Notice that the grid extends vertically and horizontally.

USING ADDED CUTS OR STARTER PIECES TO WORK BOND FOR MODULAR WALLS

Flexibility in wall lengths can be obtained by using special closure units at openings or by laying the units in the wall. These units are usually laid at the corner or next to a window or door jamb. Since laying a cut piece in the wall tends to destroy the pattern created by the bond, this technique is only used when there is no other way. If this method is used, place the unit under a window or opening so it is not very noticeable. Figure 15-7 shows several examples of how starter pieces can be laid on the corner to make the bond work when using different size masonry units.

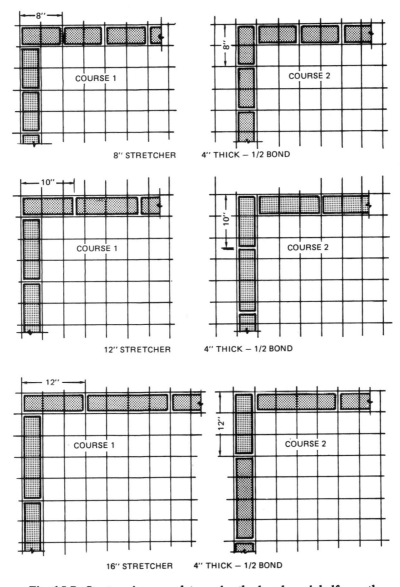

Fig. 15-7 Starter pieces used to make the bond work half over the stretcher units when using different lengths of bricks. Three examples are shown, 8-inch stretchers, 12-inch stretchers, and 16-inch stretchers. Notice how all the units are drawn to the 4-inch grid.

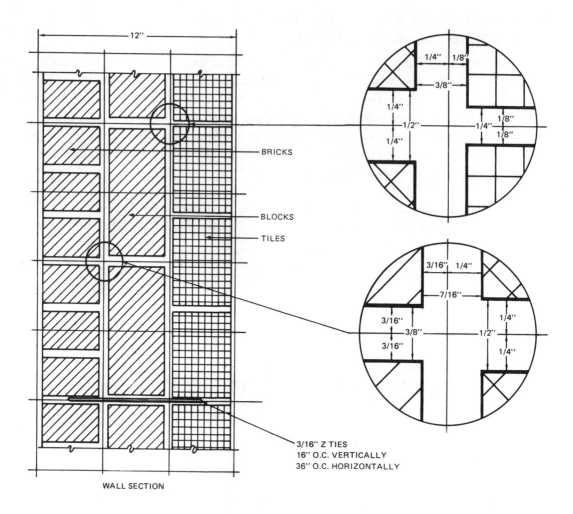

Fig. 15-8 Modular coordination of different masonry units. Notice that each of the units shown is different in size, but all are level at 16-inch intervals, to be bonded together.

HOW VARIOUS MASONRY UNITS COORDINATE TOGETHER IN THE WALL

Figure 15-8 shows how different masonry units coordinate together in the wall. Notice that although all of the mortar joint thicknesses are different, the top of the 6 courses of bricks is level with the interior tile backing and the finished glazed tile on the inside of the wall. The brick is laid with a 3/8-inch joint, the interior back up tile is laid with a 1/2-inch joint; and the inside facing tile is laid with a 1/4-inch joint. The complete coordination is shown in the enlargement; the thickness of the vertical joints between the different types of units is the average of the joint thickness used with each unit. Together the 3 units make a 12-inch wall in width.

> NOTE: If standard brick is used for the facing of the wall and backed up with concrete block, the mortar bed joints are the same (3/8 inch). This is the combination that is most frequently used for solid masonry wall construction rather than tile.

ADVANTAGES OF A MODULAR COORDINATED SYSTEM

The major advantages of a modular coordinated system for the construction industry are summarized as follows:

- All plans are drawn the same because they are all based on the 4-inch module. This greatly simplifies the understanding of modular coordination for the mason.
- Because all modular plans and products are made on a standard grid, the replacement and repair of old buildings is much easier. In the past, many different parts of a structure were difficult to obtain. This resulted in having to make them custom fit at a large expense, and caused a delay in the job.
- If revisions or changes must be made to a new building after construction has started, the process is simpler because the grid allows the architect to substitute another product or material. In this way, the job is not delayed. This is a major problem when a building has a deadline for completion. If the building is not completed on schedule it can cause the owner a considerable loss of rent.
- Increased production by the workers on the job also results from modular coordination, because modular materials can be used without expensive cutting or alterations. The laying of bricks, installing of electrical boxes, or setting of metal partitions all work together.

ACHIEVEMENT REVIEW

Select the best answer from the choices offered to complete each statement. List your choice by letter identification.

1. The base unit of the modular coordination system is called the
   a. centimeter.
   b. nominal.
   c. module.
   d. stretcher.
2. The base unit of measurement in the modular system is
   a. 2 inches.
   b. 4 inches.
   c. 6 inches.
   d. 8 inches.
3. A nominal dimension for a masonry unit means
   a. the sum of the unit plus the mortar joint.
   b. the difference between the length of the masonry unit less the mortar joint.
   c. the width of the unit.
   d. the height of the unit.
4. All masonry units listed on the modular rule bond together at a height of
   a. 4 inches.
   b. 8 inches.
   c. 12 inches.
   d. 16 inches.
5. The most frequently used joint thickness for laying face brick or concrete block is
   a. 1/4 inch.
   b. 3/8 inch.
   c. 1/2 inch.
   d. 5/8 inch.

6. The correct bed joint thickness for glazed tile is
   a. 1/4 inch.
   b. 3/8 inch.
   c. 1/2 inch.
   d. 5/8 inch.

7. The height of a course of standard brick including the mortar joint is
   a. 2 1/4 inches.
   b. 2 2/3 inches.
   c. 2 3/4 inches.
   d. 3 inches.

8. The side of a window or door is called the
   a. head.
   b. sill.
   c. jamb.
   d. face.

9. One course of glazed tile including the mortar bed joint is
   a. 5 inches.
   b. 5 1/4 inches.
   c. 5 1/3 inches.
   d. 5 3/4 inches.

10. If a 12-inch wall is built in half-bond lap, the correct size piece for the corner is
    a. 6 inches.
    b. 8 inches.
    c. 10 inches.
    d. 11 inches.

Unit 16 ESTIMATING MODULAR MASONRY MATERIALS

OBJECTIVES

After studying this unit the student will be able to

- describe some of the problems resulting from inaccurate estimates by the mason.
- work out an estimate for a brick masonry job by using basic math and the tables included in the unit.
- work out an estimate for concrete block work by using basic math and the tables supplied in the unit.

THE IMPORTANCE OF ACCURACY IN ESTIMATING

Before ordering any materials for the job, a mason must be able to estimate the amount needed. Although the estimating of large jobs is done by a professional estimator in the office, the mason is expected to be able to estimate materials to some degree.

If the amounts are greatly overestimated, many materials will be left over after the job is finished. This creates several problems. The materials must be moved to another job or stored in a yard at a high labor cost to the contractor. The materials are more subject to chips and breaks when moved from one place to another. The type of material may be the kind that cannot be readily used again — such as an unusual color of brick or tile. All of these factors subtract from the contractor's profit.

Underestimating is also a problem because this causes a shortage of materials and the work is thus delayed. Colors of bricks sometimes change from one run to the next, and the materials may not be available immediately.

In this unit, no attempt is made to estimate the total cost of the masonry work for several reasons. First, labor and material costs differ in various parts of the country. Also, material prices change frequently. The last reason is that the masonry costs are based on past performances, so careful records are kept of the production of the masons. These records compute the amount of bricks, concrete blocks, tile, etc., that can be laid under a variety of conditions.

It is important, however, that the amount of materials needed to do the job can be estimated accurately. A mason or foreman should be able to do this type of estimating. Large commercial structures should be estimated by a trained estimator. Rule-of-thumb estimating is used for most small jobs. A more efficient method of estimating for the mason is by using tables that have been worked out mathematically. These tables are based on the amount of masonry units needed for a certain number of square feet of wall and mortar mixes based on volume and weight. To serve this purpose, numerous tables are found in this unit that allow the student, with the aid of basic math, to calculate the amount of materials necessary to do a masonry job.

Since the two principle masonry materials used in construction today are brick and concrete masonry, this unit deals with the estimating of these two materials.

ESTIMATING BRICK MASONRY

Because it is simple and accurate, the most commonly used estimating procedure is the *wall-area* method. It consists simply of multiplying known quantities of materials required per square foot by the net wall area (total areas less areas of all openings).

Estimating material quantities is greatly simplified by using the modular system. For a stated nominal size, the number of modular masonry units per square foot of wall is the same regardless of the mortar joint thickness. (This is assuming that the units are to be laid with the joint thickness for which they are designed.) As stated previously, it is important to remember that there are only three standard modular joint thicknesses: 1/4 inch, 3/8 inch, and 1/2 inch. All three of these joints can also be used for collar joints. The *collar joint* is the continuous, vertical joint placed between two tiers (thicknesses) of masonry walls. In contrast, the number of nonmodular standard bricks required per square foot of wall varies according to the thickness of the mortar joint.

To follow this procedure, always determine the net quantities of all materials before adding any allowance for waste. The amount of waste varies but as a general rule, 5 percent is allowed for brick and 25 percent for mortar. These percentages can vary due to job conditions.

ESTIMATING TABLES

Several tables have been devised to make the task of estimating masonry work easier. Figures 16-1 through 16-5 are tables supplied by The Brick Institute of America that are based on estimating bricks and mortar only.

Figure 16-1 includes the net quantities of brick and mortar required to construct single-wythe walls (one brick thick) using various modular and nonmodular sized bricks. Nonmodular brick is listed at the bottom half of figure 16-1. Mortar quantities are given for both 3/8-inch and 1/2-inch mortar joints. The joints are to be full mortar joints. Brick and mortar quantities are for running (all stretcher) or stack bond containing no headers. If brick headers are to be used to bond the wall, a bond correction factor table must be used. (See figure 16-2, which shows the correct bond factors to be used when estimating a wall with headers.)

Figure 16-3 contains mortar proportions by volume and weight of portland cement, hydrated lime, and sand for mortar types M, S, N, and O. Figure 16-4 contains data on collar joints for mortar of various thicknesses. For multiwythe construction (at least three thicknesses of masonry units) collar joints are added to the bed and head joint estimates. Figure 16-5 lists the weight of materials per purchase unit for mortar.

It should be understood that the amount of moisture in sand varies. Therefore, volume measurement of sand is not always exactly accurate. The two examples on estimating in this unit assume that each cubic foot of damp, loose sand (bulked approximately 38 percent) contains 80 pounds of dry sand. This is based on the proportion specifications contained in ASTM C270.

	Brick Size — in.			Brick per 100 sq ft		Cubic Feet of Mortar per 1000 Brick	
	t	h	ℓ			3/8-in. joint	1/2-in. joint
Modular (Nominal[1])	4	2 2/3	8	675		8.1	10.3
	4	3 1/5	8	563		8.6	10.9
	4	4	8	450		9.2	11.7
	4	5 1/3	8	338		10.2	12.9
	4	2	12	600		10.8	13.7
	4	2 2/3	12	450		11.3	14.4
	4	3 1/5	12	375		11.7	14.9
	4	4	12	300		12.3	15.7
	4	5 1/3	12	225		13.4	17.1
	6	2 2/3	12	450		17.5	22.6
	6	3 1/5	12	375		18.1	23.4
	6	4	12	300		19.1	24.7
	8	4	12	300		25.9	33.6
	t	h	ℓ	3/8-in. joint	1/2-in. joint	3/8-in. joint	1/2-in. joint
Nonmodular (actual)	3	2 5/8	8 5/8	532	505	7.6	9.7
	3	2 5/8	9 5/8	481	457	8.2	11.1
	3	2 3/4	9 3/4	457	433	8.4	11.3
	3	2 1/4	10	529	500	8.2	11.1
	3 3/4	2 1/4	8	655	616	8.8	11.7
	3 3/4	2 3/4	8	551	522	9.1	12.2

[1] A nominal size or dimension is the actual dimension of the unit plus the joint thickness

Fig. 16-1 Modular and nonmodular solid brick units with mortar required for single-wythe walls in running bond. (There is no allowance for breakage or waste.)

Bond	Correction Factor[1]
Full headers every 5th course only	1/5
Full headers every 6th course only	1/6
Full headers every 7th course only	1/7
English bond (full headers every 2nd course)	1/2
Flemish bond (alternate full headers and stretchers every course)	1/3
Flemish headers every 6th course	1/18
Flemish cross bond (Flemish headers every 2nd course)	1/6
Double-stretcher, garden wall bond	1/5
Triple-stretcher, garden wall bond	1/7

[1] Note: Correction factors are applicable only to those brick which have lengths of twice their bed depths.

Fig. 16-2 Bond correction factor.

	Quantities by Weight			
	Mortar Type and Proportions by Volume			
Material	M	S	N	O
	1:¼:3	1:½:4½	1:1:6	1:2:9
Cement	31.33	20.89	15.67	10.44
Lime	3.33	4.44	6.67	8.89
Sand	80.00	80.00	80.00	80.00

Fig. 16-3 Material quantities per cubic foot of mortar.

Cubic Feet of Mortar Per 100 Sq Ft of Wall		
1/4-in. Joint	3/8-in. Joint	1/2-in. Joint
2.08	3.13	4.17

Note: Cubic feet per 1000 units =
$$\frac{10 \times \text{cubic feet per 100 square feet of wall}}{\text{number of units per square foot of wall}}$$

Fig. 16-4 Cubic feet of mortar for collar joints.

Material	Weight of Material per Purchase Unit
Portland Cement	94 lb per sack
Hydrated Lime	50 lb per sack
Sand	2000 lb per ton

Fig. 16-5 Weight of materials.

USING THE TABLES

The best way to learn how to use the tables is to study an example of an on-the-job situation. A brick veneer, 3 bedroom, single-family house with a fireplace and carport with outside storage room is to be estimated. The information needed is as follows:

- There are 1,441 square feet (net) of brick veneer wall to build.
- The brick units are 4 inches x 2 2/3 inches by 8 inches (standard brick size), laid in half running bond pattern.
- Head and bed joints are 3/8 inch in thickness.
- There are no collar joints.
- The mortar is type N (1:1:6) portland cement, hydrated lime, and masonry sand.

Solving The Problem

The square feet of wall area (1,441) is given in the problem. This is found by multiplying the length of the walls by the height of the walls. The problem can be solved by using the information which follows:

1. Net wall area (of 4-inch, single-wythe walls) is 1,441 square feet.

2. Brick quantity for 4-inch x 2 2/3-inch x 8-inch units is found in figure 16-1.

    $$\frac{1,441}{100} \times 675 = \frac{972,675}{100} = 9,726.75 \text{ or } 9,727 \text{ units*}$$

    5% waste = 9,727
                   x .05
                  486.35 = 487 units (waste)

    To find the total bricks required, add 9,727 units to 487 units (waste).

         9,727
         + 487
         10,214 units (total bricks required)

3. Mortar quantity using 3/8-inch head and bed joints is found in figure 16-1.

    $$\frac{9,727}{1,000} \times 8.1 = \frac{78688.7}{1,000} = 78.6 \text{ or } 79 \text{ cu. ft.}$$

    25% waste =   79
                   x .25
                  395
                 158
                 19.75 = 20 cu. ft. (waste)

---

*When estimating material needs, round fractions up to the next whole number.

To find the total mortar required, add 79 cubic feet to 20 cubic feet (waste).

$$\begin{array}{r} 79 \\ + \ 20 \\ \hline 99 \text{ cu. ft. (total mortar required)} \end{array}$$

4. Mortar material quantities of portland cement, hydrated lime, and sand for type N mortar (1:1:6) includes waste. See figures 16-3 and 16-5 for material quantities.

$$\frac{\text{Cu. ft. of mortar required} \times \text{lbs. of material per cubic feet required}}{\text{Weight of material per purchase unit}} = \text{Total amount of material required}$$

Portland cement:

$$\frac{99 \times 15.67}{94} = 16.5 \text{ or } 17 \text{ sacks}$$

$$\begin{array}{r} 15.67 \\ \times \ 99 \\ \hline 1{,}551.33 \end{array}$$

1,551.33 divided by 94 = 16.5 or 17 sacks

Hydrated lime:

$$\frac{99 \times 6.67}{50} = 13.2 \text{ or } 14 \text{ sacks}$$

$$\begin{array}{r} 6.67 \\ \times \ 99 \\ \hline 660.33 \end{array}$$

660.33 divided by 50 = 13.2 or 14 sacks

Sand:

$$\begin{array}{r} 99 \\ \times \ 80 \\ \hline 7{,}920 \end{array}$$

7,920 divided by 2,000 = 3.8 or 4 tons

Total materials are listed as follows:

Total brick	—	10,214 units
Portland cement	—	17 sacks
Hydrated lime	—	14 sacks
Sand	—	4 tons

Figure 16-6, page 232, shows the brick equivalent factors for modular brick as a comparison.

ESTIMATING CONCRETE MASONRY UNITS AND MORTAR

The simplest estimates for concrete masonry units are for walls that have no openings and are built of the standard 16-inch long block. Various tables have been established to help estimate concrete masonry units and mortar materials. The series of tables in figures 16-7 through 16-13 are printed in cooperation with The National Concrete Masonry Association. Refer to these tables as often as needed.

Unit Designation	Nominal Dimensions, in.	Brick Equivalents[1]
Standard Modular	4 x 2 2/3 x 8	1.0
Engineer	4 x 3 1/5 x 8	1.2
Economy or Jumbo Closure	4 x 4 x 8	1.5
Double	4 x 5 1/3 x 8	2.0
Roman	4 x 2 x 12	1.13
Norman	4 x 2 2/3 x 12	1.5
Norwegian	4 x 3 1/5 x 12	1.8
Economy 12 or Jumbo Utility	4 x 4 x 12	2.25
Triple	4 x 5 1/3 x 12	3.0
SCR	6 x 2 2/3 x 12	2.25
6-in. Norwegian	6 x 3 1/5 x 12	2.7
6-in. Jumbo	6 x 4 x 12	3.38
8-in. Jumbo	8 x 4 x 12	4.5

[1] Based on nominal face dimensions only.

Fig. 16-6 Brick equivalent factors for modular brick.

As an example, assume a basement has no windows or doors. To estimate the number of blocks needed, first determine the length and width of the walls in feet. Use figure 16-7 to find the number of blocks needed for one course. Next, find the height of the wall from the plans. The number of courses that correspond to the height are found in figure 16-8, page 234.

Multiply the number of blocks needed for one course by the number of courses. The result is the total number of concrete blocks needed to build the wall. An allowance for waste or breakage is added. There is no perfect standard allowance that fits all jobs, although 2 percent is used for most concrete block jobs.

The following example illustrates how a job is figured by this method. How many standard 8-inch x 8-inch x 16-inch blocks must be ordered for a basement measuring 26 feet x 40 feet (outside dimensions), if the walls are 10 feet high and there are no openings?

1. Examine figure 16-7. Read across from left to right and down until the two figures meet on the table. It is found that 97 blocks are needed for one course all the way around the basement.

2. Looking at figure 16-8, it is found that a wall 10 feet high requires 15 courses of blocks.

3. The number of blocks (in this case, 97) is then multiplied by the number of courses (in this case, 15), to get the total.

```
     97
   x 15
    485
    97
  1,455   (Total number of blocks needed)
```

4. Add 2% for breakage or waste.

```
   1,455
   x .02
   29.10  =  30 blocks (waste)
```

Number of Block Per Course For Solid Walls of Various Sizes																				
Size In Feet	2	4	6	8	10	12	14	16	18	20	22	24	26	28	30	32	34	36	38	40
2	4	7	10	13	16	19	22	25	28	31	34	37	40	43	46	49	52	55	58	61
4	7	10	13	16	19	22	25	28	31	34	37	40	43	46	49	52	55	58	61	64
6	10	13	16	19	22	25	28	31	34	37	40	43	46	49	52	55	58	61	64	67
8	13	16	19	22	25	28	31	34	37	40	43	46	49	52	55	58	61	64	67	70
10	16	19	22	25	28	31	34	37	40	43	46	49	52	55	58	61	64	67	70	73
12	19	22	25	28	31	34	37	40	43	46	49	52	55	58	61	64	67	70	73	76
14	22	25	28	31	34	37	40	43	46	49	52	55	58	61	64	67	70	73	76	79
16	25	28	31	34	37	40	43	46	49	52	55	58	61	64	67	70	73	76	79	82
18	28	31	34	37	40	43	46	49	52	55	58	61	64	67	70	73	76	79	82	85
20	31	34	37	40	43	46	49	52	55	58	61	64	67	70	73	76	79	82	85	88
22	34	37	40	43	46	49	52	55	58	61	64	67	70	73	76	79	82	85	88	91
24	37	40	43	46	49	52	55	58	61	64	67	70	73	76	79	82	85	88	91	94
26	40	43	46	49	52	55	58	61	64	67	70	73	76	79	82	85	88	91	94	97
28	43	46	49	52	55	58	61	64	67	70	73	76	79	82	85	88	91	94	97	100
30	46	49	52	55	58	61	64	67	70	73	76	79	82	85	88	91	94	97	100	103
32	49	52	55	58	61	64	67	70	73	76	79	82	85	88	91	94	97	100	103	106
34	52	55	58	61	64	67	70	73	76	79	82	85	88	91	94	97	100	103	106	109
36	55	58	61	64	67	70	73	76	79	82	85	88	91	94	97	100	103	106	109	112
38	58	61	64	67	70	73	76	79	82	85	88	91	94	97	100	103	106	109	112	115
40	61	64	67	70	73	76	79	82	85	88	91	94	97	100	103	106	109	112	115	118
42	64	67	70	73	76	79	82	85	88	91	94	97	100	103	106	109	112	115	118	121
44	67	70	73	76	79	82	85	88	91	94	97	100	103	106	109	112	115	118	121	124
46	70	73	76	79	82	85	88	91	94	97	100	103	106	109	112	115	118	121	124	127
48	73	76	79	82	85	88	91	94	97	100	103	106	109	112	115	118	121	124	127	130
50	76	79	82	85	88	91	94	97	100	103	106	109	112	115	118	121	124	127	130	133
52	79	82	85	88	91	95	97	100	103	106	109	112	115	118	121	124	127	130	133	136
54	82	85	88	91	94	97	100	103	106	109	112	115	118	121	124	127	130	133	136	139
56	85	88	91	94	97	100	103	106	109	112	115	118	121	124	127	130	133	136	139	142
58	88	91	94	97	100	103	106	109	112	115	118	121	124	127	130	133	136	139	142	145
60	91	94	97	100	103	106	109	112	115	118	121	124	127	130	133	136	139	142	145	148

Fig. 16-7 Standard 16-inch concrete masonry units.

5. Total number of blocks needed   1,455
    Plus breakage or waste   + 30
    Total blocks to be ordered   1,485

ALLOWANCE FOR WINDOWS, DOORS, AND OTHER OPENINGS

For more complicated structures that have windows, doors, or other openings, the number of blocks can be figured by using the tables. This is done as follows:

1. Find the length and height (in feet) of every wall, using the plans. Multiply the length times the height of each wall to obtain the gross area of the wall. Add the gross area of all walls to find the total gross wall area. List this total on a sheet of paper as *gross wall area.*

2. Using the plans, find the height and width (in feet) of each window, door, or other opening. Multiply the height times the width of each opening to get the area of the opening. Add the areas of all the openings and list this total on the paper as *openings.*

(Height of unit 7-5/8")
(Joint thickness 3/8")

Height	No. of courses
8"	1
1'4"	2
2'0"	3
2'8"	4
3'4"	5
4'0"	6
4'8"	7
5'4"	8
6'0"	9
6'8"	10
7'4"	11
8'0"	12
8'8"	13
9'4"	14
10'0"	15
10'8"	16
11'4"	17
12'0"	18
12'8"	19
13'4"	20
16'8"	25
20'0"	30
23'4"	35
26'8"	40
30'0"	45
33'4"	50

Fig. 16-8 Concrete masonry courses by height.

3. On a sheet of paper, subtract all openings from the total gross area. List the result on the paper as *net wall area*.
4. Count the number of corners in the walls. Multiply this by the wall height (in feet) and by the wall thickness (in inches). Divide the product by 12. This gives the number of square feet of corner blocks.
5. Subtract the square feet computed for corner blocks from the net wall area.
6. Check the plans for any special blocks such as lintels, sash, jamb, control joint, and square-end corner blocks. Enter the number of special blocks on the paper. Be careful that none are missed.

Nominal Height and Length of Units in Inches	Number of Units Per 100 Sq. Ft.
8 x 16	112.5
8 x 12	150.0
5 x 12	221.0
4 x 16	225.0
2 1/4 x 8*	675.0
4 x 8**	450.0
5 x 8***	340.0
2 x 12****	600.0
2 x 16****	450.0

*Modular Concrete Brick (2 1/4 x 3 5/8 x 7 5/8).
**Jumbo Concrete Brick (3 5/8 x 3 5/8 x 7 5/8).
***Double Concrete Brick (4 7/8 x 3 5/8 x 7 5/8).
****Roman Concrete Brick: (1 5/8 x 3 5/8 x 11 5/8).
(1 5/8 x 3 5/8 x 15 5/8).

Fig. 16-9 Concrete masonry requirements for 100 square feet of wall area.

7. To determine the total number of concrete masonry units, one must know the type of wall being built. Check the plans for this information. If a single-wythe wall (one masonry unit in thickness) is being built, figure 16-9 is used to find the total number of blocks. If a composite wall (wall with two or more different types of units) or a cavity wall is being built, figure 16-10, page 236, is used to find the number of blocks.

8. If a single-wythe wall is being built, select the factor from figure 16-9 which corresponds to the nominal block size being used in the wall. Multiply this factor by the net masonry area. Divide the result by 100 to get the total number of CMU's (concrete masonry units). If the wall being built is composite or multiwythe, the total number of concrete masonry units is obtained by using figure 16-10.

This procedure can be best shown by using an example.

Example

A 20-foot x 40-foot concrete block building is to be built as a storage building for a farm. The height of the wall is 10 feet. Standard 8-inch x 8-inch x 16-inch concrete blocks are to be laid. From the plans it can be seen that no special blocks are needed except corner blocks. There are 3 windows and 1 door. The openings total 57 square feet. What are the quantities of blocks needed to build the building? The following steps are taken to estimate the blocks needed.

1. The gross wall area for each wall is calculated by multiplying the length by the height. Enter this under gross wall area.

    20 ft. x 10 ft. = 200 ft.
    20 ft. x 10 ft. = 200 ft.
    40 ft. x 10 ft. = 400 ft.
    40 ft. x 10 ft. = 400 ft.
    Total   1,200 sq. ft. (gross wall area)

Wall Description	Number of Masonry Units Per 100 Sq. Ft.	
	Exterior Wythe	Interior Wythe
COMPOSITE WALLS 4-in. concrete brick plus 4-in. block:		
Masonry Bonded	772	97.0
Wire Ties	675	112.5
4-in. concrete brick plus 8-in. block:		
Masonry Bonded	868	97.0
Wire Ties	675	112.5
CAVITY WALLS 4 in. concrete brick plus 4-in block	675	112.5
4-in. block plus 4-in. block	112.5	112.5

Fig. 16-10 **Number of concrete masonry units required for 100 square feet of various composite or multiwythe walls.**

2. The figure of 57 sq. ft. for the openings is subtracted from the gross wall area. The result is listed as net wall area.

    Gross wall area   1,200 sq. ft.
    Minus openings   – 57 sq. ft.
    Net Wall Area   1,143 sq. ft.

3. The corner deduction is computed as follows:

$$\frac{(\text{Number of corners}) \times (\text{Wall height}) \times (\text{Wall thickness})}{12} = \frac{4 \times 10 \times 8}{12} = 26.7 \text{ or } 27 \text{ sq. ft.}$$

4. The corners are subtracted from the net wall area and listed as the masonry area.

    Net wall area   1,143 sq. ft.
    Corners   – 27 sq. ft.
    Masonry Area 1,116 sq. ft.

5. Refer to figure 16-9. For a unit 8 inches x 16 inches (nominal size), a factor of 112.5 units per sq. ft. is obtained. This is multiplied by the masonry area. The result is divided by 100 to get the number of concrete masonry units needed. This is listed as the total number of CMU's.

$$\frac{112.5 \times 1,116 \text{ sq. ft.}}{100} = 1,255.5 \text{ or } 1,256 \text{ CMU's}$$

6. The plan shows that 15 square-end corner blocks are required for each corner. They are laid up to a height of 8 inches for each course. Since there are 4 corners on the building, multiply 4 x 15.

    15
    x4
    60 corner blocks needed

7. The corner blocks can be deducted from the total number of blocks needed if they must be ordered separately. Usually, every third block made is a square-end corner block so it may not be necessary to order special corner blocks. Check the local area to find out which is the case.

Assuming that corner blocks must be ordered, subtract 60 from the total amount needed:

    1,256 blocks
    – 60 corner blocks
    1,196 total CMU's needed

8. It is now known that 1,196 stretcher units and 60 corners are needed for the walls. Adding 2 percent for breakage and waste gives the total number to be ordered:

	Stretchers	Corners
Units	1,196	60
Plus breakage and waste (2 percent)	+ 24	+1
Total to Order	1,220	61

ESTIMATING THE MORTAR

A mason should also be able to estimate the mortar needed for the job. The amount of mortar and the quantities of raw materials to make the mortar must be figured. Figure 16-11 shows the cubic feet of mortar needed for 100 sq. ft. of wall area in a single-wythe wall. The amount of mortar depends on the size of the block being used. Figure 16-12, page 238, shows the number of cubic feet of mortar for various composite and multiwythe walls.

Nominal Height and Length of Units in Inches	Cu. Ft. of Mortar Per 100 Sq. Ft.
8 x 16	6.0
8 x 12	7.0
5 x 12	8.5
4 x 16	9.5
2 1/4 x 8*	14.0
4 x 8**	12.0
5 x 8***	11.0
2 x 12****	15.0
2 x 16****	15.0

    *Modular Concrete Brick (2 1/4 x 3 5/8 x 7 5/8).
    **Jumbo Concrete Brick (3 5/8 x 3 5/8 x 7 5/8).
    ***Double Concrete Brick (4 7/8 x 3 5/8 x 7 5/8).
    ****Roman Concrete Brick: (1 5/8 x 3 5/8 x 11 5/8)
                        (1 5/8 x 3 5/8 x 15 5/8).

Fig. 16-11 Mortar requirements per 100 square feet of wall area. (Wall is assumed to be one masonry unit in thickness.)

238 ■ Section 4 Modular Coordination

Wall Description	Cu. Ft. of Mortar Per 100 Sq. Ft.
COMPOSITE WALLS 4-in. concrete brick plus 4-in. block:	
Masonry Bonded	18.0
Wire Ties	18.0
4-in. concrete brick plus 8-in. block:	
Masonry Bonded	13.5
Wire Ties	18.0
CAVITY WALLS 4-in. concrete brick plus 4-in. block	15.0
4-in. block plus 4-in. block	12.0

Fig. 16-12 Mortar required for 100 square feet of various composite or multiwythe walls.

In the example given for estimating block, the area was figured to be 1,116 sq. ft. In figure 16-11 it is found that for a nominal block size of 8 inches x 16 inches, 6.0 cubic feet of mortar are needed per 100 square feet of wall area. The amount needed for 1,116 square feet is, therefore:

$$\frac{1,116 \times 6}{100} = 66.9 \text{ or } 67 \text{ cubic feet of mortar}$$

The waste allowance is added to the quantity to determine the amount of mortar to be mixed. (Mortar waste is 10 percent of the calculated quantity.)

    67 cu. ft. (calculated quantity)
    +7 cu. ft. (10% waste)
    74 cu. ft. (amount of mortar to mix)

Since there are 27 cubic feet in a cubic yard, the amount of mortar to be mixed is:

$$\frac{74}{27} = 2.74 \text{ or } 2.7 \text{ cubic yards}$$

ESTIMATING THE INGREDIENTS

Once the amount of mortar to be mixed is known, the mason must determine the quantity of ingredients needed. It is first necessary to find out the type of mortar to be used. This can be found by consulting the specifications. As a rule, type N is the mortar used.

Refer to figure 16-13 to find the quantity of materials for a cubic yard of masonry mortar. The first column shows the type of mortar to be mixed. Two entries are listed for

Mortar Type, ASTM: C270	Sand, damp, loose volume	Cementitious Materials (bags of material or cubic feet)*		
		Portland Cement	Masonry Cement	Lime
Type M	1.0 cy	4.5	4.5	–
Type M	1.0 cy	7.5	–	2.0
Type S	1.0 cy	3.0	6.0	–
Type S	1.0 cy	6.0	–	3.0
Type N	1.0 cy	–	9.0	–
Type N	1.0 cy	4.5	–	4.5
Type O	1.0 cy	–	9.0	–
Type O	1.0 cy	3.0	–	6.0

*Cementitious materials are usually one cubic foot volume per bag.

Fig. 16-13 Quantities of materials for a cubic yard of concrete masonry mortar.

each mortar type. These provide for mixing the mortar with or without masonry cement. If masonry cement is selected, then the first of the two entries is used. If portland cement mortar is selected the second of the two entries is used. The mason then knows the quantities of materials for one cubic yard of mortar.

For the example discussed to this point, the mortar needed has been figured at 2.7 cubic yards. It is also type N mortar. Masonry cement is to be used in mixing the mortar. The amount of materials needed is determined as follows: In figure 16-13, it is found that for type N mortar using masonry cement, 1 cubic yard of sand and 9 bags of masonry cement are needed for each cubic yard of mortar. The required materials for 2.7 cubic yards are, therefore

2.7 x 1.0 = 2.7 cu. yds. sand
2.7 x 9.0 = 24.3 or 24 bags of masonry cement

As mentioned previously, estimating is made easier when modular masonry units are used. The tables given in this unit for concrete masonry units are based on the modular system because all concrete blocks are made to the modular grid. Once the mason learns how to use them, the tables help ease the task of estimating a masonry job. Rule-of-thumb estimating is used on many small jobs but the most accurate method is to use predetermined tables such as the ones given in this unit.

ACHIEVEMENT REVIEW

Solve the following problems by using the estimating tables given in the unit. Show all of your calculations.

Part A

The following problem is for a grocery store built of brick veneer. After deducting all openings for windows and doors, the following factors are known.

a. The net area is 1,600 square feet.
b. Brick units are 4 inches x 2 2/3 inches x 8 inches, laid in a running bond.

c. Head and bed joints are 3/8 inch thick.
d. There are no collar joints.
e. Type N portland cement mortar is to be used in a ratio of 1:1:6 mix. The materials used in the mortar are portland cement, hydrated lime, and sand.

Calculate the following material needs, based on the above information.

1. Calculate the number of bricks needed including the waste.
2. Calculate the amount of mortar needed.
3. Calculate the amount of portland cement needed.
4. Calculate the amount of hydrated lime needed.
5. Calculate the amount of sand needed.

Round off all fractions in material needs to the nearest whole figure *exceeding* the number. For example, if 19 1/3 bags of portland cement are needed, it is rounded off to 20 bags.

Part B

Solve the following problem for a concrete block building by using the tables for concrete block in the unit.

A concrete block garage is to be built. Standard 8-inch x 8-inch x 16-inch blocks are to be used. No special blocks are needed except for square-end corner blocks for the corners and window and door jambs. Each course of block lays up to 8 inches with the bed joint.

The following information is given for the building.

a. Garage outside dimensions are 24 ft. x 24 ft.
b. Height of the wall is 10 ft.
c. The two windows each measure 2 ft. x 4 ft.
d. The two garage doors each measure 8 ft. x 8 ft.
e. Type N mortar is used with masonry cement.

Use the tables provided in this unit and basic math skills to solve the problems.

1. Determine the total amount of square feet of wall area.
2. Deduct all openings and give the square feet of wall remaining.
3. Determine the number of square-end corner blocks needed for the corner and the jambs. Allow for the waste factor.
4. Determine the total number of stretcher blocks needed. Allow for the waste factor.
5. Determine the amount of cubic feet of mortar needed and change it into cubic yards.
6. Determine the amount of sand needed.
7. Determine the amount of masonry cement needed.

SUMMARY – SECTION 4

- Modular coordination is the method of applying the principles of mass production by standardizing the manufacturing and construction of building materials to a 4-inch grid.
- Plans should be drawn so that all dimensions are applications of standard modular products.
- In recent years more and more buildings are being built on the modular concept.
- The base unit of measurement in modular coordination is a grid of 4 inches, called the module.
- Over 50 percent of all bricks made today are modular in size.
- Nominal dimensions of a masonry unit mean that the sum of the actual size of the unit plus the mortar joint thickness are included in the unit's length.
- Nominal dimensions of a masonry unit on a plan are from center to center of the mortar joints.
- The three mortar joint thicknesses used to lay modular masonry units are 1/2 inch, 3/8 inch, and 1/4 inch.
- All of the masonry units shown on the modular rule are coordinated at a height of 16 inches, which allows the wall to be bonded together.
- Plans drawn on the 4-inch modular grid greatly simplify the building of the structure because of the relationship between all of the parts and materials used in that structure.
- The task of repairing and renovating old buildings is more economical if they are of modular construction because new materials fit into the overall construction.
- Increased production by workers is possible when modular products are used because the materials can be installed with a minimum of cutting or alteration.
- Flexibility in masonry wall lengths can be obtained by using special corner or closure units made to fit into the modular grid.
- One of the most important advantages of the modular system is that building products and materials can be standardized for size to fit into any building.
- Accuracy in estimating is very important since it affects the contractor's profit.
- Overestimating a job is not desirable because part of the money that would have been profit is tied up in unnecessary surplus materials.
- Labor estimates are difficult to standardize due to the different wage rates in various locations and the amount of skilled labor available.

- In estimating brick masonry the modular brick greatly simplifies the procedure.
- For a given size the number of modular masonry units per square foot of wall is the same regardless of joint thickness. By comparison, the number of nonmodular standard bricks per square foot of wall varies with the mortar joint thickness.
- Because of its simplicity and accuracy, the most widely used estimating procedure is the wall-area method. This method consists of multiplying known quantities of materials required per square foot by the net wall area (gross areas minus areas of all openings).
- Determine the net quantities of all materials before adding any allowance for waste.
- Allowance for waste as a general rule for brick masonry is 5 percent for brick quantities and 25 percent for mortar.
- Tables are available for estimating modular masonry units and mortar materials.
- When using tables for estimating concrete masonry units, 2 percent should be allowed for waste.
- Mortar waste for concrete masonry units should be estimated as 10 percent.
- All of the simple math should be rechecked for accuracy before concluding an estimate. Most mistakes are made in simple addition, subtraction, multiplication, or division.

SUMMARY ACHIEVEMENT REVIEW, SECTION 4

Complete each of the following statements referring to the material found in Section 4.

1. The method of applying the 4-inch grid when manufacturing and installing building materials is known as _____ _____.

2. The three mortar joint thicknesses used to lay modular masonry units are _____, _____, and _____.

3. All of the masonry units listed on the modular rule level off together or coordinate at a height of _____.

4. The module is the base unit of measure in the modular system. It is also called the _____.

5. The length of the masonry unit including the mortar joint thickness is called its _____.

6. The length of a standard concrete block without the mortar head joint is _____.

7. The mortar joint thickness for glazed facing tile is _____.

8. Nominal dimensions of a masonry unit shown on the plans are taken from _____.

9. Due to its simplicity and accuracy, the most often used method for estimating brickwork is called the _____ _____.

10. A standard brick size is _____.

11. When estimating mortar, fractions of a bag should be rounded off to _____.

12. When estimating mortar for brick masonry, the waste figure that should be allowed is _____.

13. When estimating bricks the waste allowance is _____.

14. The waste allowance for concrete blocks is _____.

15. The waste allowance for mortar for concrete blocks is _____.

Section 5
Glazed and Prefaced Masonry Units

Unit 17 STRUCTURAL CLAY TILE AND GLAZED BRICKS

OBJECTIVES

After studying this unit the student will be able to

- describe the advantages of structural clay tile and the series available for use.
- identify the various shapes of glazed structural facing tiles and explain how they are used in a wall.
- describe the types and sizes of glazed brick available.

HISTORY OF STRUCTURAL CLAY TILE

Structural clay tile was introduced in Europe in the early 1870s and was first made in the United States in 1875. This is relatively recent considering the ancient age of bricks.

Called hollow tile at the time, structural clay tile was soon made in a variety of shapes and sizes with a glazed or unglazed exterior face. The early uses of structural clay tile were mainly for the fireproofing of beams, columns, and interior partitions of buildings. Many tile arch floor systems were also built. The steel skeleton framework created the need for a light fireproof material to protect it. Tile filled that need. Structural tile was therefore in great demand as an inexpensive backup material and fireproofing protection material.

The use of structural clay tile reached its peak in about 1929. During World War II it was used in military construction throughout the country. In recent years, glazed tiles have been in demand for the construction of sanitary buildings such as sewage and water treatment plants, bakeries, hospitals, and food establishments. These tile walls are built either as an interior part of the outside wall, or as partitions in the structure, thus making them a structural part of the building. Structural clay tile can always support great loads as well as its own weight and is frequently used as a load-bearing wall.

MANUFACTURING STRUCTURAL CLAY TILE

Ceramic-glazed structural clay facing tile is a material which combines the hardness, beauty, and long-lasting permanence of a ceramic, glazed face with the load-bearing capacity of a burned clay masonry unit. Most of the tiles sold today are glazed.

Glazed structural tile and natural finish clay tile are generally made from light burning, de-aired fireclay which is deep mined. After the clay is ground into a fine consistency, it is remixed with water in a pug mill and extruded under very high pressure into a column of clay that is cut into various sized units. It can then be glazed, predried, and fired in a kiln at approximately 2,400 degrees Fahrenheit to make the body and glaze of the tile inseparable.

The tile is then packed in paper cartons with separators and unitized into shipping packs. It is transported by rail box cars or trucks directly to the job where it will be used. Although the tile is sold through distributors, shipping directly to the job cuts down on the amount of handling and reduces chipping and breaking of the tile face. Figure 17-1 shows a pallet of glazed tile in the protective paper wrapping. For various sizes and shapes of glazed structural tile, refer to figure 17-2, pages 246 through 249.

Fig. 17-1 Pallet of glazed structural tile in protective paper wrapping.

PROPERTIES OF GLAZED STRUCTURAL TILE

The most popular features of glazed structural tile include the following: a very dense, hard surface that does not absorb stains and is easy to clean; a minimum amount of maintenance required once laid in the wall; a permanent color that does not fade; a zero flame spread which does not support combustion, char, or give off toxic fumes as some paints do; structural strength which eliminates the need for backup support walls; and a highly sanitary finish which can be kept cleaner than other masonry units due to the hardness of the facing. ASTM C126 states standard specifications for ceramic-glazed, structural clay facing tile.

Colors

Colors vary somewhat depending on the manufacturer, but the most popular finishes for glazed structural tile are the light colors, white, creams, and speckles. Decorator colors are usually used in panels or for special effect. They are not usually laid in large walls. The colors, however, are bright and do not fade.

In recent years mat finishes have become popular due to the desire for dull, rustic, earthy tones. The cleanability of the mat finish and the gloss finish are basically the same because glazed tile is nonabsorbent.

Grades And Types

ASTM Specification C126 covers two grades and two types of glazed tile described as follows:

- Grade S   (select) for use with comparatively narrow mortar joints.

246 ■ Section 5 Glazed and Prefaced Masonry Units

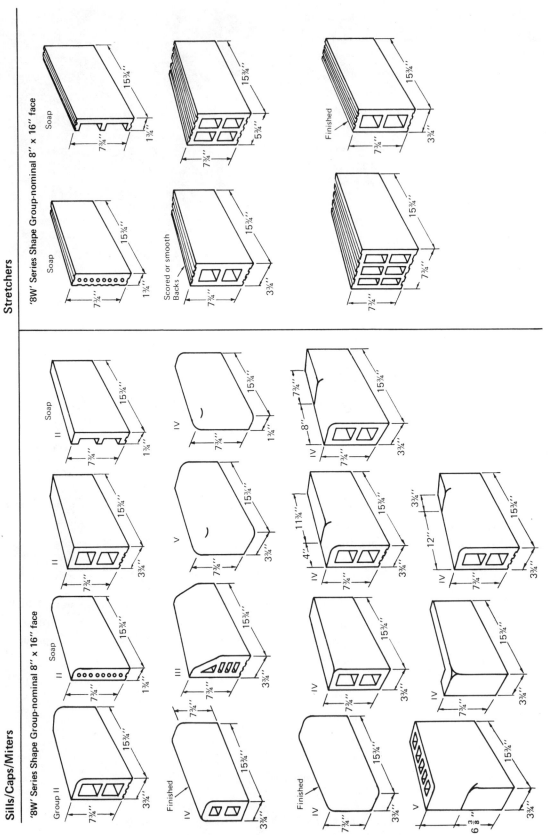

Fig. 17-2. Standard units and shapes in the 8W and 6T series of glazed structural tile. (continued)

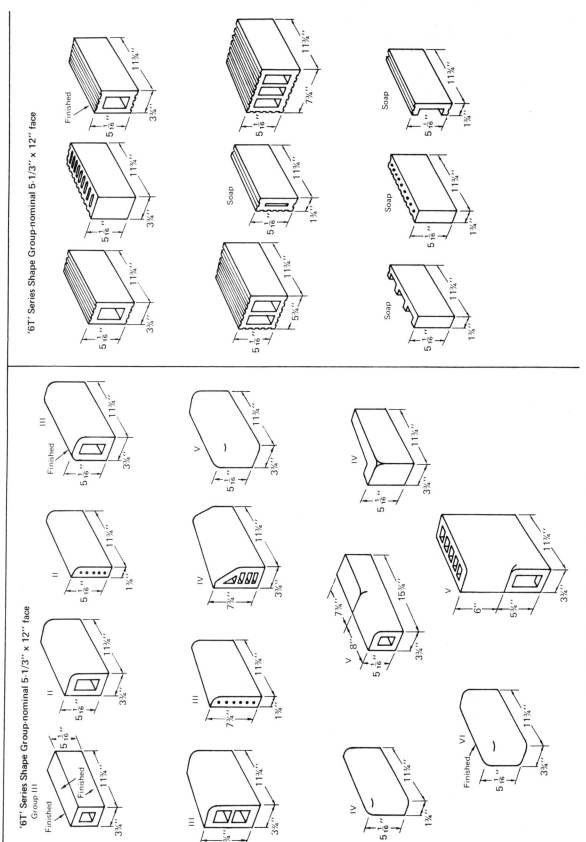

Fig. 17-2. Standard units and shapes in the 8W and 6T series of glazed structural tile. (continued)

248 ■ Section 5 Glazed and Prefaced Masonry Units

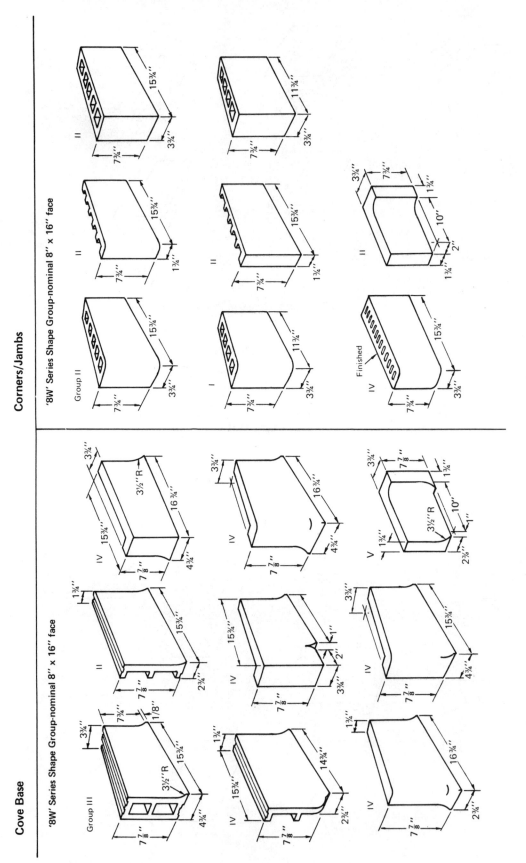

Fig. 17-2. Standard units and shapes in the 8W and 6T series of glazed structural tile. (continued)

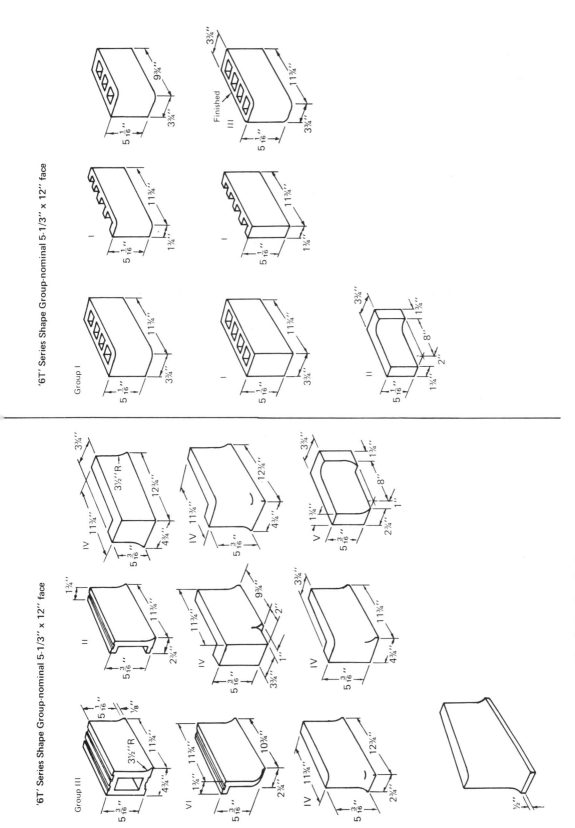

Fig. 17-2. Standard units and shapes in the 8W and 6T series of glazed structural tile.

- Grade G (ground edge) for use where variations of face dimensions must be very small.
- Type I (single-faced units) for general use where only one finished face will be exposed.
- Type II (two-faced units) for use where two opposite finished faces will be exposed.

These specifications also include limits on distortion, variations in dimensions of the tile, and performance requirements of the tile to meet minimum acceptable standards.

There is another type of structural clay facing tile which is made from either light-burning fireclay or shale and other darker burning clays. Such facing tiles are unglazed and have either a smooth or a rough textured finish. These tiles are described under ASTM C212. The specifications cover two classes of tile, standard and heavy duty, based on factors affecting the appearance of the finished wall. The two classes are described as follows:

- Smooth face tile suitable for general use in exposed exterior and interior masonry walls and partitions and adapted for use where tile must be low in absorption, easy to clean, and resistant to staining. Also used where a high degree of mechanical perfection, narrow color range, and a minimum variation in face dimensions are desired.
- Smooth or rough textured facing tile suitable for general use in exposed exterior and interior walls and partitions, and adapted for use where tile of moderate absorption, moderate variations in face dimensions and medium color range may be used, and where minor defects in surface finish, including small handling chips, are not objectionable.

Sizes

Recently, the sizes of structural clay tile offered to builders have been the 6T and 8W series. The 6T series is available with nominal face dimensions of 6 inches x 12 inches and in nominal thicknesses of 2 inches, 4 inches, 6 inches, and 8 inches. The 8W series has nominal face dimensions of 8 inches x 16 inches and is available in the same nominal thicknesses as the 6T series. The 4D series, which is shown in some of the technical and catalog information, is no longer produced as a standard in the industry. This is due to the higher in-the-wall cost and the architectural objections to using smaller units and more mortar joints.

The 8W series is generally sold with ground ends to provide a more uniform size when laying in a stack bond. The 8W series also has the same face size as a standard concrete block when laid in mortar using the specified joint. This series can be backed up with concrete block when only one side of a glazed tile wall is needed. Both the 6T and 8W series fit the modular grid and lay up to a height of 16 inches (the standard height that a wall is bonded together with the backing materials).

Shapes

Glazed structural facing tiles are made in many different shapes to serve the needs of various areas of a tile wall. Many projects can be built by using only some of the basic shapes such as stretchers, corners, closures, starters, and miters. When this is possible, it is more economical to use basic shapes because the special shapes cost more to make due to their unique function in the wall and limited use. However, for a complex wall layout, shapes such as sills, caps, lintel tile glazed on the bottom side, cove base stretchers, doorstops, and coved internal corners, are needed to complete the wall. The purpose of all the shapes is to be able to provide a wall with no internal or external corners that will catch dirt or be subject to chipping.

All-purpose combination corners and jamb units are also made with the interior coring arranged to permit ready cutting in the proper lengths needed to work the bond in the wall at these spots. Figure 17-2, pages 246 through 249, shows standard units and shapes made and recommended by the Facing Tile Institute for use with the 6T and 8W series.

One of the most frequently used shapes in glazed structural tile is the *bullnose* corner tile. It is rounded on the corner which makes it easier to keep it clean and less likely to be chipped than a square glazed corner. It is very popular when building hallways or corridors in a building because the rounded corner does not present the hard, sharp corner of a square tile. Figure 17-3 shows 8W series bullnose tile.

Fig. 17-3 8W series bullnose corner piece of glazed structural tile.

Coring And Scoring Of Glazed Tile

Types and directions of scoring and coring are optional depending on the manufacturer. In general, the manufacturer standardizes on either horizontal or vertical coring. *Coring* refers to the direction the cells run in the tile. If the coring is in a horizontal direction it is said to be *side construction*. If the coring is in a vertical position, it is said to be *end construction*. Figure 17-4 shows the two types of coring.

The purpose of scoring on the back of the tile is to make the mortar stick to the tile more readily. This also makes the tile easier for the mason to pick up, because it does not slip from the hands as easily

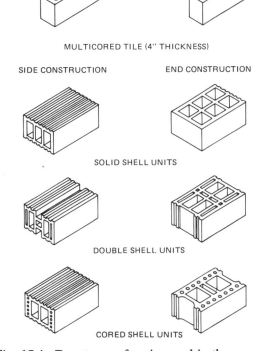

Fig. 17-4 Two types of coring used in the manufacture of glazed structural tile.

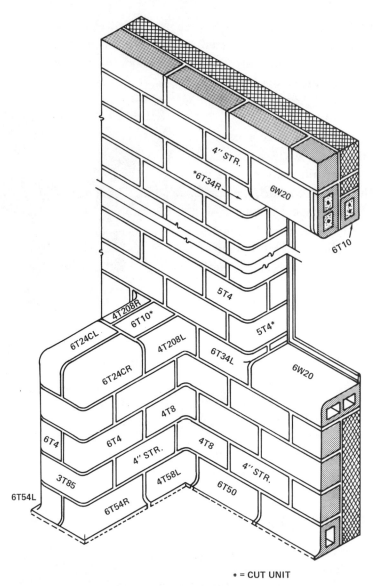

Fig. 17-5 An eight-inch double-faced wall that is built from the 6T series tile. Notice how the special shapes are used.

as a smooth back. If job conditions require that the back of a tile be of a certain type, it should be specified by the customer when ordering.

Bonds For Glazed Tile

A tile wall can be built in several bond patterns: the conventional half lap, one third and one quarter (seldom used), and the stack bond patterns. Figure 17-5 shows an application of the half-lap bond using many of the special shapes described. It is laid with the 6T series tile.

The stack bond is also called block bond or plumb bond. In the stack bond pattern, each tile is laid directly over the one underneath with all of the head joints lining up in a plumb position. When the tiles are laid in all stretchers, the bond is called a *stretcher stack bond*.

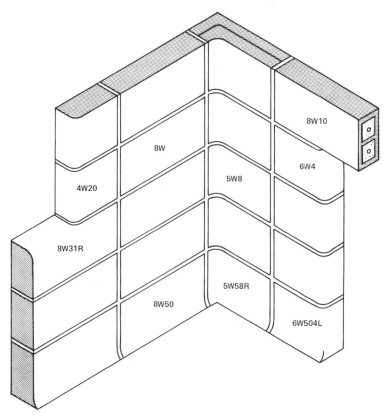

Fig. 17-6 Stretcher stack bond of glazed tile using the 8W series tile.

The stretcher stack bond is the most frequently used bond pattern. Figure 17-6 shows a stretcher stack bond using the 8W series glazed tile.

When tiles are laid in the stack bond, it is very important that all of the tiles are the same length and height. This is necessary because the continuous head and bed joints show even the slightest difference and will cause the bond to be uneven or the size of the joints to be too big or too small. Therefore, it is recommended to use glazed tile specified as select quality, with a ground edge.

Using Metal Reinforcement With Stack Bond

When structural facing tiles or any other masonry units are laid in stack bond, they do not have the overlapping of one unit over the other to distribute weight or tie them together. Even if this does not weaken the wall, it can present a problem in bonding all of the masonry work together. For this reason it is recommended to use steel reinforcement and wall ties laid horizontally in the wall and bonded into the backup work. The typical requirement is that one 3/16-inch rod of steel reinforcement, or its equivalent, for each 6 inches of wall thickness (or fraction thereof), be placed in horizontal joints 16 inches on center.

Coursing Of Glazed Tile

The table in figure 17-7, page 254, gives the horizontal and vertical coursing of the 6T and 8W series, including the mortar head and bed joints as they would appear in a glazed tile

Horizontal Coursing Table

Number of Stretchers	6T 12" Nominal Length	8W 16" Nominal Length
1/2	6"	8"
1	1'0"	1'4"
1 1/2	1'6"	2'0"
2	2'0"	2'8"
2 1/2	2'6"	3'4"
3	3'0"	4'0"
3 1/2	3'6"	4'8"
4	4'0"	5'4"
4 1/2	4'6"	6'0"
5	5'0"	6'8"
5 1/2	5'6"	7'4"
6	6'0"	8'0"
6 1/2	6'6"	8'8"
7	7'0"	9'4"
7 1/2	7'6"	10'0"
8	8'0"	10'8"
8 1/2	8'6"	11'4"
9	9'0"	12'0"
9 1/2	9'6"	12'8"
10	10'0"	13'4"
10 1/2	10'6"	14'0"
11	11'0"	14'8"
11 1/2	11'6"	15'4"
12	12'0"	16'0"
12 1/2	12'6"	16'8"
13	13'0"	17'4"
13 1/2	13'6"	18'0"
14	14'0"	18'8"
14 1/2	14'6"	19'4"
15	15'0"	10'0"
15 1/2	15'6"	20'8"
16	16'0"	21'4"
16 1/2	16'6"	22'0"
17	17'0"	22'8"
17 1/2	17'6"	23'4"
18	18'0"	24'0"
18 1/2	18'6"	24'8"
19	19'0"	25'4"
19 1/2	19'6"	26'0"
20	20'0"	26'8"
20 1/2	20'6"	27'4"
21	21'0"	28'0"
21 1/2	21'6"	28'8"
22	22'0"	29'4"
22 1/2	22'6"	30'0"
23	23'0"	30'8"
23 1/2	23'6"	31'4"
24	24'0"	32'0"
24 1/2	24'6"	32'8"
25	25'0"	33'4"
25 1/2	25'6"	34'0"
26	26'0"	34'8"
26 1/2	26'6"	35'4"
27	27'0"	36'0"
27 1/2	27'6"	36'8"
28	28'0"	37'4"

Vertical Coursing Table

Number of Courses	6T 5 1/2" Nominal Height	8W 8" Nominal Height
1	5 5/16"	8"
2	10 5/8"	1'4"
3	1'4"	2'0"
4	1'9 5/16"	2'8"
5	2'2 5/8"	3'4"
6	2'8"	4'0"
7	3'1 5/16"	4'8"
8	3'6 5/8"	5'4"
9	4'0"	6'0"
10	4'5 5/16"	6'8"
11	4'10 5/8"	7'4"
12	5'4"	8'0"
13	5'9 5/16"	8'8"
14	6'2 5/8"	9'4"
15	6'8"	10'0"
16	7'1 5/16"	10'8"
17	7'6 5/8"	11'4"
18	8'0"	12'0"
19	8'5 5/16"	12'8"
20	8'10 5/8"	13'4"
21	9'4"	14'0"
22	9'9 5/16"	14'8"
23	10'2 5/8"	15'4"
24	10'8"	16'0"
25	11'1 5/16"	16'8"
26	11'6 5/8"	17'4"
27	12'0"	18'0"
28	12'5 5/16"	18'8"
29	12'10 5/8"	19'4"
30	13'4"	20'0"
31	13'9 5/16"	20'8"
32	14'2 5/8"	21'4"
33	14'8"	22'0"
34	15'1 5/16"	22'8"
35	15'6 5/8"	23'4"
36	16'0"	24'0"
37	16'5 5/16"	24'8"
38	16'10 5/8"	25'4"
39	17'4"	26'0"
40	17'9 5/16"	26'8"
41	18'2 5/8"	27'4"
42	18'8"	28'0"
43	19'1 5/16"	28'8"
44	19'6 5/8"	29'4"
45	20'0"	30'0"
46	20'5 5/16"	30'8"
47	20'10 5/8"	31'4"
48	21'4"	32'0"
49	21'9 5/16"	32'8"
50	22'2 5/8"	33'4"

Fig. 17-7 Horizontal and vertical coursing tables for the 6T and 8W series of glazed structural tile.

wall. The mason's modular rule can also be used to mark off courses of both series of tile. The 6T series is represented on the mason's modular rule as number 3 and the 8W series is represented as number 2 on the rule.

Noise-Absorbent Glazed Tile

Special tiles have been designed and manufactured to meet the demands for noise control. This has been done by making a glazed tile that has a series of holes in the face and a factory-inserted, chemically resistant, fiberglass pad in the coring of the tile. These tiles are called *SCR acoustile®*. They are very effective in reducing noise while still having the advantages of a glazed structural clay tile. Figure 17-8 shows a sound-absorbing glazed structural tile.

CERAMIC GLAZED FACING BRICKS

Glazed bricks are very similar to tile except that they are made in the size of brick units. It is strongly recommended that the glazed brick manufacturer be consulted before using the product on the exterior. Some units are not as durable as others. Grade EW bricks, for example, are intended for use where a high and uniform degree of resistance to frost action and disintegration by weathering is desired.

Glazed bricks are made from clay shale, fireclay, or a mixture of both. They are fired in a kiln much the same as glazed tile.

Types

Three types of glazed facing bricks are available. Type GBX is for general use in exposed exterior and interior walls and partitions where a high degree of mechanical perfection, narrow color range, and minimum variation in size are required. Type GBS is for general use in exterior exposed and interior masonry walls and partitions where wider color ranges and greater differences in sizes are allowed. Type GBA is for general use in exposed exterior and interior masonry walls and partitions. This brick is manufactured to produce characteristic architectural effects resulting from nonuniformity in the size, color, and texture of the individual units.

Fig. 17-8 A glazed structural tile that has holes in the face and fiberglass inside the coring to absorb sound.

Sizes

They are made in different sizes to fit the needs of designers and architects. The standard size is 3 3/4 inches x 2 1/4 inches x 8 inches. Modular size is 3 5/8 inches x 2 1/4 inches x 8 inches. Oversize (jumbo) brick is 3 3/4 inches x 2 3/4 inches x 8 inches. A utility size is 3 5/8 inches x 3 5/8 inches x 11 5/8 inches. (These dimensions are from Facing Tile Institute Specifications, September 1973.) Figure 17-9 shows a group of glazed brick sizes available.

Features And Uses

All of the types of glazed bricks mentioned are resistant to cracking and spalling (chipping) when subjected to heat or cold. The surface does not stain under normal conditions and the glazed finish gives maximum resistance to most chemical actions.

These bricks present a very hard, ceramic-glazed surface when laid in the wall. They offer an attractive appearance along with the qualities found in a glazed surface.

Glazed bricks are used for the exterior of office buildings, hallways and corridors, restaurants, and schools. They are highly resistant to air pollution which deposits residue on many of the masonry faces of structures. Figure 17-10 shows the use of glazed bricks in a restaurant.

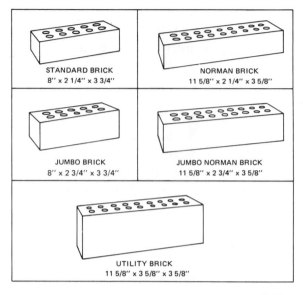

Fig. 17-9 Different sizes of glazed bricks available.

Fig. 17-10 Glazed brick wall in a restaurant. Notice the high gloss of the brick.

Colors

As with glazed structural tile, there are many different colors of glazed bricks available. However, the most popular colors for glazed bricks are the light, neutral colors and bricks with light specks, because they do not go out of style. The manufacturer should be consulted for current colors listed before any material is ordered because colors are sometimes changed.

WORKMANSHIP

Good workmanship is essential when laying glazed structural tiles or glazed bricks. Both units have fragile faces, so when laying them in the wall, extra care should be used. Glazed structural tiles are recommended for use in walls on the interior of a building, while ceramic glazed bricks are used when a glazed surface is desired on the interior or exterior of a building. *Always check the manufacturer's recommendation of the product before installing the product in a building.*

Mortar joints should be well filled and all work should be level and plumb. Chipped or cracked tiles should not be installed in the wall where they can be seen. Clean all mortar stains from the work as soon as possible to reduce problem cleaning at the end of the job.

ACHIEVEMENT REVIEW

Select the best answer from the choices offered to complete each statement. List your choice by letter identification.

1. Structural clay tile was first introduced in the United States in the year
    a. 1806.
    b. 1850.
    c. 1875.
    d. 1890.

2. When first introduced, the main use of structural clay tile was for
    a. the facing of the exterior of buildings.
    b. bathrooms and service areas.
    c. fireproofing.
    d. decorative purposes.

3. Structural clay tiles are made from
    a. ceramic materials.
    b. fireclay.
    c. shale.
    d. portland cement.

4. The most popular colors of glazed structural tile are the
    a. darker shades.
    b. bright shades.
    c. light neutral shades.
    d. pastels.

5. A glazed tile that absorbs sound is called
    a. ceramic tile.
    b. bullnose tile.
    c. universal tile.
    d. SCR acoustile ®.

6. Two grades of glazed structural facing tile are
    a. S and G.
    b. D and S.
    c. C and E.
    d. F and L.

7. In recent years, the manufacture of glazed structural facing tile has been reduced to only two series. These series are
    a. 4W and 5T.
    b. 6T and 8W.
    c. 8T and 9W.
    d. 10T and 12W.

8. Glazed structural tiles are sold with ground edges because
    a. the mortar bonds to the edges better.
    b. the color of the tile does not chip from the edges as quickly.
    c. they are more exact in size.
    d. they match the color of the mortar joints more evenly.

9. A special glazed structural tile that has a rounded corner for use at corners, windows, and door jambs is called a
   a. starter.
   b. lintel tile.
   c. universal.
   d. bullnose.

10. When glazed tiles are laid so that all of the vertical head joints are plumb over each other it is called a
    a. stack bond.
    b. half lap.
    c. quarter bond.
    d. running bond.

11. On a mason's modular rule, the 8W series tile is noted as number
    a. 2.
    b. 3.
    c. 6.
    d. 8.

12. The most important problem encountered by the mason in laying glazed brick walls is
    a. keeping the mortar joints the same thickness.
    b. preventing water from entering the wall.
    c. stopping the colors in the brick from fading.
    d. keeping materials on the job for the mason.

Unit 18 LAYING GLAZED STRUCTURAL TILE AND GLAZED BRICKS

OBJECTIVES

After studying this unit the student will be able to

- explain the different kinds of mortar used for laying and pointing glazed structural tile.
- describe the layout and building of a glazed tile wall.
- identify some of the problems associated with laying glazed bricks.

CARE OF GLAZED STRUCTURAL TILE BEFORE LAYING

Glazed tile is a masonry unit that needs no further treatment after it has been laid, tooled, and cleaned. Most of the damage to tiles occurs when unloading them from the box cars and stacking them on the job. Care in handling saves money for the contractor because chipped tile may have to be cut out and replaced. Corners and special shapes are more likely to chip or break due to the nature of the special design.

Glazed tile and brick are delivered to the job in protective paper cartons. The materials should not be removed from the cartons until ready to be laid in the wall. Special shapes such as doorstops, universal pieces at windowsills, caps, and so forth are shipped in cardboard cartons packed with straw. These should be checked as soon as they arrive so that if they are damaged, a claim can be made to the railroad or trucking company. If a long time passes before the tiles are checked, the claim may not be honored.

Glazed tile should not be stacked under scaffolds where waste materials or masonry units on the scaffold can fall on them. Welding splatter can also ruin a glazed masonry unit if it hits the face.

MORTAR FOR GLAZED TILE

As a rule, the mortar used to lay glazed tile is always specified. Tile mortar must meet the compressive strength requirements of the job and be matched with the color of the tile being laid. The mix is approved by the architect after a sample is inspected on the job site. Colored mortars (white or off-white) are the most frequently used. White portland cement mortar, hydrated lime, and white sand are very popular when an all-white joint is required.

For an all-purpose general mortar, type N portland cement-lime mortar is used. This mix is 1 part portland cement, 1 part hydrated lime, and 6 parts sand. Regardless of the type of mortar used, the mortar joints should be well filled and free from any object that can stain through to the surface.

Consistency Of The Mix

Mortar for glazed tile should be a little stiffer and richer than regular mortar because the tile does not set as quickly as bricks or concrete blocks. In addition, the mortar should not be mixed more than two hours before using or it loses its strength. Mortar also should not be mixed and loaded in the mortar pans just before lunch and left to set. Good management by the foreman on the job eliminates this problem. It is recommended not to temper mortar for glazed tile more than once, and then only with water lost from evaporation. This is because tempering more than once affects the bond strength and elasticity of the mortar.

The use of low-strength mortar with poor adhesive bond results in an inferior wall. A mason's work output is greatly affected if the mortar is not the proper consistency or if it does not stick to the tile when applied.

Special Epoxy Mortar For Glazed Tile

A special epoxy mortar can be pointed in mortar joints for a denser sanitary finish. It adds a longer life to new glazed tile walls and is easier to keep clean. In structures such as milk plants, laboratories, bakeries, and chemical plants, epoxy mortar provides a joint that is nontoxic, unaffected by bacteria, safe for use around foods, low in absorption, and stainproof.

Epoxy mortar consists of two or three components such as a resin, hardener, and powder. It has a working life of approximately 45 minutes after being mixed at 75 degrees F. It sets and hardens in 16 hours and is used only for pointing in the mortar joints of the structural clay tile, not the setting bed.

Applying The Mortar

Glazed tile is usually laid in mortar with a joint width of 1/4 inch. As soon as the mortar can hold the weight of the tile, it is raked out to a depth of 1/4 inch. Mortar stains on the face of the tile are wiped off with a cloth or burlap bag. The wall is then allowed to cure at least 24 hours before pointing with epoxy mortar.

It is important that epoxy mortar be mixed according to the manufacturer's directions. A mechanical mixer with an agitator is recommended. The mortar is then applied to the face of the wall with a rubber-faced trowel. The joint can then be tooled or a rubber finish can be applied if desired. At the end of the pointing process, be sure that all mortar is removed from the face of the tile before it hardens.

Cleaning Up

Clean up immediately afterward by washing down with warm water and a sponge. (Add several drops of household detergent per bucket.) Clean the surface of the wall with a circular motion, changing the water when it is dirty. Epoxy mortar should never be allowed to remain on the face of the wall for more than 45 minutes.

LAYING GLAZED TILE

Glazed tiles are made very accurate in length so that a rule or steel tape can be used for marking the bond on the floor over a distance. Checkpoints should be marked on the base about every three tiles to prevent the bond from getting ahead or behind as the mason is laying out the units. The 6T series stretcher is made 11 3/4 inches long, so that it measures 12 inches with the mortar head joint. When laying out the bond with a steel tape or rule, any multiples of 12 inches will work bond or whole tiles. For example, assume a wall between two concrete columns is 14 feet long. This will work 14 of the 6T series tile perfectly, with no cutting, except for the halves on every other course if the bond is running or all stretcher bond (half lap). Figure 18-1 shows a 6T series tile laid out between two concrete columns.

If the 8W series is used, the bond is laid out in multiples of 16 inches, because the tile is 15 3/4 inches and with the 1/4-inch head joint, it measures 16 inches. Since the 8W series is a larger tile, less tiles are needed than with the 6T series. Figure 18-2 shows the 8W series tile laid out between two columns.

Starter Pieces

When there is a corner return on one or both ends of the wall, a starter piece must be used on the corner every course. This is because the glazed end of the tile is not wide enough to lay out over the tile below on the corner and break half bond.

Figure 18-3, page 262, shows a layout of two courses of 6T series tile from the corner to the door jamb. The corner piece is cut 9 3/4 inches long with the masonry saw. This is

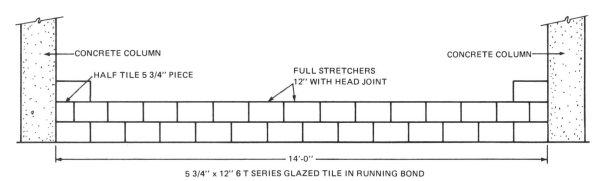

Fig. 18-1 Elevation view of 6T series glazed tile wall laid out between two columns.

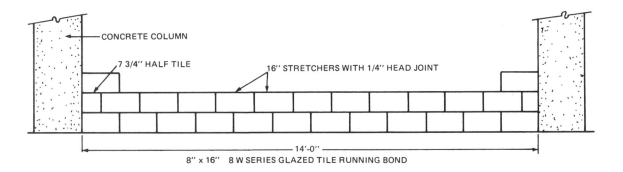

Fig. 18-2 Elevation view of 8W series glazed tile wall laid out between two concrete columns.

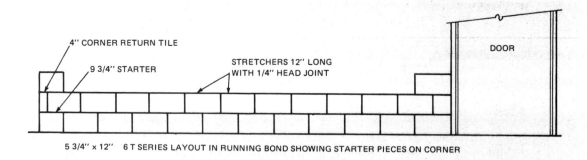

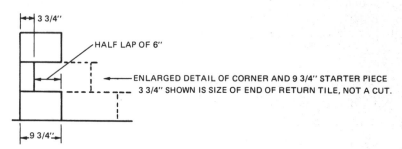

Fig. 18-3 Elevation view of 6T series glazed tile wall that has a corner on one end and a door on the other. Notice the arrangement of the starter piece on the corner.

done so that the 4-inch corner return tile, which will be laid on the course on top of it, will bond half over the one beneath. The difference is 2 inches because the 4-inch corner return tile is 2 inches short of the 6 inches required to bond exactly half over the tile beneath. The 6T series starter pieces are shown in figure 18-3.

Figure 18-4 shows the starter pieces needed for the 8W series tile for the corner. The 8W series tile has a 4-inch glazed tile end including the mortar head joint. Therefore, the corner starter piece for 8W tile must be 11 3/4 inches long.

Keeping Pieces Against A Window Or Door Jamb

If there must be a piece of tile cut in the wall, it is a good practice to locate it against a window or door jamb. In this manner it is not obvious that there is a piece in the wall. Cuts should never be put in the center of the wall unless there is no other choice. If cuts must be put in the center of the wall, cut the tiles and lay them under a large window. It is a sign of good workmanship when no pieces or cuts of glazed tile are in the wall except at corners or jambs. The use of good judgment in the initial layout of the tile saves cutting with the saw and increases the total efficiency of the job.

Cove Base Tile

If cove base tiles, figure 18-5, are being laid on the first course, care should be taken to make sure that the lip of the cove at the base of the tile is in perfect alignment. If misaligned, the floor tile will not fit against it correctly. Alignment can be checked by straightedging the lip of the cove base tile with the plumb rule.

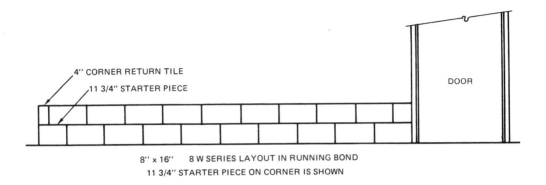

Fig. 18-4 Elevation view of 8W series glazed tile showing the arrangements of the bond at the corner and the starter piece.

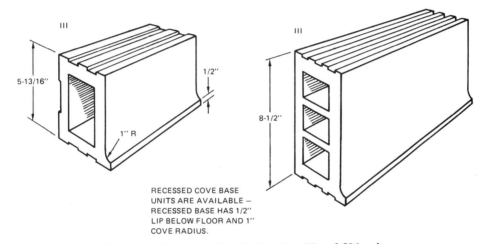

Fig. 18-5 Cove base glazed tile units, 6T and 8W series.

Cove base tile or regular stretcher tile is laid out by laying one tile at each end of the wall, then leveling and plumbing the unit. A line is attached and pulled tight from left to right. It may be necessary to fasten the line to an unmovable object on each end of the wall, and to fasten a trig to the line so the strain of the line does not pull the tile loose. The course is then run through, establishing the bond. The tiles are tooled and wiped off with a cloth.

It is a good idea to cover the projected bottom lip of the tile with sand, or a tile wrapper with mortar laid on top of it to prevent the edge from breaking as the wall is built. Cutting out and replacing cove base tile is very time consuming and costly. Each individual tile on the first course should be plumbed, even though the tiles are being laid to a line, especially if the wall is built in stack bond.

Building The Corner Or Lead

The corner or lead is always built up ahead of the line so it does not pull loose when the line is attached and tightened. The exception to this is if a door, window jamb, or concrete column is against the end of the wall. The line can be attached to one of these and pulled tight. Then a tile is laid against the jamb and the line is held to the tile with a trig. This practice is economical when it can be done.

When building the lead, the tile should be leveled with care so the face is not chipped as it is laid true. Keep the level in the center of the tile and settle the level into position gently either with a hammer or trowel. Figure 18-6 shows a mason leveling 8W series glazed structural facing tile. Steel reinforcement is essential when laying stack bond tile to ensure that the work is tied together properly.

Plumbing The Tile

When laying tile in a stack bond, it is important to plumb the vertical mortar joints (figure 18-7), as well as the face of the wall (figure 18-8). The select tile in the 8W series has a ground edge for accuracy and any deviation is very noticeable. Great care should be taken in the plumbing process to protect the face of the tile from damage. It is also very important to tie the lead to the existing work if it is a double wall. This is done by building in all wall ties fully with mortar.

Striking The Mortar Joints

The mortar bed and head joints are usually struck with a concave joint; others are raked out. Joint finish types are stated in the job specifications. A plastic or glass tool should be used so it does not cause black marks in the joints. The mortar joints should not be struck, however, until they are thumbprint hard or the tile can sink or fall out of plumb. Figure 18-9 shows a mason tooling the mortar joints on 8W series glazed structural tile.

Fig. 18-6 Mason leveling an 8W series glazed structural facing tile.

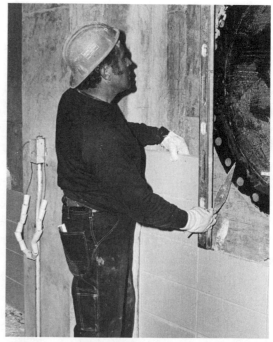

Fig. 18-7 Mason plumbing the vertical head joint in a stack bond wall.

Unit 18 Laying Glazed Structural Tile and Glazed Bricks ■ 265

Fig. 18-8 Mason plumbing the face of a glazed structural tile wall.

Fig. 18-9 Mason tooling the mortar joints.

Wiping Off The Wall

After building the lead and tooling the joints, they should be gently wiped off with a clean cloth or burlap bag, figure 18-10. Care should be taken that the cloth or bag is clean, because oil or foreign substances are absorbed into the mortar joint finish. Do not use a dirty cloth or bag to wipe off a tile wall. This precaution is especially important if white portland cement mortar is being used.

If the tile is SCR acoustile®, every hole must be cleaned of mortar or the tile will not be effective in reducing sound.

Laying Tile To The Line

After the corner or leads are built, the line is attached and the wall is built. Glazed tiles are laid the same as other masonry units, except the mason must take special care not to chip or damage the edges. Figure 18-11, page 266, shows a mason laying a course of tile to the line.

When crossing over a window or door frame, the top of the tile that is to be laid over the door may be cut out. A length of steel rods is then inserted, and the cavity is

Fig. 18-10 Wiping off the glazed structural tile wall with a clean burlap bag to remove all mortar stains.

Fig. 18-11 Laying glazed tile to the line. The tiles shown are being used in a sewage plant to line the interior walls for sanitary and noise control purposes.

Fig. 18-12 This glazed tile is being laid over the head of an opening. The tile are cut out for the insertion of steel rods and concrete to form a lintel or bond beam.

filled with grout for greater strength. This is known as a *bond beam* type of lintel. Figure 18-12 shows a course of cut tile being laid across the top of the door frame. After the tile has set, the wood frame is removed and the lintel is self-supporting.

Final Cleaning Of Glazed Tile

After the wall has cured, it is cleaned by using warm water and soap. A fiber brush can be used to scrub the wall. The wall should then be rinsed thoroughly to remove all of the soap. It is not recommended to use acid or metal scrapers to clean tile walls because the ceramic face can be damaged. A wood paddle usually removes any particles of hardened mortar.

LAYING GLAZED BRICKS

When laying any masonry wall, good construction principles should be followed. It is especially important that glazed bricks be protected from the penetration of water into the wall. Waterproofing such as flashing under windowsills and over openings, weep holes installed in the mortar joints at grade line, and dampproof courses at the base of the wall will help prevent water from getting into the wall.

Coping should have proper drips built in or provided by flashing. Regular (unglazed) brick allows moisture to escape from evaporation through the face of the unit. The glaze on the face of glazed units prevents this loss of moisture by evaporation.

The cavities in cavity walls must be kept clean of mortar droppings and masonry waste at the conclusion of building the wall. Drainage must also be provided to allow the water to escape from cavity walls built of glazed bricks. Glazed bricks are impervious to water (do not allow moisture to pass through the face of the brick). Therefore, in order for water to be able to escape from a completed wall, the above protective measures must be followed. Glazed bricks are not recommended for coping, caps, or other horizontal surfaces.

Glazed bricks should also be kept dry until ready to lay in the wall. The protective paper wrappings should remain on the bricks until the mason is ready to lay the bricks in the mortar. Covering the stacks with canvas or plastic helps keep the brick dry.

Handling

Due to the fragile nature of the ceramic glaze on the face of the brick, care must be taken in stacking or moving the bricks on the job. Dropping the unit or handling it roughly results in chippage or breakage.

Expansion Joints

Expansion joints should be built into the wall to allow for movement, as shown on the plans. Joints should be kept completely clean of mortar for their full width. The size and location of expansion joints are the responsibility of the architect and design engineer. The location of the joints should be closely followed by the mason.

Expansion joints should be filled with a nonhardening, permanently elastic material designed to prevent water entry. Horizontal control joints and bond breaks should also be included where shown on the plans.

Mortar

Mortar for glazed bricks is made of portland cement and lime. The types used are N, S, O, or M, of ASTM C270 or equal strength mortar as stated by the job specifications.

Tooling The Mortar Joints

Tool the mortar joints with a glass or plastic jointer to prevent rust or black marks from appearing in the mortar joints. This should not be done until the mortar joints are thumbprint hard.

PROCEDURES FOR LAYING GLAZED BRICKS

Glazed bricks are laid in the wall in almost the same way as regular bricks. There are, however, some important practices the mason must know. The rate of absorption of the mortar is slower with a glazed brick than with regular fired clay brick. Glazed brick may have less suction than regular brick. Therefore, never wet glazed units and be sure to store them in a dry place. The unit should be as dry as possible and, as mentioned before, not removed from the paper wrappers until ready to lay in the wall. Since the mortar does not set as fast, the mason must develop a lighter touch when laying the brick in the mortar. The mortar must be of the proper consistency so the bricks do not sink or bleed too much. However, if the mortar is too stiff and the brick must be tapped into the mortar, it may cause the wall to bulge or fall out of plumb. Once this happens, it is very difficult to correct. This is because of the low absorption rate of glazed bricks. When the bricks are first laid in the mortar bed joint, they must be level and plumb without using too much pressure.

Building From One Scaffold Height To The Next

Enough time must be allowed for glazed brickwork to set properly before the next section of wall is built on top of work below. Often, the brickwork is not given sufficient time to set when a large building is constructed using a swinging scaffold. The scaffold is put up and the masons start to lay the next section of wall before the wall below sets enough to hold the weight without sagging or falling out of plumb. The rule of thumb is to try to work around the building and let the wall set until the next day. This practice is not always possible when many masons are working on a structure, but should be followed for best results. It also requires more planning by the foreman.

The work should be tied to the substructure by placing wall ties periodically as the work progresses. If this is not done, the brickwork can pull away, making it necessary to rebuild the wall. This applies to any masonry work when facing a skeleton-type structure such as steel or concrete.

Rubbing The Wall At The End Of The Striking Process

After striking the mortar joints, the wall should not be bagged or cleaned with a cloth until the joints have set enough not to smear or rub out. It is recommended to always bag glazed brick walls to reduce the amount of cleaning necessary. Use very little pressure when rubbing the wall with a burlap bag or cloth so the wall is not disturbed.

The important point to remember when laying units in the mortar bed is to use as little pressure as possible to keep the wall level and plumb. In the trade this is called developing a "light touch". A light touch is acquired through experience and practice with the tool skills.

CUTTING GLAZED TILE AND BRICKS

Because the units are hard, yet have a fragile face, it is always recommended to cut them with a wet masonry saw equipped with a diamond blade. Eye protection and the proper work clothing should be worn when cutting to protect the worker from accidents.

Due to the hardness of the masonry units, they do not absorb much water from the cutting process. Therefore, there is no problem when laying the units immediately afterwards. Wipe off the excess water before laying the unit in the mortar bed.

NOTE: The preceding information on glazed structural facing tile and glazed bricks does not cover every possible situation that will occur. The situations discussed are the most frequent problems experienced when laying or building walls with glazed tile and bricks.

ACHIEVEMENT REVIEW

Select the best answer from the choices offered to complete each statement. List your choice by letter identification.

1. An all-purpose mortar for laying glazed tile is type
   a. M.
   b. N.
   c. O.
   d. S.

2. When mixing mortar for laying glazed tile it should be mixed ahead of time not more than
   a. 1/2 hour.
   b. 1 hour.
   c. 2 hours.
   d. 3 hours.

3. A special mortar is made for the pointing of mortar joints for glazed tile. This mortar is very hard and has a low absorption level. This type of mortar is called
   a. high-lime mortar.
   b. caulking mortar.
   c. epoxy mortar.
   d. masonry cement mortar.

4. The correct length for the starter piece on the corner for 6T series glazed tile is
   a. 4 3/4 inches.
   b. 6 3/4 inches.
   c. 9 3/4 inches.
   d. 11 3/4 inches.

5. The correct length for the starter piece on the corner for 8W series tile is
   a. 6 3/4 inches.
   b. 9 3/4 inches.
   c. 11 3/4 inches.
   d. 14 3/4 inches.

6. A popular bond used in glazed tile work has all of the tile laid over one another with all of the vertical head joints in a plumb position with each other. This type of bond is called a
   a. Flemish bond.
   b. running bond.
   c. 1/4 running bond.
   d. stack bond.

7. The most frequently used joint finish for glazed tile work is
   a. raked out.
   b. concave.
   c. grapevine.
   d. slicker.

8. Glazed structural tile walls should be cleaned with
   a. a muriatic acid solution.
   b. scouring powders.
   c. warm water and soap.
   d. a vinegar solution.

9. A major problem in laying glazed bricks is that the brick is impervious. Impervious, as used in relation to glazed brick, means
   a. the strength of the unit.
   b. the color of the bricks.
   c. the bricks do not allow moisture to pass through the wall.
   d. a difference in the size of the bricks.

10. Mortar for glazed brickwork must be mixed to the proper consistency because
    a. of the low absorption rate of glazed bricks.
    b. the bricks are smeared too much on the face side.
    c. the mortar joints are weakened.
    d. too much mortar is dropped from the wall.

11. Glazed bricks should always be cut with a
    a. tile hammer.
    b. brick set chisel.
    c. trowel blade.
    d. masonry saw.

Unit 19 PREFACED CONCRETE MASONRY UNITS

OBJECTIVES

After studying this unit the student will be able to

- describe the types and sizes of prefaced concrete masonry units available.
- explain the key factors to consider when laying and cleaning prefaced concrete masonry units.
- list the advantages of a prefaced block wall.

DEVELOPMENT OF PREFACED CONCRETE MASONRY UNITS

Prefaced concrete masonry units, more commonly called prefaced blocks, were developed to meet the need for a masonry unit that had a tile-like face on one side and concrete block face on the opposite side. This was accomplished by applying a thermoplastic resin finish to the face of a lightweight concrete block. The finish or facing is 1/8 inch thick and is resistant to most chemicals, abrasions, and impacts. This thermoplastic resin finish can be applied to one side of a concrete block, figure 19-1, or to both sides of the block.

The factory finish is easy to keep clean, resists marking under normal conditions, and furnishes a durable masonry unit. These units can serve as a wall having a concrete block finish on one side and a tile-like face on the opposite side. The plastic tile-like finish is very hard and durable.

Concrete blocks used for this process must conform to ASTM specifications for load-bearing and non-load-bearing units. The particular fire rating can be obtained by reading the manufacturers' brochures on their products.

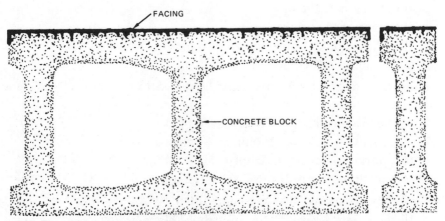

Fig. 19-1 Top view of a prefaced concrete block showing how the facing is bonded to the block.

Unit 19 Prefaced Concrete Masonry Units ■ 271

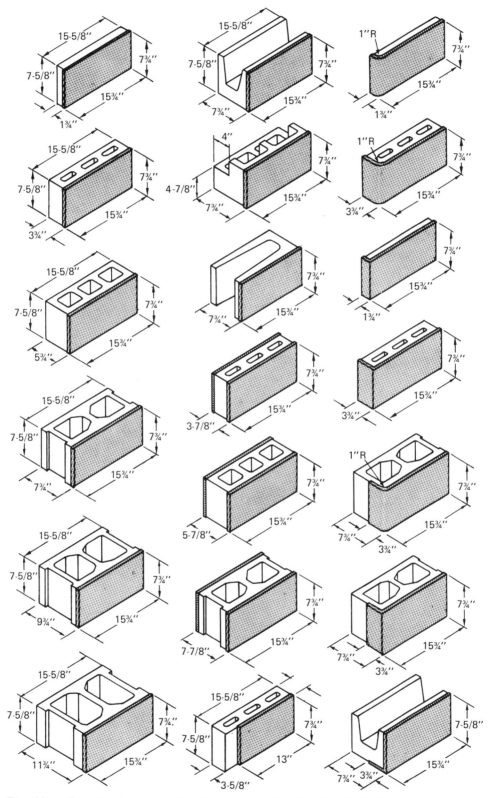

Fig. 19-2 Functional series prefaced masonry units, showing the field stretcher units available.

SIZES

Block face sizes are 8 inches x 16 inches, 8 inches x 8 inches, and 4 inches x 16 inches. Prefaced concrete block thicknesses are 2 inches, 4 inches, 6 inches, 8 inches, 10 inches, and 12 inches. The 8-inch x 16-inch block is considered standard size and is the most used.

TYPES

There are three types of prefaced block faces available: functional, scored, and design series.

Functional Series

The functional series can be supplied with either 2- or 3-core blocks, and open or closed ends depending on local block manufacturing practices. Special right or left shapes must be specified when ordering. The blocks are available in nominal 5-inch heights, and 6-inch and 8-inch thicknesses.

The functional series has a regular, smooth tile-like face and is used when no special design is required. The blocks come in different shapes, figures 19-2, page 271, and 19-3, to fit the specific job.

Scored Series

The scored series takes its name from the design of the scoring in the face of the block. This series gives a choice of scales and patterns plus the economy of a standard 8-inch x 16-inch block. The scoring can be used vertically or horizontally to provide simple or unusual patterns for special effects in scale and design.

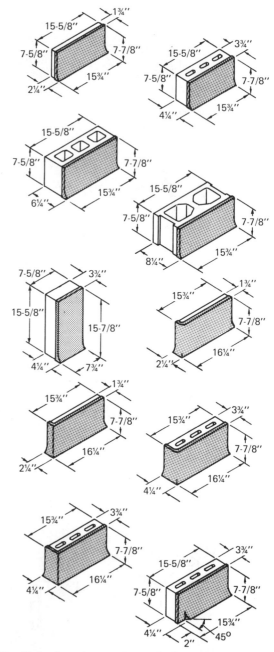

Fig. 19-3 Cove base units available in the functional series of prefaced masonry units.

The scored joints are grouted or pointed with mortar after installation. For double-faced scored walls, it is recommended to use two individual units in thickness.

Design Series

The design series adds a three-dimensional appearance to the wall. It includes a wide selection of sculptured shapes and forms that provide shadows and highlights in soft shading or very bold contrasts. This series can be combined with the scored series to provide an attractive wall, as shown in figure 19-4.

The design series supplied a creative approach for the architect and designer in the construction of prefaced block walls. Figure 19-5 shows a prefaced block wall built of the design series in stack bond. The design of the face accents the shadows falling on the wall and presents an appealing finished wall. All 8-inch x 16-inch face units are available in bond beam and brick bonding types, in addition to the standard stretchers.

Cove Base Units

The cove base unit is designed to provide a modular sized sanitary base unit which can be installed without a floor recess. Figure 19-6, page 274, shows how the unit appears laid in the wall.

COLORS

Prefaced blocks are available in many different colors. Color charts should be obtained from manufacturers or dealers when ordering the blocks since colors may vary depending on the company. Colors range from dark to light; all, however, have the sparkle and brilliance of a glazed surface.

USES OF PREFACED MASONRY UNITS

Prefaced concrete masonry units are used for schools, restaurants, beverage plants, hospitals, power stations, sewage treatment plants, water purification plants, commercial

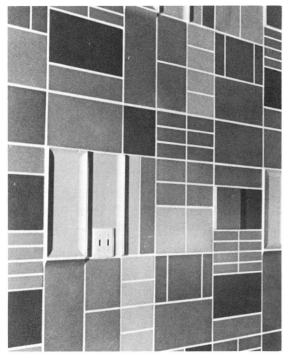

Fig. 19-4 A prefaced masonry unit wall that combines the scored and design series.

Fig. 19-5 Design series prefaced masonry units laid in a vertical position.

buildings, and any other structure in which a prefaced structural unit is desired. The most important advantage of prefaced blocks is that when a single masonry unit wall is required, a tile-like surface is showing on one side and a standard concrete block finish appears on the opposite side.

CARE IN HANDLING UNITS

When prefaced blocks are delivered to the job site, they should be carefully stacked and protected from the elements with tarpaulins or sheets of plastic. They should not be moved more than necessary due to the possibility of shipping the face or edges of the units. Leave the unit in its protective paper carton until it is to be laid in the wall. It is also important to stock a wall with just enough prefaced blocks to build that wall. Moving the blocks to build another wall increases the risk of chipping the units.

INSTALLING UNITS IN THE WALL

A standard type of mortar, such as type N or equal masonry cement mortar,

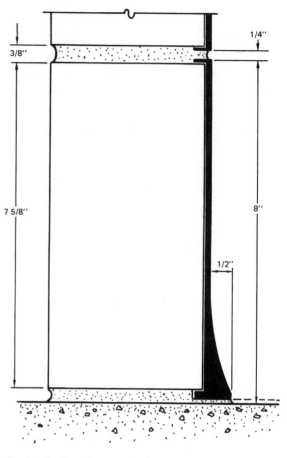

Fig. 19-6 Cove base unit that is made to be installed without a floor recess.

can be used to lay prefaced block. Mix the mortar a little thinner than for glazed structural tile because the main body of the unit is concrete block which absorbs water from the mortar at a faster rate. The joint can be raked out and repointed later, if desirable, with a white or special mortar depending on the job specifications.

WORKMANSHIP

All units should be laid in a neat manner, level, and plumb. All mortar head and bed joints should be well filled with mortar to obtain maximum strength. If stack bonds are used, all vertical head joints should be in a plumb, vertical position. This is not difficult to do because prefaced concrete blocks are very accurate in height and length due to the method of curing.

Methods of laying prefaced blocks are similar to the methods for laying other masonry units. One difference, however, is that the mason must be very careful in handling the units to prevent chipping the corners or sides, since there is a lip of facing that extends slightly over the edges of the block.

The units are modular in size. They are made to be laid in a 1/4-inch mortar bed joint and a 1/4-inch head joint.

STRIKING THE MORTAR JOINTS

For best results it is recommended to strike the mortar joints with a glass or plastic tool to prevent black marks from appearing in the mortar joints or on the face of the block next to the joint. (Metal jointers should not be used because they can cause a black mark in a white joint if the mortar sets too long.)

If being repointed or grouted, all mortar joints should be dampened with water and tuck-pointed flush with a thin bladed tool and then struck with the glass or plastic tool. (To *tuck-point* a mortar joint is to fill the joint with mortar using a tool such as a slicker.) It is never advisable to smear the entire wall with grout and wipe it off later.

CUTTING THE UNITS

All cuts of prefaced blocks should be made with a masonry saw, preferably one having a diamond blade lubricated with water. It is a good practice to cut prefaced blocks before laying them in the wall so the block part of the unit has time to dry out. When laying a prefaced block on the saw table, caution should be taken that no chipped pieces from former cuts are laying underneath because these chips can damage the face of the block.

> **Masons operating the masonry saw should wear eye protection at all times and stand on a wooden platform.**

At the completion of the cut it is also a good practice to return the cut blocks into the paper wrapper to transport them back to the mason working on the wall. This is especially important if the cut blocks are being moved on a wheelbarrow.

CLEANING THE WORK

Wipe off all work as it progresses. Any excess mortar should be wiped off with a clean cloth or burlap bag. When the cloth is beginning to become caked with mortar, discard it and use a clean cloth. Muriatic acid solutions are not recommended, so care should be taken to keep the work as clean as possible during construction. Instead of a muriatic acid solution, a masonry cleaning compound can be used. Use cleaning compounds in accordance with the manufacturer's directions.

All walls should be rinsed with clean water under pressure from a hose, and then damp dried with a cloth. Do not use steel wool or abrasive compounds on prefaced block.

ADVANTAGES OF PREFACED BLOCK

Listed are some of the advantages of building masonry walls of prefaced blocks.

- They come in a wide range of colors.
- They have a tile-like masonry face.
- The units are the same size, allowing a precise stack bond pattern.
- Flat ends permit the use of soldier coursing either in stack bond or half lap.

- Large cores or cells in the block allow the use of granular fill to increase fire resistance or insulation of the wall. This makes it easy to reinforce the wall with metal joint reinforcement.
- The facing is highly resistant to stains and dirt.
- The blocks meet USDA and OSHA requirements for sanitary walls.
- The units provide fire ratings depending on the thickness of the block (1, 2, 3, or 4 hours).
- Through-wall, load-bearing units eliminate expensive backup walls.
- Units are available throughout the United States.
- There is a minimum amount of maintenance and upkeep.

ACHIEVEMENT REVIEW

Select the best answer from the choices offered to complete each statement. List your choice by letter identification.

1. The facing on a prefaced concrete block is
   a. 1/16 inch thick.
   b. 1/8 inch thick.
   c. 3/16 inch thick.
   d. 1/4 inch thick.

2. When a prefaced block that has a tile-like face with no special design is wanted, the series used is the
   a. design series.
   b. functional series.
   c. soldier series.
   d. scored series.

3. The prefaced blocks that give a three-dimensional appearance are
   a. scored.
   b. functional.
   c. design.
   d. stack bond.

4. The standard height and length of a prefaced block is
   a. 8 inches x 16 inches.
   b. 8 inches x 8 inches.
   c. 4 inches x 12 inches.
   d. 12 inches x 12 inches.

5. The most important advantage a prefaced block offers that no other masonry unit offers is
   a. a fire rating that causes a lower insurance rate.
   b. colors that do not fade or go out of style.
   c. a design in the finish that creates a beautiful wall.
   d. a tile-like finish on one side and a concrete block surface on the opposite side.

6. Prefaced concrete masonry units are made to be laid with a mortar bed and head joint thickness of
   a. 1/8 inch.
   b. 1/4 inch.
   c. 3/8 inch.
   d. 1/2 inch.

7. For best results it is recommended to tool the mortar joints of prefaced concrete blocks with
   a. stainless steel strikers.
   b. copper strikers.
   c. glass or plastic strikers.
   d. wood strikers.

8. Prefaced concrete blocks should be cut with a
   a. tile hammer.
   b. wet masonry saw.
   c. dry masonry saw.
   d. very fine steel chisel.

9. The cleaning method that should never be used when cleaning prefaced concrete masonry units is
   a. soap and water.
   b. manufacturer's cleaning compounds.
   c. muriatic acid.
   d. a cloth and burlap bag.

SUMMARY – SECTION 5

- Structural clay tile was introduced in the United States in 1875.
- Unglazed structural tiles were mostly used for the building of partition walls and for fireproofing structural parts of buildings.
- Structural tiles are made from light-burning de-aired fireclay fired in a kiln.
- Glazed structural tiles have a glazed finish applied to the tile. The tiles are fired in a kiln to make the finish inseparable from the tile.
- Glazed tiles are made in different colors but the light, neutral shades are the most popular.
- Glazed structural tiles are available in two grades, S (select) and G (ground edge).
- The only two series of glazed structural tile made now are the 6T and the 8W series.
- Different shapes are made in both series to satisfy the particular needs of a glazed structural tile wall.
- The 6T and 8W series are modular in size.
- Coring in glazed tile runs vertically and horizontally.
- The most popular bonds for glazed tile are running and stack bonds.
- Glazed bricks can be laid inside or outside with no special treatment.
- Glazed bricks are made in different sizes and types.
- Glazed bricks are available in different colors and are highly resistant to moisture.
- Well-filled mortar joints should be used when laying glazed tile and bricks.

- Glazed or prefaced masonry units of any type should be left in the protective paper cartons until they are laid in the mortar joints.

- Type N mortar is recommended as an all-purpose mortar for the laying of glazed structural tile.

- White portland cement mortar and epoxy mortars are used when necessary or when special effects are desired.

- When building corners of the 6T and 8W series glazed tile, starter pieces are needed to make the bond work correctly on the corner.

- It is good workmanship to keep all pieces or cuts of glazed tile either against a jamb or under a window or opening.

- When laying glazed or prefaced masonry units, the mason must be careful not to chip the face when leveling or plumbing the unit in the wall.

- Glazed masonry units and prefaced blocks should be wiped off with a cloth or clean burlap bag after the mortar joints are tooled to remove all mortar stains.

- Final cleaning of glazed structural tile should be done with soap and water, using pressure from a hose.

- It is very important that exterior glazed brick walls have drains, such as weep holes, to allow any moisture to drain from within the wall. This is necessary because of the hardness of the unit.

- Expansion joints should be used in glazed brick walls to allow for expansion and contraction.

- Glazed brick walls should be allowed to set before the next section of wall is built on top, due to the low absorption rate of the unit.

- All glazed or prefaced masonry units should be cut with a masonry saw, preferably one with a wet diamond blade.

- Eye protection must be worn by the saw operator whenever cutting any units.

- Prefaced blocks are made by applying a thermoplastic resin finish to a concrete block, and firing the unit in a kiln until the block and finish are inseparable.

- There are three major types of face finishes for prefaced blocks: functional, scored, and design series.

- The functional series is the standard prefaced block face; the scored series has a scored variation of lines on the face that can be used to form different designs; and the design series has a three-dimensional or sculptured face that gives an appearance of depth to the finished wall.

- Prefaced blocks, like glazed tile, are available in a variety of colors.

- The most important advantage of a prefaced block is that a single-thickness unit can provide a tile-like finish on one side of a wall and a concrete masonry surface on the opposite side. This is very economical since a single wall serves two purposes.
- The standard size of a prefaced block unit (8 inches x 16 inches) is large in contrast to other prefaced or glazed units, which makes prefaced blocks a very high-production masonry unit.
- Mortar joints for prefaced blocks should be struck with either a glass or plastic tool to prevent black marks in the mortar joints.
- Final cleaning of prefaced blocks should be done with a cleaning compound recommended by the manufacturer. Never use muriatic acid.

SECTION SUMMARY REVIEW, SECTION 5

Part A

Complete each of the following statements, referring to material found in Section 5.

1. Structural clay and natural finish clay tiles are made from _____ _____.

2. Glazed tiles are packed in paper cartons to prevent _____ _____.

3. Glazed structural tiles are available in a variety of colors but the most popular are _____.

4. Special noise-absorbent, glazed tiles are made with holes spaced in the face. Inside the cell of each tile there is a layer of _____ to absorb sound.

5. Glazed tiles that have a glazed finish on both sides are known as _____.

6. Glazed tiles that have a nominal face measurement of 6 inches x 12 inches are called _____ series.

7. Glazed tiles that have a nominal face measurement of 8 inches x 16 inches are called _____ series.

8. Coring in a glazed tile that runs in a horizontal direction is called _____.

9. When laying a stack bond, glazed tile wall, it is recommended to always use _____ edges.

10. If white mortar is to be used in glazed tile walls on the job, before giving final approval to the mason contractor the architect usually inspects a _____.

Section 5 Glazed and Prefaced Masonry Units

Part B

Column A contains statements associated with glazed and prefaced masonry units. Column B lists terms. Select the correct term from B and match it with the proper statement in Column A. Review Section 5 if necessary.

Column A

1. special mortar used to point glazed tile joints when a dense, sanitary finish is desired
2. standard mortar joint thickness for glazed tile
3. thickness of facing on prefaced blocks
4. starter corner piece for the 6T series
5. starter corner piece for the 8W series
6. special glazed tile used when the tile finish must meet the floor
7. standard glazed tile unit
8. glazed tile bond in which all mortar head joints are in a plumb, vertical position
9. reinforced tile course over a window or door
10. recommended cleaner for glazed tile
11. a term meaning that moisture will not pass through
12. elastic materials built into the masonry wall to control movement
13. prefaced concrete block made with indented lines in the face so that it can be grouted with mortar to resemble mortar joints
14. prefaced concrete block made with a standard tile-like face
15. prefaced concrete block with a sculptured face
16. striking tool used for tooling prefaced block mortar joints

Column B

a. stretcher
b. stack bond
c. half lap
d. soap and water
e. impervious
f. epoxy
g. cove base
h. expansion joint
i. header
j. muriatic acid
k. sulfuric acid
l. bond beam lintel
m. design series
n. 1/4 inch
o. 11 3/4 inches
p. 9 3/4 inches
q. glass convex jointer
r. functional series
s. stainless steel
t. scored series
u. 1/8 inch
v. 1/2 inch
w. 3/8 inch
x. rake out
y. bullnose

Section 6
Arches

Unit 20 DEVELOPMENT OF ARCHES

OBJECTIVES

After studying this unit the student will be able to

- describe the major types of masonry arches used today.
- explain the difference between a minor and major arch.
- describe the theory and function of an arch.
- define the terms associated with arch construction.

An *arch* is a section of masonry work that spans an opening and supports not only its own weight, but also the weight of the masonry wall above. With a curved arch, the more weight bearing on the arch, the more thrust is exerted outwards to the sides of the wall.

BRIEF HISTORY

Arches date back many centuries. Chinese and Egyptian civilizations used arches before the Christian era, but it was the Romans who expanded the use of the arch in the construction of canals and waterways called *viaducts* that carried water to the cities. This was the semicircular arch, which is still known as the *Roman arch.*

Arches have appealed to builders for centuries because of the various shapes, types, and designs that can be built. The arch's beauty is expressed in the many forms which are used for balance, proportion, scale, rhythm, and character. Figure 20-1, page 282, shows a series of semicircular brick arches.

The structural advantage of the arch over a level beam or lintel is that when supporting a load, the pressure is mostly compressive (within) and to the sides, instead of straight down. For this reason, the arch frequently provides a very efficient way of supporting masonry work over an opening.

Arches were also used for building stone bridges, especially in the Eastern part of the United States. Figure 20-2, page 282, shows a stone bridge which uses the Roman arch for support.

Concrete and steel lintels have now greatly replaced the masonry arch due to a savings in installation time. Arches are still, however, used in many buildings for architectural effect and design. Examples of arches can be found in churches, banks, educational institutions, and many homes.

TYPES OF ARCHES

Many arch forms have been developed ranging from the jack arch, through the circular, elliptical, parabolic, and others, to the high-rise Gothic or pointed arch. Figure 20-3 shows some of the most frequently used masonry arches today.

Looking at figure 20-3, it can be seen that arches are constructed with various curvatures. One of the most beautiful arches is the Gothic arch. It was used in the past in many European cathedrals. Figure 20-4, page 284, shows Gothic arches in a church.

CLASSES OF ARCHES

The two general classes of arches are minor and major. The classifications are based on span and the load the arch can support.

Minor arches have a maximum span of 6 feet and a maximum equivalent load of about 1,000 lbs. per linear square foot of span. Minor arches are usually used in masonry walls as lintels over openings.

Major arches have spans or loadings whose dimensions start where the maximum stops for minor arches. In general, a minor arch is relatively short and a major arch is long.

Fig. 20-1 These semicircular arches were designed by Thomas Jefferson and constructed during the period in which he lived.

Fig. 20-2 A stone bridge that uses the semicircular Roman arch for support of the masonry over the stream.

ARCH CONSTRUCTION

Due to the nature of its construction, the arch exerts not only a downward pressure, but also a strong sideways pressure or a tendency to spread. This is called the *thrust* of an arch. In order for the arch to remain strong and not "give", the thrust must be resisted by abutments, buttresses, or the strength of the wall itself. How the arch is designed is directly related to the amount of thrust it must resist.

ARCH TERMINOLOGY

To understand how arches are constructed, one must know the various terms associated with arches. Figure 20-5, page 284, shows the various parts of an arch. Refer to figures 20-3 and 20-5 when studying the list of arch terms on page 285.

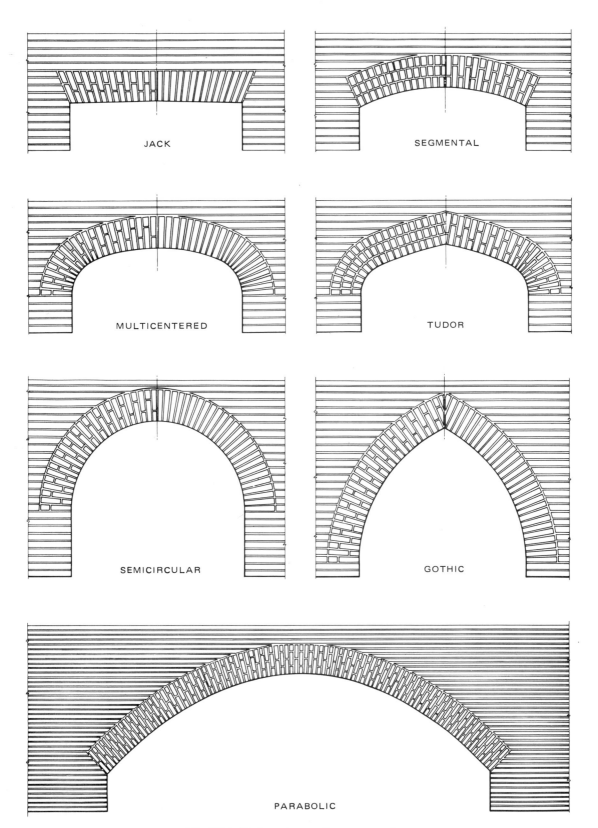

Fig. 20-3 Various types of masonry arches.

284 ■ Section 6 Arches

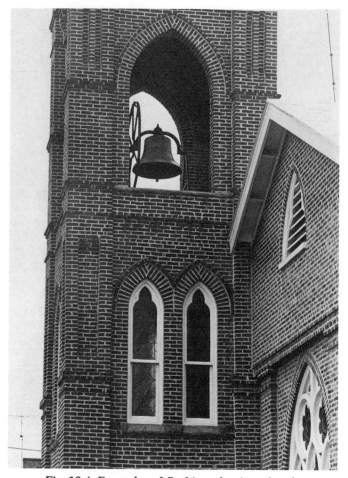

Fig. 20-4 Examples of Gothic arches in a church.

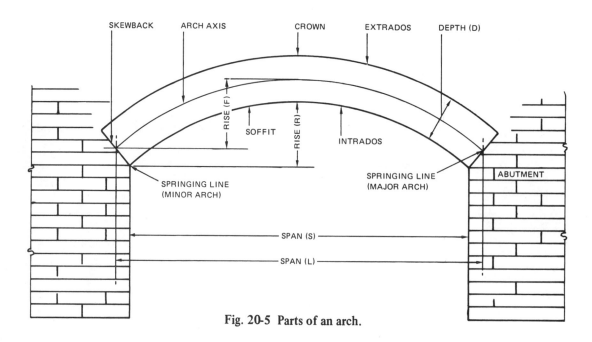

Fig. 20-5 Parts of an arch.

abutment — the skewback of the arch and the masonry which supports it.

arch axis — the median line of the arch ring.

camber — the relatively small rise of a jack arch.

constant cross-section arch — an arch that has a constant depth and thickness throughout the arch.

crown — the apex (center) of the arch ring. In symmetrical arches the crown is at midspan.

depth (see "d" in figure 20-5) — the dimension which is perpendicular to the tangent of the axis. The depth of a jack arch is its greatest vertical dimension. Depth as shown in the figure refers to height.

extrados — the convex curve which bounds the upper limits of the arch.

fixed arch — an arch that has the skewback fixed in position and inclination (slope). Plain masonry arches are, by the nature of their construction, fixed arches.

Gothic or pointed arch — an arch with relatively high rises, the sides of which consist of arcs of circles, the centers of which are at the level of the springing line. The Gothic arch is often referred to as a drop, equilateral, or lancet arch (depending upon whether the spacings of the centers are less than, equal to, or greater than the clear span).

intrados — the concave curve which bounds the lower limit (the entire curvature) of an arch (see figure 20-5). The difference between the soffit and the intrados is that the intrados is linear (length), while the soffit is a surface, in this case the bottom surface of the arch.

jack arch (see figure 20-6) — a flat arch that can carry a load but should be supported on steel angle irons if the opening is more than 2 feet long. It is the weakest of all arches.

multicentered arch — an arch that has a curve consisting of several arcs of circles which are usually tangent at their intersections.

rise (see "r" in figure 20-5) — the maximum height of the arch soffit above the level of its springing line. The rise of a major parabolic arch (shown as "f" in figure 20-5) is the maximum height of the arch axis above the springing line.

segmental arch (see figure 20-6) — an arch that has a circular curve that is less than a semicircle.

skewback — the inclined (sloping) masonry units which the beginning arch bricks rest against at each end of the opening.

soffit — the undersurface or bottom of the arch.

span — the horizontal distance between abutments (masonry jambs on each side of an opening). For minor arch calculations, the clear span of the opening is used (shown as "S" in figure 20-5). For a major parabolic arch, the span is the distance between the ends of the arch axis at the skewback (shown as "L" in figure 20-5).

springing line — for minor arches, the line where the skewback cuts the soffit. For major parabolic arches, the term commonly refers to the intersection of the arch axis with the skewback (see springing line in figure 20-5).

286 ■ Section 6 Arches

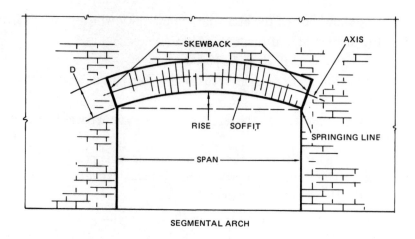

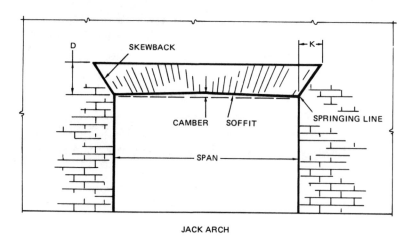

Fig. 20-6 Parts of the segmental and jack arches (minor arches).

tudor arch — a pointed four-center arch with medium rise-to-span ratio (shown in figure 20-3).

voussoir — one of the wedge-shaped masonry units which form the arch ring. An example is a brick in a jack arch.

keystone — the center masonry unit or stone in an arch that is cut to a taper on each side from the top to the bottom of the arch. It locks the other pieces in place and helps prevent the arch from collapsing.

APPLYING ARCH TERMINOLOGY TO THE CONSTRUCTION OF A MINOR ARCH

The rise of a segmental arch should not be less than 1 inch per foot of the span and the skewback should be at right angles to the arch axis. In the jack arch the camber should be 1/8 inch per foot of span and the inclination of the skewback should be 1/2 inch per foot of span for each 4 inches of arch depth. Both arches, shown in figure 20-6, are minor arches used over the head of a window or door.

Fig. 20-7 Bonded jack arch over a window in a house.

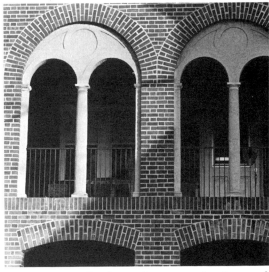

Fig. 20-8 Applications of two popular forms of arches. The arch in the bottom part of the figure is a segmental arch, and the arch in the top part of the figure is a semicircular arch.

USES OF ARCHES IN STRUCTURES TODAY

The most commonly used arch types today are the semicircular and the jack arch. They are used more for ornamental purposes than to support a load. Minor arches should not be expected to carry heavy, concentrated loads.

The Gothic arch is still used mainly for religious institutions. The segmental arch is used to obtain a slightly arched effect in buildings in which ornate brickwork is desired. Figures 20-7 and 20-8 are examples of three popular arches used in masonry work.

The building of arches is considered one of the more difficult tasks of the mason. Close attention must be given to correct bonding and workmanship. Units 21 and 22 deal with the laying out and building of various arches.

ACHIEVEMENT REVIEW

Part A

Select the best answer from the choices offered to complete each statement. List your choice by letter identification.

1. The small rise found in the bottom of a jack arch is called the
   a. skewback.
   b. crown.
   c. camber.
   d. springing line.

2. An arch that is flat and supported by an angle iron is known as a
   a. Gothic arch.
   b. segmental arch.
   c. jack arch.
   d. parabolic arch.

3. The wedge-shaped masonry units that make up the arch ring are known as the
   a. keystones.
   b. soffits.
   c. skewbacks.
   d. voussoirs.
4. An arch that is shorter in length than 6 feet is classified as a
   a. major arch.
   b. minor arch.
   c. multicentered arch.
   d. constant cross-section arch.
5. The underside of an arch is known as its
   a. crown.
   b. depth.
   c. fall.
   d. soffit.
6. The inclined masonry units that are cut on an angle on each side of a jack arch and which the arch bricks are laid against are called the
   a. axis.
   b. skewbacks.
   c. intrados.
   d. extrados.
7. The most popular arch built in churches is the
   a. semicircular.
   b. Gothic.
   c. tudor.
   d. parabolic.
8. Of all the types of arches constructed, the one considered to be the weakest is the
   a. jack.
   b. semicircular.
   c. tudor.
   d. Gothic.

Part B

Identify the various parts of the arch indicated by the letters in figure 20-9. Refer back to the unit if necessary.

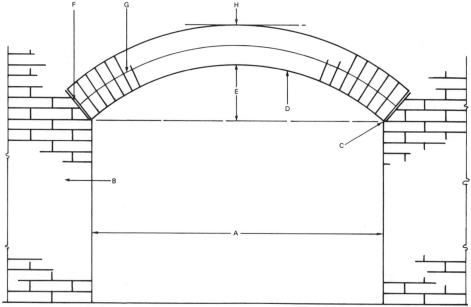

Fig. 20-9 Illustration for achievement review, Part B.

Unit 21 CONSTRUCTION OF A SEMICIRCULAR ARCH

OBJECTIVES

After studying this unit the student will be able to

- describe how wood arch forms are constructed for semicircular arches.
- explain how the arch bricks are spaced on the arch form before laying them on mortar.
- describe the different types of brick positions used in the construction of semicircular arches.

INTRODUCTION

If properly constructed, a brick arch will support a heavy load such as a masonry wall. The arch can support this load mainly because of its curved shape. Although some arches, such as jack arches, have a flat shape, they can also support a load if the arch is designed properly. Semicircular arches are considered to be the strongest of all masonry arches because the pressures and overhead loads applied are divided equally over the entire area of the arch.

Successful masonry work depends on correct layout and following good working practices. This is especially important in the construction of arches. Constructing arches requires careful planning and understanding of the tool skills needed to lay the bricks accurately and neatly.

Many architects specify arches in the construction of certain brick structures such as banks, churches, office buildings, and educational institutions. The arch is mostly used today for design and beauty in a brick building, not as a main supporting member, as in the past.

The building of brick arches is a challenge to the mason's ability and is considered to be a difficult, but satisfying, task. It is one of the more creative parts of the masonry trade and is required of all journeyman masons. This unit deals only with the construction of semicircular arches since they are the most popular arches used today.

WOOD ARCH FORMS

Before a semicircular arch is constructed, a curved wood form must be built to support the arch until the arch has cured. This wood center arch form is usually constructed by the carpenter using information given on the plans.

For arches up to six feet long, a good construction grade of 3/4-inch plywood is excellent for the front and back of the arch form. The two pieces are cut to the specified curvature. Then, 2-inch x 4-inch wood pieces are cut and spaced in between them as spreaders

Fig. 21-1 Wood arch form constructed from plywood. A strip of 1/4-inch plywood is nailed over the top for the arch bricks to lay on. Notice the 2-inch x 4-inch bracing around the inside of the curvature.

which serve to hold the form together. A piece of 1/4-inch plywood is cut and fitted over the top, or over the part of the arch form that the bricks will lay on. The 1/4-inch plywood bends easily over the curved form without breaking. A sufficient number of 2-inch x 4-inch pieces should be spaced between the front and back to prevent the 1/4-inch plywood from sagging. Figure 21-1 shows a wood center arch form constructed from plywood.

The purpose of the arch form is to support the dead load of the arch and wall above the arch until the masonry has cured and is able to carry the weight. The length of time to leave the arch form in place depends on the size of the arch, the weight it must support, and the curing conditions. For minor arches, three days is usually enough time if the temperature is not freezing and the mortar is setting correctly. Major arches, which are larger, should not be disturbed nor have the forms removed before ten days have passed, if conditions are the same. These allowances are not absolute rules. The most important considerations are the prevailing weather conditions and the setting time of the mortar.

DETERMINING THE CORRECT CURVATURE FOR A SEMICIRCULAR ARCH

Although the carpenter usually constructs the arch form, sometimes the mason is asked to construct it. This is especially likely on a small job or when the repair of old work involves replacing arches that have deteriorated.

In order to lay out the curvature of an arch, the span and rise must be known. These dimensions are usually given on the plan. It is also necessary to find the radius in order to mark the curve of an arch on the wood form. A simple geometric method of finding the radius is shown in figure 21-2. Multiplying one-half of span A by itself and dividing the result by rise B gives distance C. To obtain the radius add B and C, and then divide the answer by 2. The following example illustrates this method.

Example: Find the radius of an arch that has a 4-foot (48-inch) span and a 12-inch rise. Solution: Take one-half of the span and multiply it by itself (24 inches times 24 inches, which equals 576 square inches). Divide 576 by the rise (12 inches), and get 48 inches. Then add together 48 inches and the rise (12 inches), to get 60 inches. Divide 60 inches by 2, which equals 30 inches, the radius.

The radius of an arch is 1/2 of the springing line. (The *springing line* is a line across the span of an arch passing through the points where the arch intersects with the skewbacks.) To correct the optical illusion of flatness in semicircular arches, the *radial center point* (exact center of the circle) is usually raised one or two inches above the springing line.

SETTING THE FORM IN PLACE

After the arch form is constructed and brick piers are built to the height where the arch is to be laid over the opening, the arch form is set in place. A pair of legs is made from 2-inch x 6-inch or 2-inch x 8-inch lumber (depending on the depth of the arch). The legs are cut about 1 inch shorter than the actual height of the opening.

The legs are set in place, plumb and level. Four wood wedges are then laid on top of the legs, two on each. The arch center form is set on top of the wedges and adjusted into the proper position by tightening up on the wedges. Care must be taken during the process, so that when it is finished, the arch form is at the correct height, level and plumb with the face of the brickwork. Appropriate braces can be fastened to the form to hold the arch center securely

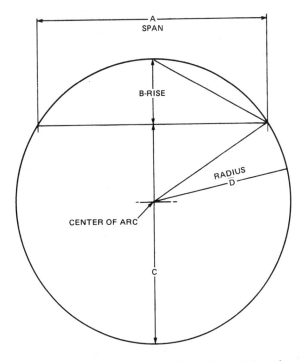

Fig. 21-2 Geometric method of determining the radius. The radius is necessary to scribe the curve of a semicircular arch.

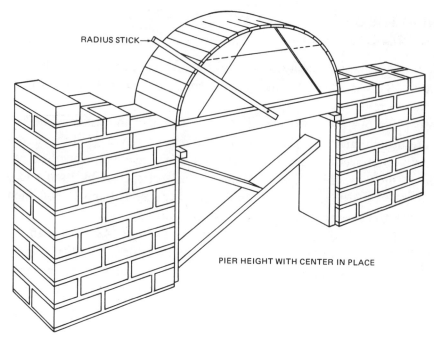

Fig. 21-3 Wood center arch form set in place between two brick piers, ready for the arch bricks to be laid on it.

in place. Figure 21-3 shows a wood arch form for a semicircular arch, braced in position, ready for the bricks to be laid on the form.

MARKING OFF THE ARCH BRICKS ON THE FACE OF THE FORM

There are two different methods of allowing for mortar joints in the construction of brick arches. The first involves the use of special arch bricks obtained from the manufacturer. For the second method, the bricks are cut with a masonry saw to a specific angle to achieve uniform mortar joints. In the first method, standard bricks of uniform shape are used and the joint thickness is varied to obtain the desired curvature. The special arch bricks can be cut on the job with a power masonry saw or they can be ordered from the brick manufacturer as a special order. If ordered from the manufacturer, the arch bricks will be sent separately and carefully marked.

The brick spacing is marked off with a mason's rule. This is done by bending the rule or modular steel tape to fit the curvature of the bottom of the arch form and marking on the face of the form with a sharp pencil. The course counter rule, commonly called a spacing rule, is used most often because it allows the mason to evenly adjust any differences in the joint thickness.

Although the spacing of the brick courses are marked off on the bottom of the arch form, it is difficult to lay and keep the top of the arch bricks at the correct angle. To simplify this step, attach a radius stick to the radial center point with a finishing nail. As the arch is being laid, swing the radius stick and it will indicate the exact position of the top of the arch brick, thus eliminating all of the guesswork. To be of value, the radius stick must extend to or past the top of the arch. Refer to figure 21-3 to see the placement of the radius stick.

LAYING THE ARCH BRICKS ON THE WOOD CENTER ARCH FORM

Before laying bricks on the wood arch form, clean out any chips or nails protruding from the construction of the carpentry work. Brush off the form with a fine brush to remove all chips.

Construction of an arch is always begun at the two ends or piers, and laid up to the center or key of the arch. The bottom mortar joint that sets on the arch form should be slightly angled inward. This is because it must be cut out with a chisel to a depth of 1/2 inch after the arch form is removed and repointed. Whether the key in the arch is brick or stone, it should be laid so that it extends the same distance over the center of the form on each side.

Select the best edge of the brick (the one that is straightest and unchipped) to set on the form in order to obtain a neat appearance. The mortar should be mixed to the proper consistency so that great pressure is not needed to make the joint squeeze out as the bricks are laid. Do not tap too hard on an arch because this will knock the previously laid bricks out of position and weaken the bond.

BONDED ARCH WITH SOLDIER COURSE

Arch bricks can be laid in several positions for semicircular arches. If the arch is a bonded soldier course arch, figure 21-4, the number of courses needed to build the arch should be counted before any bricks are laid. If the number of courses is even, the sides should start differently (one side with a half brick and the other side with a whole brick). This works out correctly when the center of the arch is reached with the bond. A bonded soldier course is a minimum of 12 inches, or a brick and a half, in height.

ROWLOCK OR HEADER ARCHES

The most popular bond arrangement for semicircular brick arches is either rowlocks, figure 21-5, page 294, or header course rings. The use of two or three rings (in height) is recommended for more strength. The shorter height of the rowlock or header bricks allows them to be turned on a sharper radius. This results in better curvature with smaller mortar joints. In addition to the mortar joints being smaller (which is desirable), the rowlock or header gives a through-the-wall tie since each brick is 8 inches in length.

Fig. 21-4 Bonded soldier course semicircular arch with a stone keystone in the center.

294 ■ Section 6 Arches

Fig. 21-5 A series of brick arches with rowlock rings and a brick keystone in the center.

WORKMANSHIP

The mortar joints should all be the same size and as small as possible (consistent with the wall) to give a neat, attractive appearance. A small joint such as 1/4 inch to 3/8 inch is less likely than a larger mortar joint to shrink and absorb water. The mortar joints at the intrados (bottom) and the mortar joints at the extrados (top) should be about the same size. This is not difficult to do with the rowlock or header arch, but it is more difficult with the soldier arch because the bricks must be cut to a special tapered shape.

Once the arch bricks are laid and set with the mortar, they should not be moved, otherwise the bond strength of the mortar joint is broken. This causes a loss of adhesion, allowing water to penetrate through the wall. If a brick must be moved after the mortar has set, fresh mortar should be applied and the brick relaid.

The mortar joints should be tooled as soon as they are thumbprint hard and still pliable. The arch should be brushed at the completion of the work. It is also a good practice to have a line attached on the bottom and center of the arch as a guide to make sure the arch does not bulge out of position. A straightedge such as a plumb rule can be used on a small arch to keep the center of the arch from bulging out. This should always be done as a last check after the arch is completed.

Parging the back of the arch with mortar helps to prevent water from leaking through from the face of the wall. It also helps to strengthen the arch and make it more secure.

TYING IN THE ARCH WITH THE BACKING WORK

If an arch does not equal the full width of the masonry wall, it should be tied or bonded into the backing work with metal wall ties and well mortared. The best type to use is the coated wall tie with a galvanized or copper finish. The wall ties should be imbedded at least half the thickness of the width of the brick in the facing of the arch and the backing materials. Every 3 or 4 courses in height should be enough for a good tie. Wire ties allow the mortar to bond more fully. Corrugated ties are used in veneer.

Fig. 21-6 Plugging chisel.

REMOVING THE ARCH FORM AND REPOINTING THE MORTAR JOINTS

After the arch has cured as previously mentioned, the arch form should be carefully removed. This is done by gently driving out the wedges, holding the form in place. When doing this, care should be taken not to chip the bottom of the arch brick.

Since the bottom of the arch rests completely on the wood form, the mortar joints must be cut out with a special chisel, called a *plugging chisel*, used for repointing joints with mortar. The plugging chisel, figure 21-6, is made with an angled blade on the end so it does not chip the edge of the bricks, but cuts out only the mortar joint. The special tapered blade cleans mortar from the joints easily without binding.

After removing the form, cut out the bottom of the mortar joints to a depth of 1/2 inch. Then, with a brush and bucket of water, dampen the mortar joints immediately before repointing with mortar. This is done to prevent the mortar from drying out too rapidly and curing improperly. The same mortar mix that was originally used to build the arch should be used for repointing. Type M or S mortar is usually recommended for building arches.

Some important points to remember when installing a semicircular arch follow:

- Select bricks that are approximately the same size before laying any of the bricks on the form.
- Choose bricks free of chips or cracks.
- Select bricks that match well with the color of the structure.
- Use dry bricks because they are easier to hold in place on the arch form.
- Use well-filled mortar joints. Mix the mortar to a thickness that squeezes out to all edges without the use of undue pressure when laying.
- Mark off all brick spacings on the arch form before laying any of the bricks.
- Work from each edge at the piers up to the crown (top) of the arch.
- If the arch has a keystone (as those shown in figure 21-7, page 296) this should be marked off first. Then the arch bricks must be laid off between the keystone and the bottom of the arch.

- Always try to make the top of the arch work with the stretcher course of bricks in the wall. It is considered better workmanship to have the stretcher course of brick work over the top of an arch without having any splits of bricks.

- Use a range line across the bottom and center of the arch to make sure the arch has not bulged out of position.

- When finished, straightedge the center of the arch with a plumb rule.

- Provide enough wall ties to tie the arch and backing work together.

- Parge the back of the arch to prevent water from leaking through the wall, and to help strengthen the arch.

- Do not remove the arch forms until the arch has cured enough to carry its own weight without sagging.

Fig. 21-7 Series of semicircular arches. Notice that each arch has a keystone in the center.

- Cut out the mortar joints at the bottom (soffit) of the arch with a plugging chisel and repoint with mortar.

- Observe good safety practices by wearing eye protection when mixing mortar or cutting masonry units for arches.

The building of arches by the mason requires good workmanship in addition to an understanding of the principles of arch construction. The practices and techniques in this unit should be studied and learned to aid the student in the construction of semicircular arches.

ACHIEVEMENT REVIEW

Select the best answer from the choices offered to complete each statement. List your choice by letter identification.

1. The strongest of all arches is the
   - a. jack arch.
   - b. Gothic arch.
   - c. semicircular arch.
   - d. parabolic arch.

2. Arch center forms are usually built of
   - a. steel.
   - b. plastic.
   - c. fiberglass.
   - d. wood.

3. The most difficult brick position to build a semicircular arch, retaining all of the head joints the same size, is a
   a. rowlock.
   b. soldier.
   c. header.
   d. stretcher.

4. When building a semicircular arch, the radial center point should be raised above the springing line about
   a. 1 or 2 inches.
   b. 3 or 4 inches.
   c. 5 or 6 inches.
   d. 7 or 8 inches.

5. The top of the arch bricks can be maintained at the proper curvature by using a
   a. mason's rule.
   b. mason's line.
   c. radius stick.
   d. form board.

6. The use of rowlock or header rings rather than soldiers is recommended for semicircular arches. The main reason for this is that
   a. rowlocks or headers present a more pleasing appearance.
   b. rowlock or header bricks are easier to lay on the arch form.
   c. the mortar head joints remain more uniform in size.
   d. it is more economical to lay the rowlock or header bricks than to lay a soldier course.

7. Once the arch bricks have been laid in the mortar joint and have set, they should not be moved. The main reason for this is that if an arch brick is moved,
   a. the design will be changed.
   b. the bond of mortar with the brick is lost.
   c. the arch form may be bumped out of place.
   d. the brick becomes smeared with mortar and requires cleaning before relaying.

8. After the arch has cured and the form is removed, the mortar joints on the bottom of the arch should be cut out with the
   a. hammer.
   b. brick set chisel.
   c. plugging chisel.
   d. blade of a trowel.

9. To make sure the mortar that is pointed in the bottom of the arch bricks stays in place, the joints should be cut to a depth of
   a. 1/8 inch.
   b. 1/4 inch.
   c. 1/2 inch.
   d. 3/4 inch.

PROJECT 14: LAYING OUT AND CONSTRUCTING A TWO-ROWLOCK SEMICIRCULAR ARCH

OBJECTIVE

- The student will lay out and construct a double-rowlock arch on a previously constructed form. This arch is typical of the type used over single windows.

EQUIPMENT, TOOLS, AND SUPPLIES

mortar pan or board
mixing tools
mason's trowel
brick hammer
plumb rule
ball of nylon line
2-foot framing square
line pin and nail
convex jointer
chalk box
brush
plugging chisel
brick, lime, and sand for mortar, to be estimated from the plan by the student
one 2-inch x 8-inch, 4-foot long board, for legs and bracing boards
arch center form to be constructed by the carpenter or student

SUGGESTIONS

- Use 3/8-inch mortar head joints spacing for layout.
- Lay the courses of brick to No. 6 on the modular rule.
- Stock all materials approximately 2 feet back from the wall line.
- Select dry bricks for the project.
- Choose arch bricks of the same size and free from chips.
- Mix only enough mortar ahead that can be used in one hour because building an arch is slow work.
- Strike the mortar joints as soon as they are thumbprint hard.
- Wear eye protection whenever mixing mortar or cutting bricks.
- Recheck all jambs before attempting to set the arch form in place.

PROCEDURE

1. Dry bond the first course of the project as shown in the plan, figure 21-8.
2. Set the arch center form in place over the dry bonded bricks to be sure it fits and then remove it.
3. Lay out the first course in mortar.
4. Build the piers to a height of 8 courses, level with each other.
5. Cut a pair of wood legs from the 2-inch x 8-inch board to the height of the pier minus enough room for placement of wood wedges.
6. Level the arch form on the legs with the wood wedges so that the radial center point is 1 inch above the springing point.
7. Brace the form into position.
8. The height of the arch center form plus the 2 rowlocks should be even with the brick coursing.
9. Attach a radius stick to the radial center point with a small nail.
10. With the mason's spacing rule, mark off the coursing around the bottom of the arch form.
11. Lay the rowlock arch bricks from the edges of the piers to the center of the arch.

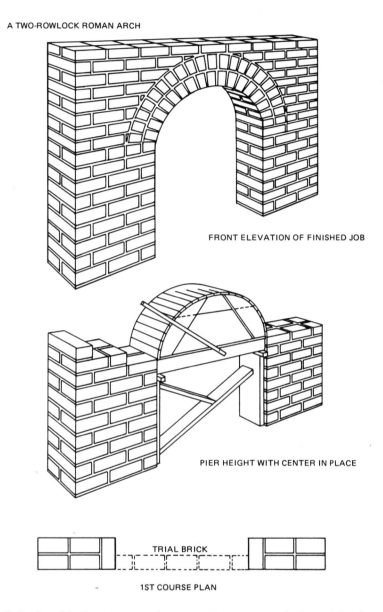

Fig. 21-8 Project 14. Laying out and constructing a two-rowlock semicircular arch.

12. Swing the radius stick in line with the marks on the bottom of the arch form to check the top of the rowlock curvature.
13. As the arch ring is laid, continue building the piers up and cut the bricks on the correct angle to fit against the arch ring.
14. Check the alignment of the arch with the piers using a line attached across the face of the project.
15. Continue building the project until it is completed as shown on the plans in figure 21-8.
16. After the project has cured, remove the arch center form and point up the bottom (soffit) of the arch.
17. Have the instructor evaluate the project.

Unit 22 CONSTRUCTION OF A JACK ARCH

OBJECTIVES

After studying this unit the student will be able to

- name the two types of jack arches.
- describe how to lay out a brick jack arch.
- explain how a jack arch is installed by the mason.

INTRODUCTION

A jack arch is a flat arch capable of supporting a load imposed on it over an opening. Because of its lack of curvature, it is the weakest of all the arches. It is supported on steel angle irons if the opening is more than two feet long.

Workmanship has a great deal to do with the appearance and durability of the finished jack arch. The arch should be constructed so that the mortar head joints are the same width for the entire height of the arch. To accomplish this, special bricks must be molded at the brick plant according to the plan specification for the inclination of the arch, or the arch bricks must be cut on the job to a pattern and rubbed perfect with a stone.

The top and bottom of the arch bricks or voussoirs must be in line with the angle iron at the bottom and the line at the top of the arch. The bottom and top of the arch bricks must be cut to the correct angle so they are level with the angle iron and with the line since the bricks are laid on a slant. These are cut with the masonry saw.

A jack arch that is built correctly gives the appearance of sagging in the center. This is only an optical illusion but can be corrected by having a slight camber (upwards curvature) in the angle iron. The longer the arch, the more it will appear to be sagging. Three lines are usually used in the construction of a jack arch. Once the wall is built and the skewback of the arch is installed, a line is attached at the bottom of the arch and across the face. Another line is used at the top edge to keep the top of the arch in line and as a guide for making the angle cut for the top of the bricks. Lastly, a range line is used across the center of the arch to make sure the arch does not bulge out of position.

TYPES OF JACK ARCHES

There are two main types of jack arches. One is called common, the other, bonded.

A *common jack arch,* figure 22-1, is composed of only one full brick in height, laid in a soldier position. It is the simplest type of jack arch and requires the least amount of work to install. A common jack arch serves to decorate the house or structure.

A *bonded jack arch,* figure 22-2, is laid out and built the same way as a common jack arch except that the bonded arch is higher and is usually either a brick and a half or two

Fig. 22-1 Common jack arch over a window.

Fig. 22-2 Bonded jack arch over a window.

bricks in height. To make the arch stronger, it is bonded by staggering the mortar joints the same as in a brick wall, only the arch bricks are laid in a vertical position. Since the height of the bonded arch is greater, the brick skewback on each side is five courses in height. The bricks are cut parallel to the angle iron and to the line at the top of the arch.

Bonded jack arches can also have a keystone in the center for architectural effect and design, figure 22-3. If the arch has a keystone, the stone should be centered in the arch and marked on the frame allowing for the mortar joints on each side. The bricks are then laid to that point.

LAYING OUT A JACK ARCH FROM THE RADIAL CENTER POINT

Skewbacks

Before laying a jack arch on the angle iron the inclination (angle) of the skewbacks must be determined. This is done by locating the radial center of the arch, which is the center point at the bottom of the arch opening. Once this point is located, drive a small nail at the center point and attach a line, figure 22-4, page 302.

The line is then held from the radial center to the edge of the opening at the point where the jack arch starts on the

Fig. 22-3 A bonded jack arch with a keystone in the center. This type of jack arch is used on buildings that have limestone trim.

angle iron. The line can be attached to a wood board and the brick skewbacks cut to the correct angle. The skewbacks are either cut with a sharp brick chisel or cut with the brick saw for accuracy.

MARKING OFF SPACING OF THE ARCH BRICKS

To make sure that the arch bricks work even and full bricks, the spacing of each brick must be marked off on the bottom of the angle iron and the top of the arch. Refer to figure 22-5 while reading the procedure below.

First, lay a wood board (2 inches x 4 inches or 2 inches x 6 inches) on top of the skewback on each side of the arch so that it spans the opening. This board is used to mark the top of the arch bricks.

Next, attach the line to the radial center Hold the line tightly from the center of the arch. Mark each course of arch brick on the top of the board, and on the angle iron until the center of the arch is reached. Remember that the center of the arch should be the center of the keystone. Mark in the same manner from opposite sides until the arch is completely marked off. It is important that both marks at the top and bottom of the arch be marked to keep the correct angle.

LAYING THE JACK ARCH IN MORTAR

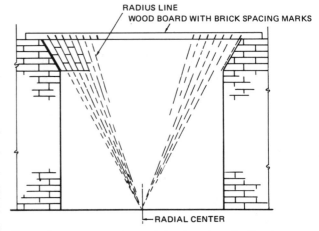

Fig. 22-4 A line is attached from the radial center to the skewback of the jack arch. The line is fastened to a board spanning the opening above the arch and the skewbacks are cut to the line.

Fig. 22-5 Spacing the arch bricks using the line from the radial center. The line is swung to the center of the arch, marking off the individual bricks.

After the jack arch has been laid out on the board and angle iron, attach the three lines in the way previously described. The mortar should be mixed so it is stiff enough to squeeze out when laid, without too much tapping on the bricks. Use solid mortar joints on all bricks.

Start laying the arch bricks from the ends to the center of the arch. The spacing marks on the board and on the angle iron must be checked often or the inclination may be lost. It is also a good practice to recheck the radius line at intervals as the work progresses. Check the angle of the arch bricks to make sure the correct angle is maintained.

A relatively small mortar joint should be used between the arch bricks because it gives a better appearance and allows the work to set up more quickly. Large mortar joints detract from the overall appearance and therefore should not be used in the construction of arches.

When the last bricks are laid in the center of the arch (key) make sure to butter both ends of the bricks in place. (*Buttering* is using a trowel to apply mortar on a masonry unit, to form a joint.) Also butter both ends of the keystone to make sure the mortar joint is well filled and a complete bond is established.

Tool the mortar joints as soon as they are thumbprint hard and then brush the work. Be sure to look under the bottom of the arch and point up any holes neatly with mortar when the work is finished. A slicker does a nice job of pointing. The jack arch is the easiest of all arches to construct, but good workmanship should be used at all times to obtain the best results.

ACHIEVEMENT REVIEW

Select the best answer from the choices offered to complete each statement. List your choice by letter identification.

1. A jack arch should be supported on angle irons if the span is more than
    a. 2 feet.
    b. 3 feet.
    c. 4 feet.
    d. 5 feet.

2. The layout line for a jack arch is taken from the
    a. springing line.
    b. radial center.
    c. center of the angle iron.
    d. top of the arch.

3. The jack arch is laid from
    a. the center to the jambs.
    b. each end to the center.
    c. one end to the other end of the arch.
    d. the center to the ends.

4. A jack arch is marked off on the top and bottom of the arch in order to
    a. maintain the exact coursing of each brick.
    b. correct the curvature.
    c. find the keystone.
    d. make it easier for the mason to see the marks.

5. The last brick laid in the center of the arch is called the
    a. crown.
    b. keystone.
    c. extrados.
    d. inclination.

6. The radial center of a jack arch is always taken from
    a. the end of the arch.
    b. the top of the arch.
    c. the bottom of the arch.
    d. halfway up the opening and in the center of the arch.

7. The position that jack arches are laid in is called
    a. rowlock.
    b. stretcher.
    c. soldier.
    d. header.

PROJECT 15: CONSTRUCTING A COMMON JACK ARCH

OBJECTIVE

- The student will use standard bricks to lay out and construct a common jack arch from a radial center. This kind of jack arch is typical of the type used in a house where a simple arch is desired for design.

EQUIPMENT, TOOLS, AND SUPPLIES

mortar pan or board
mixing tools
mason's trowel
plumb rule
mason's spacing rule and
   modular rule
chalk line
mason's hammer
ball of nylon line
2 corner blocks
line pins and nails
convex sled runner striker
brick set chisel
brush

pencil
square
supply of bricks, hydrated lime, and sand for the mortar — to be estimated by the student from the project plan, figure 22-6. Bricks are standard face bricks.
one 8-inch x 8-inch x 16-inch concrete block to put the level in when it is not in use
two wood 2-inch x 6-inch boards, approximately 56 inches long
three 3 1/2-inch x 3 1/2-inch angle irons, each 56 inches long

SUGGESTIONS

- Use solid joints for all of the brickwork.
- Lay out the first course with 3/8-inch head joints.
- All brick coursing is to be No. 6 on the modular rule.
- Do not mix more mortar than can be used in a 1-hour period.
- Keep all materials at least 2 feet away from the wall line for working space.
- Use eye protection whenever cutting bricks or when mixing mortar.
- Strike the mortar joints when they are thumbprint hard.
- Make all cuts with the brick set chisel for accuracy.
- Select straight bricks that are free from chips to be used for the arch.

PROCEDURE

1. Mix the mortar for the project.
2. Lay out the project with the chalk line according to the plan in figure 22-6, and dry bond the first course.
3. First build the piers on each side as shown on the plan. Use No. 6 on the modular rule.

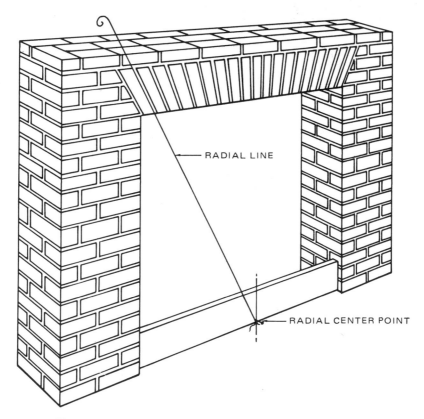

Fig. 22-6 Project 15: Constructing a common jack arch.

4. Set the angle iron in place when the height of the piers is reached. Divide the bearing on each side of the pier equally. This project requires 3 angle irons because the wall is 12 inches wide.
5. Rack up a lead on both sides of the project to a height of 3 courses, but do not build all the way out to the jamb.
6. Cut the radius board to fit between the piers at the front bottom of the project. The radius board is a 2-inch x 6-inch piece of framing lumber.
7. Find the center of the radius board and attach a small nail at the bottom center.
8. Attach a line to the radial center point and draw it up to the inside of the piers to determine the skewbacks.
9. To the other 2-inch x 6-inch piece laid on top of the project, attach the skewback line at the top of the arch line with a nail.
10. Cut the skewbacks to the line and lay in position.
11. By moving the radius line across the project to the center and using the spacing rule, mark off the individual bricks, both on the top and on the angle iron with a pencil. (Use a 1/4-inch mortar joint on the spacing rule. This may differ according to the thickness of the face bricks available.)

12. Make sure that the keystone (key brick) is laid out so it is in the center of the arch.
13. Attach a line at the top and bottom of the arch and proceed to lay the jack arch from each end to the middle. Cut the top and bottom of all arch bricks so they lay level.
14. Check the inclination of the arch periodically as the arch is laid with the radius line.
15. Continue laying the arch until it is completed.
16. Lay the jack arch on one side of the wall only. Lay stretcher courses in the middle and back of the wall because the project is 12 inches in thickness.
17. Strike the mortar joints and point up the joints on the bottom of the arch.
18. Continue to lay the project as shown on the plan in Figure 22-6, until the course over the arch is laid and the backing is completed.
19. Strike all work and have the instructor evaluate the project.

PROJECT 16: CONSTRUCTING A BONDED JACK ARCH

OBJECTIVE

- The student will lay out and construct a bonded jack arch of standard bricks. The angles at the top and bottom of the arch should be cut with the masonry saw or brick set chisel.

  This type of bonded jack arch is typical of the kind used on a colonial building such as a bank, school, or educational institution. Bonded jack arches are laid over windows and door openings mainly for design purposes, not to support a load.

EQUIPMENT, TOOLS, AND SUPPLIES

mixing tools
mortar pan or board
mason's trowel
plumb rule
mason's spacing rule and modular rule
chalk line
mason's line
ball of nylon line
2 corner blocks
several line pins and nails
convex sled runner striking iron
brick set chisel
brush

pencil
square
supply of standard bricks, hydrated lime, and sand for the mortar to be estimated by the student from the plan in figure 22-7.
three 3 1/2-inch x 3 1/2-inch angle irons, 56 inches long, for the opening
two wood 2-inch x 6-inch pieces of framing lumber, 56 inches long
one 8-inch x 8-inch x 16-inch concrete block to store the level in when not in use
masonry saw with a diamond blade (if available)

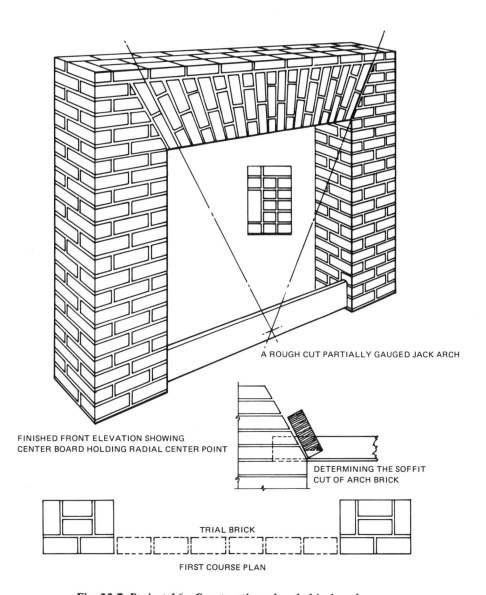

Fig. 22-7 Project 16: **Constructing a bonded jack arch.**

SUGGESTIONS

- Select good, straight bricks for the arch.
- All bed joint coursing should be No. 6 on the modular rule.
- Lay out the head joints for the first course with 3/8-inch mortar joints.
- Do not mix more mortar than can be used in one hour.
- This project can be built by a team of two students.
- Strike all mortar joints as soon as they are thumbprint hard.
- Observe good safety practices at all times and wear eye protection whenever mixing mortar or cutting bricks.

PROCEDURE

1. Mix the mortar for the project to the consistency desired.
2. Lay out the project with the chalk line according to the plan in figure 22-7, page 307.
3. Dry bond the first course of bricks as shown on the plan.
4. Build the brick piers up to the angle iron height. Keep the bricks level with each other.
5. Set the angle irons in place dividing the bearing evenly on both sides. (Three angle irons are needed because the wall is 12 inches wide.)
6. Rack up a lead in mortar on each side of the pier 5 courses high above the lintel height.
7. Locate the radial center point. Cut and lay the radius 2-inch x 6-inch board, in line with the front of the project.
8. Attach a line to the radial center point and draw it up to the top of the arch at the pier line.
9. Lay the other 2-inch x 6-inch board on top of the project so it spans the arch opening and is in line with the face of the project.
10. Holding the radius line in position, attach it to a nail in the board at the point where the skewback bricks must be cut.
11. Cut the skewbacks on each side to the line with either the brick set or masonry saw (if available).
12. By moving the radius line toward the center of the arch and using the spacing rule, mark off the individual arch bricks on the angle iron and the edge of the board. (Remember to keep the brick key in the center of the arch while the spacing is being laid out.)
13. Attach a line at the top and bottom of the arch as a guide. One may also be attached in the center as a range line.
14. Set a brick on top of the angle iron dry (without mortar) at the ends and mark the angle of the bottom cut with a pencil. Cut the angle with the saw or chisel. Rub the bottom cut smooth with a rubbing stone.
15. Repeat step 14 for the top of the arch since it must be cut to fit level with the line.
16. Cutting the bricks as shown in figure 22-7, proceed to lay the arch on the angle iron toward the center from both sides.
17. Periodically check the inclined angle of the arch with the radius line.
18. Lay the brick key last and double-joint or butter the edges of the bricks and brick key in place for a solid joint.
19. Strike up the mortar joints and point up under the bottom (soffit) of the arch.
20. Continue to build until the project is completed.
21. Recheck the work to be sure it is level and plumb. Have the instructor evaluate the project.

SUMMARY – SECTION 6

- In the past, arches were used for supporting the weight of masonry work above openings; now they are used more for design and architectural effects.
- The semicircular arch is called the Roman arch, named for the builders of the Roman era.
- The strongest of all arches is the semicircular arch.
- The Gothic arch is often used in building churches.
- The two major classifications of arches are minor and major.
- The theory of the arch is that the upward curve resists downward pressure by exerting pressure to the sides rather than straight down.
- The intrados is the bottom curvature and the extrados is the top curvature of an arch.
- A jack arch is a flat arch supported by angle irons.
- The center of an arch is called the key.
- The individual bricks in an arch are called its voussoirs.
- Arch construction is considered to be one of the more difficult tasks of the masonry trade.
- Semicircular arches are constructed on a wood arch form.
- The curvature of a semicircular arch is determined by using geometry.
- The installation of a semicircular arch can be simplified by using a radius stick attached to the center of the arch form.
- Each arch brick should be marked on the arch form before any of the bricks are laid.
- The most popular brick positions for a semicircular arch are rowlocks or headers.
- The mortar joints in an arch should be kept to a minimum width, such as 1/4 inch.
- Three lines should be used when installing an arch: one at the bottom, one at the middle, and one at the top.
- Parging the back of an arch makes it stronger and tends to stop moisture from penetrating.
- A masonry arch should be tied into the backing part of the wall using wall or masonry ties.
- Arch forms should not be removed until the masonry work has cured.
- When cutting out the bottom of an arch which is resting on a form for repointing, it is important to dampen the cut out portion with water before repointing.

- Well-filled mortar joints should always be used in the construction of brick arches.
- The building of masonry arches is one of the most difficult tasks for the mason and requires more advanced tool skills.

SUMMARY ACHIEVEMENT REVIEW, SECTION 6

Complete the following statements referring to the material found in Section 6.

1. A masonry arch that is constructed with a half-circle is a _____ _____ arch.
2. The advantage of an arch over a beam or lintel is its ability to withstand a load under downward pressure. This type of pressure is called _____ _____.
3. The simplest type of arch is the _____.
4. An arch that forms a quarter of a circle is a _____ arch.
5. The two general classifications of arches are _____ and _____.
6. The pressure exerted outward to the sides of a semicircular arch under a load is called its _____.
7. The stretcher bricks cut on an inclined angle on each side of a jack arch are called _____.
8. An arch that has a curve consisting of several arcs of circles which are usually tangent at their intersections is a _____ arch.
9. The bottom of an arch is called the _____.
10. The line where the skewback intersects with the soffit is the _____ _____.
11. Bricks that form the arch ring are known as the _____.
12. The upward curvature in the bottom of an angle iron used for installing a jack arch is called the _____.
13. The top center of the curvature of a semicircular arch is the _____ _____.
14. Bricks that are cut to fit against the curvature of a semicircular arch are called _____.
15. A pointed four-center arch with a medium rise-to-span ratio is a _____ arch.
16. Semicircular arches are constructed on forms built of _____.
17. Before a form for an arch can be built, it is necessary to locate a certain point to attach a line to. This point is known as the _____.

18. To aid in laying out the correct curvature of the arch when marking off the individual courses, a _____ is attached to the arch form.

19. Arches are built from both ends to the _____.

20. Due to its length, the most difficult brick position for the construction of semicircular arches is the _____.

21. When building a jack arch it is important to tie the arch bricks into the backing materials by installing _____.

22. An arch brick must not be moved after it has initially set because the bond strength _____.

23. A special chisel is used for cutting out mortar from the soffit of a semicircular arch after the form has been removed. This chisel is called a _____.

24. The arch that is usually associated with churches is the _____.

25. The distance between piers or abutments the arch crosses is known as the _____.

26. A jack arch which consists of two bricks in vertical height and in which the mortar joints are staggered for strength is called a _____.

27. Individual bricks are marked off across the angle iron with the aid of a _____ rule.

28. The top of a jack arch can be regulated for brick spacing by marking each brick with a pencil on a _____.

29. The mortar joints should be struck with a striking tool as soon as they are _____.

30. When a masonry arch has a stone-shaped wedge installed in the center to hold the arch in place, this piece is called a _____.

Section 7
Stonemasonry

Unit 23 DEVELOPMENT OF STONEMASONRY

OBJECTIVES

After studying this unit the student will be able to
- explain the development of stonemasonry from its origin to the present.
- list the properties of stone as a building material.
- describe the two major types of stonemasonry.

Stonemasonry is the art of cutting and laying stone either dry or in mortar. The work may be the building of a fireplace, bridge, wall for a home, piers, or similar construction. The arrangement of the stone in a wall, combined with the finishing of the mortar joints by the mason, can create many attractive patterns and bring out the best qualities of the stone.

HISTORY OF STONEMASONRY

It is difficult to pinpoint exactly when the first stonemasonry was built, but records and ancient stonework that is still standing prove that the trade is thousands of years old. Some of the earliest recorded stonework is found in Egypt and India. It is still a mystery how the ancient stonemasons cut and moved the huge stones into position when building the pyramids without the help of power equipment.

Some of the most beautiful examples of stonework are in the temples and government buildings in Rome, Italy and in Greece. Stonemasonry was also practiced in the western hemisphere in places such as Mexico. Figure 23-1 shows an example of a Mayan temple built of stone approximately 2,500 years ago.

Arches and columns were also used in the Mayan civilization, requiring advanced

Fig. 23-1 This Mayan temple was built in Mexico approximately 2,500 years ago. The stonework is still in relatively good condition.

masonry skills in the cutting and shaping of stone with metal chisels. Figure 23-2 shows circular stone columns with flat bondstone laid on top to support the roof members. The columns not only supported the roof but also added beauty to the structure.

The mortar used to lay the stonework is still intact. Formulas for the mixes have since been lost, but it is known that the mortar did contain lime and a cementing ingredient. Many of the buildings also included carvings that reflected the craftsmanship of the Mayan stonemasons as carvers as well as builders.

The stonemasonry of today uses bond arrangements that are basically the same as those used centuries ago. The most important changes are advances in cutting, sawing, and finishing techniques (such as grinding and polishing), and in the transporting of materials to the job site.

Fig. 23-2 Mayan stone columns with flat bondstone on top.

The stonework of the past was dependent mainly on the use of manpower. Today, stonework requires less manpower due to the use of power equipment for setting large stones.

Most of the stonework done now by the mason involves setting limestone trim and sills. The laying of larger stone blocks is usually the job of a mason specializing in that particular area (generally called a stone setter). The cutting and laying of natural stone is still one of the most creative and demanding of all the masonry skills. In the past, stonemasons specialized only in cutting and laying stone. Now the mason may be called on to lay and build stonework along with the laying of bricks, concrete blocks, and glazed structural tile.

The following units in this section deal with the setting of limestone trim and sills in masonry structures, and the cutting and laying of building stone in masonry walls.

PROPERTIES OF STONE

To better understand why stone is selected as a building material, some of its important properties should be studied.

Durability

As previously mentioned, stone is the most durable of all masonry building units. This is because of the hardness, or density, of stone. The density allows a smaller amount of moisture to penetrate, so there is little deterioration of the stone. There are exceptions, of course, such as sandstone and limestone, but they are still longer lasting than the more commonly used masonry units. The chemical composition and physical structure of stone also contribute to its density or hardness. The harder the stone, the longer it lasts and the more durable it will be. Examples of very durable stone types are marble, granite, and slate.

Uniformity

The uniformity of stone is influenced by the bedrock in which it is found and the way the layers run in the earth. The colors of *sedimentary stone* (stone formed by the sediments, or deposits, of minerals) depend on the minerals present in the rock. Iron is the mineral which is most often found naturally in stone. It also causes coloring in stone. The soaking or bleeding of water through stone deposits in the earth over a period of time causes the minerals in the stone to change color. Underground caverns provide examples of some of the colors that result from this process.

Strength

The strength of stonemasonry work depends mainly on the density of the stone, the mortar being used, and the quality of workmanship. The stone varies in weight according to its hardness. Marble, granite, and dolomite limestone are examples of heavy stone. All are very hard and dense. The denser the stone, the more capable it is of supporting a load and resisting the effects of weathering. The bond arrangement used or the way the stone interlocks in the mortar joints also increases the strength of a stone wall.

A properly mixed portland cement/lime based mortar creates a strong bond between the stone and the mortar. Good workmanship by the mason (such as filling all holes in the joints, laying the work when the temperature is above freezing, and carrying a good bond) also increases the overall strength of the wall. Laying a stone on its broadest edge distributes its weight more efficiently.

TWO MAJOR TYPES OF STONEMASONRY

The laying of stonemasonry can be classified into two main types: rubblework and ashlar stonework.

Rubblework

Rubblework includes rough rubble (unsquared) stonework and roughly squared stonework. It is laid up with field or rough quarry stone. Hand dressing and rough squaring are done on the job by the mason with chisels and hammers.

Rubble stonework is built of irregular stone of different sizes and shapes. The edges can be left rounded (unsquared), or as they come from the quarry or field, and laid in the mortar joint. This type of rubblework is called *rough rubble,* and the mortar joints are either raked out and brushed or smoothed with the trowel. Figure 23-3 shows a rough rubble stone wall.

Roughly squared stonemasonry is squared on the edges by the mason. Mortar

Fig. 23-3 Unsquared rough rubble stone wall. Notice the absence of any particular pattern bond.

Fig. 23-4 Roughly squared rubble stone wall laid with a rake joint. Notice how the edges of the stone are highlighted by the raked joint.

joint thicknesses vary. The top of the stone is in a relatively level position when laid. There is no specific bond pattern. Each stone bonds over the stone below it and no vertical head joints should be directly over another joint. Figure 23-4 shows an example of a roughly squared stone wall.

Ashlar Stonework

Ashlar stonework is cut to the dimensions shown on the shop drawings. The stone is cut, dressed, and finished to job requirements at the mill and shipped to the job in a finished condition. Ashlar stonework is laid in a specified mortar joint thickness (joint thickness should not vary). The bond can be a repeated pattern or a varied bond called *random ashlar*. One of the most popular methods of bonding ashlar stonework is by laying two smaller stones against one stone (called the *two-against-one* pattern, figure 23-5).

Fig. 23-5 Example of ashlar cut stonework

Fig. 23-6 Coursed ashlar stone wall being built. Notice that the stones are of different lengths, but are of the same height and are laid in succeeding courses.

Ashlar stone walls consisting of stones of the same height but of different lengths can also be built. This type of ashlar is called *coursed ashlar,* figure 23-6. Because of the rather porous, softer composition of Indiana limestone and sandstone, they are preferred for coursed ashlar work.

Both rubblework and ashlar stonemasonry can be laid either random or coursed. Various types of face jointing used in rubblework and ashlar stonework are shown in figure 23-7.

KINDS OF STONE USED IN STONE WALLS

Some of the most commonly used types of stone for the building of stone walls are limestone, (figure 23-8, page 318), granite, sandstone, marble, quartzite, assorted mountain stone, and fieldstone. The types of stone available vary in different parts of the country.

The problem with using fieldstone for building homes is that the width of the stone is very irregular and the stone must be cut before it can be laid. For a veneer home, the tolerance from the framing to the outside wall is rather small, usually 4 to 6 inches. This makes building with fieldstone very slow. The thin bed edge of the fieldstone makes it difficult to hold in the mortar. (A regular stone wall is usually thicker in width and most stone can be worked in with a minimum of cutting or fitting.)

SETTING LARGE STONE

Not all stones laid or set by the mason are small or can be handled by being lifted into place by manpower. Many government buildings, libraries, churches, and colleges are built

Unit 23 Development of Stonemasonry ■ 317

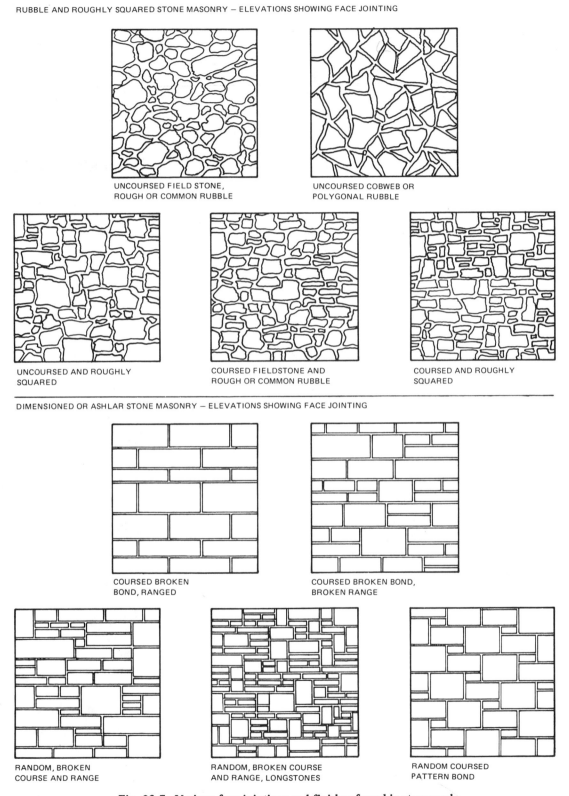

Fig. 23-7 Various face jointings and finishes found in stonework.

Fig. 23-8 Roughly squared stone entrance pier.

Fig. 23-9 Masons setting large cut ashlar stones on a building. The crane next to the building is used to lift heavy stones to the stone setter.

with very large facing stone. The mason, in this case, must also be an expert in leverage, and be able to guide the stone into position with the use of large power equipment such as cranes or lifts. The stones are cut, polished, numbered, and ground smooth at the quarry and then shipped to the job site. Figure 23-9 shows masons setting large cut ashlar stones into place on a building. Notice the crane that is used to lift the stone from the ground to the wall.

Fig. 23-10 Ashlar stone church with brick ornamental work in the front. This illustrates how stonework and brickwork can be blended together in a structure.

THE FUTURE OF STONEMASONRY

In the past, the stonemason specialized only in the cutting and laying of stone in mortar. The present trend is for the average mason who also lays bricks, concrete block, and glazed structural tile to learn some of the skills needed in laying and setting stone in mortar. The cutting and laying of rubblework and ashlar work can supplement the income of the mason who wants to be able to do more than just lay bricks and concrete blocks.

The practice of laying stonework is thus another important part of the masonry trade. Stonemasonry has lasted for thousands of years and will continue to endure, because it is the most durable of all masonry work. The building of stone walls reflects the creative abilities and craftmanship of the mason. Figure 23-10 shows how brick and ashlar stone can be combined in a structure.

ACHIEVEMENT REVIEW

Select the best answer from the choices offered to complete each statement. List your choice by letter identification.

1. The mortars used in the construction of ancient stonework contained
   a. lime.
   b. portland cement.
   c. masonry cement.
   d. high-strength additives.

2. The density of a stone means its
   a. color.
   b. hardness.
   c. chemical composition.
   d. mineral content.

3. The most common mineral found naturally in stone is
   a. copper.
   b. silver.
   c. magnesium.
   d. iron.

4. The type of stonemasonry in which the stone is cut into a squared shape at the mill and shipped to the job is called
   a. random rubble.
   b. coursed rubble.
   c. ashlar.
   d. rough rubble.

5. Stonework should always be laid on its
   a. end.
   b. bed side.
   c. face side.
   d. cut side.

6. When ashlar stones are cut and laid the same height on each course but are of different lengths, they are called
   a. random ashlar.
   b. rubble ashlar.
   c. coursed ashlar.
   d. bonded ashlar.

7. A popular bond pattern used in the building of ashlar stone walls is
   a. two against one.
   b. three against one.
   c. four against one.
   d. five against one.

8. The most important change in stonework in modern times is the
   a. use of larger stone.
   b. bond which the stones are laid in.
   c. type of chisels used by the mason.
   d. use of power equipment.

9. The coarsest type of stonemasonry laid by the mason is
   a. ashlar.
   b. rubble.
   c. veneer.
   d. limestone.

Unit 24 CUT LIMESTONE — COMPOSITION, TYPES, AND INSTALLATION

OBJECTIVES

After studying this unit the student will be able to
- describe how limestone is quarried, classified, and stored on the job.
- list uses of limestone in buildings.
- explain methods of lifting and setting limestone.

COMPOSITION

Limestone used in building construction is a product of the earth. It consists of shells and shell fragments that are cemented together by calcite. The particles which form the stone are oval or egg shape, therefore, the term *oolitic* is used when describing this type of stone. Limestone is a highly preferred building material when it is necessary to have a stone that can be cut or fitted with a minimum of effort and expense for the sills or trim of a building.

LOCATION OF STONE DEPOSITS

Most of the oolitic limestone used in buildings in the United States is found in Indiana. It is therefore known as Indiana limestone. Bedford, Indiana is the center of the limestone industry. Limestone is also found in other areas of the country, such as dolomite limestone from Minnesota.

REMOVING LIMESTONE FROM THE EARTH

The first step in quarrying limestone is the removal of the overburden. This overlaying limestone-dolomite formation is drilled and blasted free using black powder. Black powder is slow acting and does not crack the underlying select limestone, which is called Salem limestone. These loosened chunks are bulldozed and removed, exposing the top floor of the good limestone.

Machines are then brought into the quarry. A chisel located on the machine chips a cut 10 feet deep into the relatively soft stone. This cut is called a *boundary cut*. After the boundary cuts are finished, a canal cut is made lengthwise through the center of the stone. Other key cuts are then made to divide the stone into blocks or squares.

The limestone is then cut free with wire saws and wood wedges that are tightened into the stone with sledge hammer blows until the cuts split along the bottom line. The cut sections are generally 60 feet in length. These sections can be graded and broken into smaller blocks by redrilling holes and splitting the blocks further apart with wedges.

The stone blocks are removed from the quarry, placed on trucks and taken to a central stacking yard or mill. They can be processed immediately or left to cure before being cut to the size desired. Figure 24-1, page 322, shows how the large chunks of cut limestone are loaded on trucks for shipment to the mill.

322 ■ Section 7 Stonemasonry

Fig. 24-1 Loading large sections of Indiana limestone from the quarry into trucks for transporting to the mill.

CUTTING THE LIMESTONE AT THE MILL TO JOB SPECIFICATIONS

Limestone facing requires that each piece have the exact dimensions that the architect specifies for a particular job. Each piece of stone has a ticket accompanying it which describes exactly what is to be done from the time it goes into the mill until it is ready to be shipped to the job site.

The stone is cut in the mill using gang saws. The blades have diamond segments on the edges and are lubricated with water to give a smooth cut. Special pieces or shapes that have a beveled face or curve are placed on a planer made especially for shaping stone. The stone is then drilled, smoothed, and stored, ready for shipment. These heavy stones are handled during cutting and smoothing processes by specialized overhead equipment that moves on tracks called *travelers*.

SPECIAL CARVING OF LIMESTONE

Some limestone is specially carved by trained carvers at the mill according to the architect's specifications and plans. This carved limestone can be very ornamental and it reflects the specialized skill of the stone carver. Stone carvers are not masons; they are artists. The tools used by the stone carver are chisels and hand-size air hammers. Figure 24-2 shows the decorative part of the outside of a limestone structure.

SHIPMENT TO THE JOB

The limestone is shipped to the job site by either truck or railroad in a protective covering of wood or straw. Pieces are marked with numbers according to the building plans, indicating the location where they are to be set. As soon as they are unloaded on the job site, they should be stored off the ground on nonstaining skids, or lumber. They should be protected by a covering of waterproof materials such as plastic sheets or tarpaulins. Care should also be taken to make sure that the limestone is stored in a manner that does not subject it to anything dropping on it from the building that is being constructed. The stone should be arranged so that the pieces to be used first according to the plans are easy to reach to avoid extra moving.

GRADES, COLORS, AND FINISHES

The grades of Indiana limestone are classified by the fineness of the grain: fine, medium, and coarse. The range is from the finest to the coarsest. When a sample is desired, three are shown — the maximum, the average, and the minimum degree of fineness for the grade specified.

The colors of limestone vary from light cream to dark blue-gray. Indiana limestone is available in a wide variety of standard machine and hand-tooled surface finishes. New and special textures are constantly being developed. They are supplied in a variety of interesting surface finishes. Most limestone, however, has a rather fine finish and a light cream color.

Fig. 24-2 Stone carver at work carving limestone with the aid of chisels and air hammers.

USES OF CUT LIMESTONE

Limestone is generally used for the construction of masonry buildings that are intended for commercial or public use. Buildings are not usually built completely of limestone due to the cost and length of time needed to construct the building. Cut limestone is rarely used in house construction except for sills or special trim.

Limestone Sills And Coping

Sills are required under windows, doors, and other openings. Limestone sills and coping are usually selected when a structure has a limestone trim or facing. Coping is used to cap off walls and prevent water from entering at the top of walls.

Slip Sills. Before setting any stone sills or coping, the brickwork must be built up to receive them. The most common type of limestone windowsill is the slip sill.

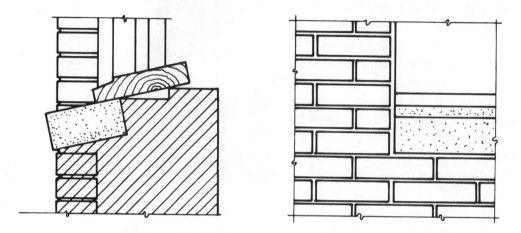

Fig. 24-3 Limestone slip sill. This type of sill is the same width as the opening minus two mortar joints.

The *slip sill* gets its name from the fact that it is inserted or slipped in the opening and laid in the mortar bed joint to the desired height. The length of the slip sill is two mortar head joints (approximately 1 inch) shorter than the actual opening.

The slip sill is usually not inserted until the window frame is ready to be set. This is to reduce the chance of breaking or staining the sill. The slip sill is a plain cut stone that can be made with a forward slope on top to turn the water, or made level with the sill sloped on the mortar bed joint to accomplish the same purpose. Figure 24-3 shows a slip sill set in place in a brick wall.

Lug Sills. Lug sills produce a better job than slip sills because they are built into the wall on both ends and the brick jambs start on the end of the sill. One problem with using lug sills, however, is that if the sill is damaged during the building of the walls above the window height, the lug sill must be cut out and replaced. This is difficult to do because the brick jambs rest on the sill end. Figure 24-4 shows a lug sill installed in a brick wall. The lug sill and the slip sill both have a water drip on the bottom front edge so the water does not

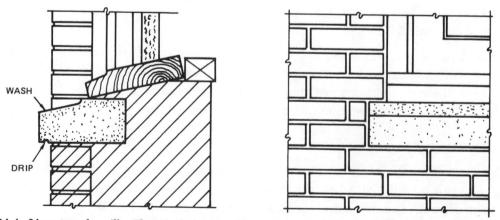

Fig. 24-4 Limestone lug sill. The edge of each brick jamb rests on the end of the sill, locking the sill into place in the wall.

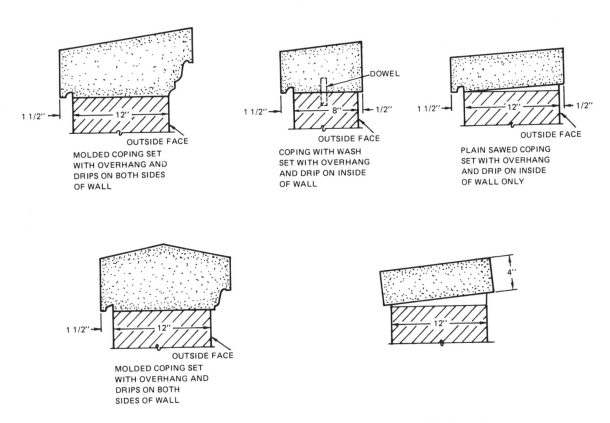

Fig. 24-5 Various types of limestone coping used to cap off brick walls.

run down the wall. Immediately after installation, a protective wood covering should be built over the sill to prevent it from being broken or stained.

Coping. Limestone coping, figure 24-5, is made in different shapes and forms. Coping is essential in capping off a wall to stop water from entering. Stone coping does this job very well because the stone is solid (except at the head joints).

Coping should always be set in a full bed of nonstaining mortar. A small leak allows water to penetrate into the wall below. Flashing, dowels, and steel masonry anchors are usually recommended under coping. Expansion joints should be built into the vertical mortar joints if the wall is long.

Belt Courses

A course of masonry units of solid stone that encircles the building is called a *belt course*. It is used to set off the appearance of the building and sometimes as a bondstone to distribute the weight throughout the wall.

Belt courses also act as *water tables* (stone that has a slope to turn aside water) when constructing a building that is more than one story in height. The use of a belt course allows the mason to set back on each story height and continue smoothly with the next story.

Trim

Limestone trim is highly preferred at corners of buildings, keystones in arches, cornerstones that have the date the structure was built, around windows and doorways, and other places where stone is used to add to the appearance of the building. The architect designs limestone freely in brick buildings when a break in the mass of brickwork is desired.

Interior Decoration

Vestibules, rotundas, entrances of public buildings, and interior window and door trim are built of a smooth, light-color limestone that matches the exterior of the building. This type of construction is often used in libraries, colleges and universities, banks, civic centers, and similar institutions. Limestone provides a durable, hard surface which can be maintained with a minimum of effort.

SHOP DRAWINGS

When the stone is delivered to the job site, the supplier also provides drawings that coincide with the architect's drawings. Such *shop drawings* show in detail the sizes, sections, and dimensions of all stones. The drawings also show the arrangement of joints, bonding, anchoring, and any other necessary details, such as expansion joints, if the walls are very long.

Each stone indicated on the drawings will have the corresponding code number marked on the back or bed of the stone with a nonstaining paint. Anchors that will be installed to help hold stone to the structure are also shown on the drawings, as well as the space for the anchor cut in the stone.

TYPES OF MORTAR TO USE IN SETTING THE STONE

Indiana limestone should always be set in a nonstaining mortar. The following is a recommended nonstaining mortar mix from the Indiana Limestone Company.

- Nonstaining portland cement/lime mortar — composed of one part masonry cement (ASTM C150), one part hydrated lime, and six parts sand.
- Nonstaining masonry cement mortar — composed of one part masonry cement (ASTM C91), and three parts sand.

The sand used should be fine and free from pebbles. Mortar should not be retempered for the setting of limestone. White sand is often used in the mortar mix. It is further recommended that the back of all limestone be parged with nonstaining mortar to prevent stains and penetration of moisture.

SETTING TOOLS AND EQUIPMENT

Hoists

Small stones being set by the mason are usually handled easily by using lengths of pipe or a pinch bar. When lifting and setting larger stones that cannot be handled by manpower alone, a lifting device is used. The most frequently used device for jobs that do not require a crane or heavy equipment is a gin pole.

The *gin pole,* shown in figure 24-6, consists of a heavy-built wood or metal pole that is held in position by guy wires and cables. At the end of the cable, either a block and tackle

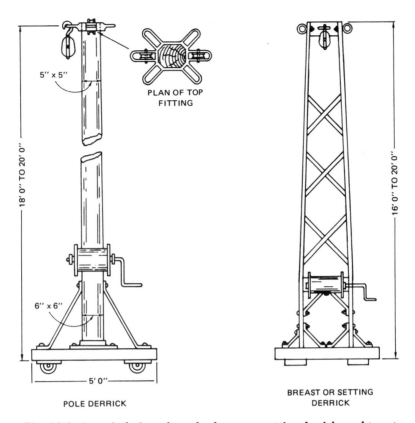

Fig. 24-6 A typical gin pole and a breast or setting derrick used to set heavy stone when a crane is not available. They can be mounted on a roof and moved around as needed.

or a hook fastened to the cable on a drum is used for attaching the stone. When the crank is turned on the bottom of the gin pole, the stone is raised to the height of the wall and swung onto the mortar bed.

A stronger arrangement for lifting stone, used on the roof of a structure when heavy coping stone is to be set, is called a *breast or setting derrick* (shown with the gin pole in figure 24-6). This type of derrick is usually mounted on wheels so it can be moved around on the roof in order to handle heavier stones than a regular gin pole. Stones that are larger than those shown in figure 24-6 are set by a motorized crane.

Slings

Gripping the stone tightly and lifting it to the wall where it is to be set requires a strong, nonslip attachment. Heavy-duty fiber slings that hold the stone on both ends are used since they do not chip or mar the surface of the stone. The main problem with slings is that wood wedges must be placed on the wall to hold the weight of the stone until the sling can be removed. The wedges must be dampened with water so they are not too difficult to remove after the stone is in position. The size of the wedge depends on the size of the stone and the thickness of the mortar bed joint. Once the stone is leveled and plumbed in place, a pinch bar or crowbar is used to pry up the stone, and the wood wedges are removed from under the stone. Care should be taken when removing the wedges that the face is not chipped by the pinch bar. Belt slings are also used with large equipment such as motorized cranes because they are easy to hook up to the crane.

Lewis Pins

Another method of gripping stone for lifting if the belt sling is not being used is the use of *lewis pins*. These pins are made of a pair of heavy steel pins from 5/8 inch to 1 inch in diameter with a steel ring attached to one end.

Holes and sinkage for the pins are cut in the stone at the mill before it is shipped to the job. Stones under 4 inches in thickness that have lewis pin holes cut into them should be approved by the architect before drilling. The holes are drilled on a 60 degree angle opposite to the force of the pull when being lifted.

The pins are inserted in the holes at each end of the stone. A chain is passed through the center ring that is connected by two smaller rings to the holes in the end of the pins. As the stone is lifted, equal pressure is exerted on both pins at an angle making the connection slipproof. Figure 24-7 shows three types of lewis pin attachments used in setting limestone: 3-leg lewis pin, knuckle lewis pin, and standard lewis pin (most often used). The standard lewis pin is the simplest because it requires the drilling of only two holes.

Anchors And Dowels

When limestone is set as a facing and backed up by other types of masonry units such as concrete block or bricks, it is advisable to have anchor and dowel holes cut into the stone. These holes receive metal ties that bond the stone to the masonry backing.

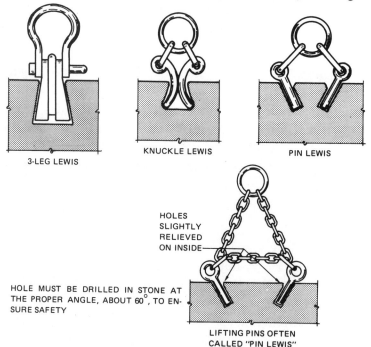

Fig. 24-7 The three different types of lewis pins used for setting limestone trim. The illustration at the lower right shows how the pins are hooked to the ring and chain for lifting.

The most common type of anchor is a thin, flat, galvanized stainless steel strap, one end of which has a 1-inch turned down edge which fits into the hole in the stone. On the opposite end, which turns into the backing materials, the turndown is about 2 inches long. Figure 24-8 shows a typical steel anchor used for setting limestone trim and sills.

Dowels are round steel pins about 1/2 inch or 3/4 inch in diameter. They are made of brass or galvanized steel. Their length depends on the height of the stone. They are mortared in the top of the masonry wall, usually in the center, and are projected upward approximately 1 1/2 to 2 inches. These pins are not installed in the top of the masonry wall until the stone is ready to be set because it would be very difficult to line them up correctly with the holes in the stones. After the dowel pins have been set and are mortared in place in the top of the wall, the stone is set down over the pins, lining up the pins with the holes in the stone. The stone is then tapped firmly in place in the mortar bed. Using dowel pins is important because it stops the stone from shifting or moving out of place once it is laid on the wall. Figure 24-9, page 330, shows steel dowel pins inserted in the wall and stone on a coping wall.

HAND TOOLS REQUIRED TO SET LIMESTONE

With the exception of a few chisels and hammers, most of the tools used to cut and set limestones are included in the standard set of masonry tools. As previously mentioned, several pinch bars or crowbars are needed to move the stone into position on the mortar bed.

Bush Hammer

A special hammer called a *bush hammer*, figure 24-10, page 330, is used for dressing a piece of limestone that has been cut or trimmed on the job. It is a very strong, dense alloy forged hammer with milled teeth on the head. The bush hammer weighs approximately 4 pounds. The teeth in the head of the hammer grind the cut edge to a fine finish without breaking the stone.

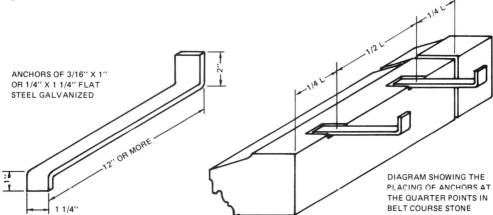

Fig. 24-8 Typical flat steel anchor. One end fits into a slot in the stonework. The opposite end fits into the masonry backing. Notice the turned edge on the anchor.

330 ■ Section 7 Stonemasonry

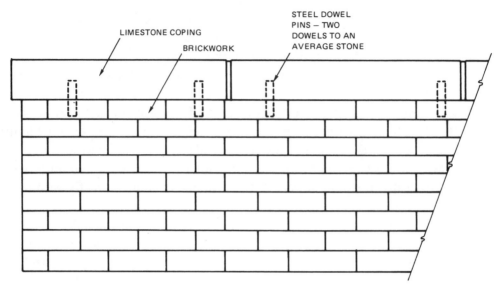

Fig. 24-9 Steel dowel pins are mortared in both the stone and the top of the wall. The hole in the stone is cut or predrilled at the mill and the hole in the top course of brickwork is cut by the mason.

Tooth Chisel

Tooth chisels, figure 24-11, are used for fine cutting of limestone or sandstone. The blade sometimes has a series of teeth. Because they may be damaged beyond repair, bush hammers and tooth chisels are not recommended for hard stone.

Slicker

The point of the mortar bed joints is almost always made with a flat-bladed steel tool called a *slicker.* This tool is used because the joint should have a flat finish or be as flush as possible with the surface of the stone. The slicker that works best is approximately 1/4 inch wide on one end and 3/8 inch wide on the opposite end. Other combinations of slicker sizes are available, however the size mentioned (which is illustrated in figure 24-12), is a good all-around slicker.

GENERAL SETTING PRACTICES FOR LIMESTONE

Safety

Safety is of prime importance when handling and setting limestone. Before any of the stones are lifted and set in the mortar bed joints, the work area should be made safe for the mason by the removal of all obstacles. If working off of a footboard (board laid on a couple of blocks), care should be taken that the board is not against the wall. The board should be laid firmly on the blocks so it cannot move when the stone is lifted and set. Injuries can result from improperly placed footboards that shift or give way while the stone is being lifted up to the wall.

Fig. 24-10 Bush hammer used for finishing off a cut or dressing limestone.

Fig. 24-12 Flat-bladed slicker used for pointing or striking the mortar joints under limestone trim.

Fig. 24-11 Tooth chisel used for cutting and dressing limestone.

If using a gin pole or derrick, make sure it is securely braced into position before attempting to lift the stone. The slings should be fitted against the stone in a way that keeps the stone fairly level during the lifting process.

If lewis pins are being used, the holes in the stone should first be examined to make sure they are correct and firm and without cracks. Insert the lewis pins in the holes and see if they fit properly. If the pins fit correctly, take up the tension on the stone slowly with a winch and check to see if the pins are holding securely. As the stone is being lifted it should be held steady without swinging. This is done by holding the stone with either the hand or a length of rope attached to the cable. The stone is thus under some control while it is in the air. The cable should not be tugged or jerked because this can dislodge the stone from the pins.

> **Never stand directly under a stone while it is being lifted by a crane or derrick.**

Figure 24-13, page 332, shows a crane lowering a large section of limestone into position. Notice that no masons are under the stone.

Applying The Mortar Joints

Spread the mortar on the wall with the trowel without furrowing. Install the dowels or anchors (if needed) and insert the dampened wood wedges in the mortar joints. Allow the wedge to protrude enough so it can be easily removed later.

Setting The Stone On The Mortar Bed Joint

After setting the stone on the mortar bed either by means of a crane or by hand, the stone should be settled firmly into the mortar bed. This is done by using a length of 2-inch x 4-inch wood laid on the stone horizontally and tapped down carefully with the striking hammer. Do not lay out the 2 x 4 on the very exterior face edge of the stone because it might chip the face.

> **Never beat a stone to position it in its bed without the aid of a wood protective board.**

When the stone is leveled or plumbed, it is done by forcing the wedges in or backing them out (whichever is necessary). Never use a hammer to drive in the wedges on the face

Fig. 24-13 Setting a heavy piece of limestone facing in place with a crane and holding device. Notice that no masons are standing under the stone, thus preventing an accident if the stone breaks loose.

side of the stone because the face of the stone may chip. Instead, pry up the stone with the pinch bar gently and either push the wedges in or slide them out. The bed joints under the stone should be pointed in firmly with mortar with the slicker. Using the slicker, force all the mortar back as far as possible to form a solid joint. The head joints are not pointed with the slicker at this time. Wedges should not be removed until the mortar has set enough to hold the weight of the stone.

Parge the back of the stone with the mortar that was used for setting the stone. This is important to prevent the staining through of mortar used in backing up other masonry units.

Pointing And Filling The Head Joints

If the stone is set flush with the wall and the bricks are laid on top of it (such as in a bondstone), mortar can either be pointed while the stone is being set or raked out to a depth of 3/4 inch and pointed later. Before pointing in any mortar, however, dampen the mortar joint area with water to decrease the drying time of the mortar. This provides for a better bond of the mortar to the stone and allows the mortar to cure properly. Wood wedges should also be dampened with water before removing. This eliminates chipping of the face of the stone.

All cornice, coping, projecting belt courses, steps, and platforms should be set with unfilled head joints. The exterior edges and bottom of the vertical head joints should then be packed with plumber's oakum or styrofoam filler rope. A strip of building paper should be laid on the top of the stone at each side of the joint in order to prevent excessive smearing when the grout is poured. After wetting the ends of the stone thoroughly, fill the joint from above with mortar grout to within 3/4 inch of the top. After the grout has set, remove the filler caulking and point up the face of the head joint with the slicker and mortar. This type of joint is known as a *poured joint* and is considered the most effective way to fill in a head joint in limestone trim or sills. Although the mortar grout is thin enough to be poured in the mortar joints, it is important that the correct proportions are followed when mixing the grout. The grout must not be made too weak.

Cut limestone should not be set, nor should joints be pointed, during freezing weather. A temporary, heated enclosure is recommended if the work is to be done during freezing conditions.

Cleaning The Limestone

Limestone should be kept as clean as possible while being stored on the job and during setting. This is a protective measure that greatly decreases the amount of cleaning necessary.

To hand clean the limestone after curing use fiber brushes, soap powder, and clean water, completely rinsed off at the completion of the cleaning. Acids attack the limestone and may cause marks and spots; therefore they should not be used.

If there is a problem spot that cannot be removed by the hand cleaning process, steam cleaning can be used. This, however, should only be done on advice from the stone company and with the permission of the architect. The steam method has also been effective in cleaning old stonework that has become darkened from air pollution settling into the pores of the stone face. Sandblasting is not recommended because it opens the pores of the stone and causes a rough appearance.

PROTECTION OF FINISHED WORK

Tops of walls should be covered with a nonstaining covering, such as plastic sheets, at times when the work is not in progress or if it is left overnight. This is especially important if bad weather is expected. Another problem may be caused by roofing. If a built-up tar roof is being applied, the stone must be protected from any spillage or dripping of tar on the stone. It is much more economical to cover the stone than to have to remove tar later.

Limestone sills and stone jambs around doorways or openings should have protective boxed wood frames built around them which are held securely in place during the construction process. Either 1/2-inch or 3/4-inch plywood can be used. This is usually the job of the carpenter, but is the mason's responsibility until the job is completed. Boxed protective coverings around the face of the stone prevent staining of the face and chipping of the corners. As a rule, this is stated in the specifications by the architect and may be the responsibility of both the general contractor and the mason. It should be remembered, however, that a covering of wood for stonework should never be constructed if there is anything in or on it that can stain the finish of the stone.

PATCHING AND REPAIRING

Cut limestone is patched and repaired only if it is absolutely necessary. It may be necessary to patch a stone to prevent having to cut out and replace an expensive stone. A surface break or defect can be repaired by first cutting a 1 1/2-inch recess in the stone. Then a piece of stone that matches as closely as possible can be cut and fitted into the cavity. The piece is cemented into place either with a rich mortar grout or a cement mix that matches the stone finish, which is supplied by the original stone supplier. After the piece has set, it can be rubbed down with a rubbing stone so it blends in with the face of the stone. If the face of the stone is chipped, it may be possible to dress down the entire length of the face enough to remove the damaged portion. This is sometimes necessary when a stone has been burred along the edge. Any repairs of stonework should be done carefully to make sure the repaired area cures properly.

ACHIEVEMENT REVIEW

Column A contains statements associated with limestone. Column B lists terms. Select the correct term from Column B and match it with the proper statement in Column A.

Column A	Column B
1. building material composed of shell fragments held together by calcite	a. gin pole
2. stone used to cap off the top of a wall	b. coping
3. type of mortar used in setting limestone	c. slip sill
4. wood or metal pole device used with a block and tackle or hook to set medium weight limestone	d. lug sill
	e. nonstaining
	f. lewis pins
	g. dowels
5. the most popular colors of Indiana limestone	h. plywood box
	i. air
6. materials used in the front of the head joint of limestone to stop the grout from running out when pouring a joint	j. soap powders
	k. oolitic
	l. boundary
7. chisel cuts that are made 10 inches deep at the quarry	m. Salem
	n. anchors
8. type of hammer used by the carver	o. Bedford, Indiana
9. rounded steel pins that are inserted at an angle in limestone for lifting purposes	p. pitching chisel
	q. limestone
10. chisel with a series of teeth, used in dressing limestone	r. shop drawings
	s. water table

11. location of limestone deposit
12. term used to describe Indiana limestone
13. type of limestone sill that is made the same size as the opening
14. type of limestone sill on which the masonry jamb rests on each end of the sill
15. ornamental stone course that encircles the entire building
16. projecting stone used in a building at each story height, made on a slope to deflect water away from the joints
17. plans supplied by the stone company to the architect
18. rounded steel pins made of brass or galvanized metal, used to anchor coping stone to the top of the wall
19. metal finishing tool used to point a flat mortar joint under stone
20. metal fastener used to tie limestone trim to the masonry backing work
21. materials that protect limestone after it is installed in the wall
22. materials used to clean finished limestone
23. chisel with a beveled blade to cut stone
24. the underlying select limestone found in the quarry

t. belt course
u. light cream and dark blue-gray
v. plumber's oakum
w. slicker
x. toothed chisel

Unit 25 CUTTING AND LAYING STONE

OBJECTIVES

After studying this unit the student will be able to

- list special tools used for cutting and laying stone.
- describe how rubble is laid and tooled.
- explain how roughly squared stones are cut, laid, and tooled.
- describe the cutting and laying of flagstone.

INTRODUCTION

The cutting and laying of rubble and roughly squared stone is a creative yet demanding masonry task. The mason should be able to "size up" a stone by examining it on the pile and then decide where it will fit into the wall with a minimum of cutting.

Rubble stone can be stone from an old fence row, stone from a field, or rough quarry stone that has been blasted from a rock formation. The stone should not weigh more than two masons can handle or lift by hand without straining. A variety of sizes adds to the character of a random or roughly cut stone wall.

ESTIMATING STONE

Stone is sold by weight, usually in tons. To determine the amount of stone needed for a job, the weight per square foot or cubic foot is found. This depends on the wall thickness of the stone being used on the job.

Rubble or roughly squared stone can be estimated by finding the square feet area of the wall, or by the old method of measurement called a perch.

Using the square feet method, find the square feet of wall area by multiplying the height of the wall by the length of the wall. The result is the total square feet of the wall. This figure can then be converted into cubic feet by multiplying by the thickness.

In the perch method, a perch of stone is 16 feet 6 inches long, 1 foot high, and 1 foot 6 inches in thickness. This equals 24 3/4 cubic feet. In some localities 16 1/2 cubic feet or 22 cubic feet are used as a perch. Although the perch method is still used in some places, the mason generally works with square and cubic feet of wall rather than the perch.

BUILDING STONEWORK TO A LINE

A stone wall is not perfectly plumb, but rather plumb from bump to bump. To do this, the mason attaches a line from the bottom of the first course of stone, on the corner, in a plumb position, to the point where the wall will end at the top. This is done by driving in a nail at the top and fastening the line to that nail, as shown in figure 25-1. The stone is then laid basically to the plumb line. The opposite corner is built the same way. After the corners are built, a range line is drawn from one corner to the other for the wall line. The range line does not have to be at any precise level but it should be approximately the same height at each end of the wall. This line is not used to find the level of the stonework but to keep the wall in alignment or plumb.

The tooling of the mortar joints is directly related to the finished appearance of the stonework. It is important that all holes be pointed full with mortar regardless of the type of pointing or joint finish being used. The information which follows in this unit deals with the cutting, laying, and finishing of rubble, roughly squared stone, and flagstone.

GENERAL PRACTICES TO FOLLOW IN DOING STONEWORK

There are a number of suggested practices to follow in laying rubble or roughly squared stonework.

- The stone wall should be started on a properly proportioned footing and installed below the frost line.

- Each stone should be laid on its broadest end (bed) for the best weight distribution and so that the stone is held in the wall more securely. It is a poor practice to lay stone and then drive stone chips in under the stone to hold it in position. This is sometimes necessary but should not be the general practice for laying stonework.

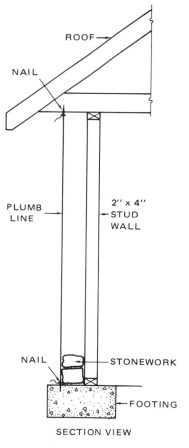

Fig. 25-1 Stonework is laid to a plumb line attached from the footing under the first course to the soffit of the building.

- Any dirt or foreign materials should be removed from the bed side of the stone before laying the stone in mortar to ensure a good bond.

- Generally, larger stones should be in the lower courses of a wall. The size of the stones should gradually become smaller towards the top of the wall. This is because the heavier stone is more stable and easier to lay on the bottom of the wall. Large stones can be used occasionally throughout the wall as bonding stones or for design purposes.

- The mortar joint thickness of rubble and roughly squared stone should not exceed 2 inches. Two-inch thicknesses are generally used in rounded rubble stonework. Large mortar joints may shrink away from the edges, appearing as a hairline crack. The addition of hydrated lime to the mix is a great help in preventing shrinkage in stonework.

- The use of small filler or chips of stone (called points) should be held to a minimum in this type of stonework. It is difficult to keep them in the mortar joint and they do not give a pleasing appearance.

- Never attempt to fill in a large amount of mortar behind a stone wall until the wall has set or cured, for it may fall out of place or collapse.

- Always wear eye protection when cutting any stone. Also, when cutting roughly squared stone, it is a good practice to wear a pair of heavy gloves. There is always the danger that a stone chip may fly off and cause injury.
- If a stone must be removed after it has been set with the mortar joint, fresh mortar should be applied to the wall after the stone is removed. Merely repointing the face of the mortar joint will not form a good bond.

MORTAR FOR LAYING STONE

A good stone mortar for rubble and roughly squared stonework is strong but elastic. Using only portland cement and sand in a rich mix is a mistake commonly made in preparing stone mortar. This mixture creates a very strong mortar, however it is brittle and does not stick to the stone. The addition of hydrated lime to the mix gives the mortar workability and provides a good bond with the stone.

A recommended mix for general stonework is the following: type N mortar in the proportions of 1 part portland cement to 1 part hydrated lime to 6 parts sand. If the sand is coarse, the number of parts of sand can be decreased to 5 rather than 6. Type N mortar is recommended for all exterior walls, except retaining walls.

A richer mix should be used for a stone wall that must withstand severe frost, or a wall that is under great pressure. This richer mix consists of 1 part portland cement to 1/2 part hydrated lime to 3 parts sand.

TOOLS NEEDED BY THE STONEMASON

The tools the mason needs when working with rubble and roughly squared stone include a group of different masonry hammers, a variety of chisels, and various types of jointing tools. Other basic tools such as the level, lines, and trowel are the same as used for regular masonry work.

There are several different types of hammers used for cutting and dressing stone. The heaviest is a sledge hammer which can be single- or double-faced. It is used for breaking large stone into smaller pieces. It is also good for breaking stone out of an outcropping or splitting rocks when they are blasted from the quarry. Sledge hammers vary in weight from 10 to 16 pounds and should be chosen according to the needs of the job.

The face hammer is larger than a brick hammer. It is shaped like a small sledge hammer except that it has one blade tapered to an edge. This type of hammer is also called a stone axe. A stone axe weighs approximately 4 to 6 pounds. It is used for dressing rough projecting points of stone before cutting the stone with the chisel. The stone axe is also used for settling stone into the mortar bed.

The mash hammer is used with the chisel for cutting or breaking medium-sized stone into pieces that can be cut with a chisel. A mash hammer usually weighs from 2 to 4 pounds. It has a square, flat head on each end.

A small bush hammer is used for trimming softer stone and dressing the edges or small projections of a stone. See figure 25-2 for an illustration of the face, mash, and small bush hammers. In addition to these hammers, a standard brick hammer is handy for fine trimming or chipping.

CHISELS

The mason uses a variety of chisels to square and dress rubble and roughly squared stone. The most frequently used chisels are the pitching chisel and the point chisel, or an assortment of small cutting chisels. Due to the hardness of rubble and roughly cut stone, the chisels rapidly become dull. Therefore, a supply of each type described is needed.

The pitching chisel is a heavy-duty type with a slightly beveled edge. Pitching chisels are available in different widths. The average width of the blade is 2 inches. When this chisel is dull, it should be sharpened by a blacksmith.

The end of the point chisel used for dressing the stone is drawn to a point. A point chisel is used to cut or chip off small knobs of stone that are too small to be cut with the pitching chisel. The point chisel is especially useful in dressing a square-cut cornerstone.

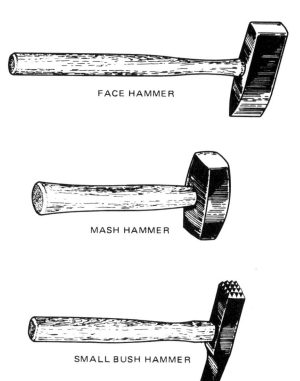

Fig. 25-2 Three hammers that the stonemason uses most often.

The last type of chisel to be discussed is the plain-bladed cutting chisel. It is used for all-purpose cutting. This chisel is good for cutting off edges or small areas of stone when fitting the stone in the wall. Figure 25-3 shows the four most common stone chisels.

LAYING RUBBLE STONE

Random rubble is the simplest type of stonework. The stones are laid in basically the same condition as they come from the field or rock outcropping, with a minimum amount of dressing. Random rubble is built with irregular stones. If the areas between the stones are large, they can be filled in with smaller pieces of stone and mortar.

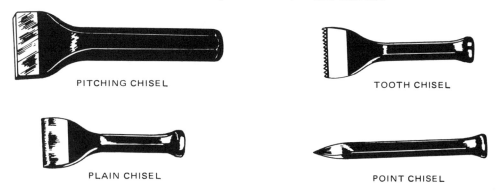

Fig. 25-3 Four chisels used most often for stonework.

Tooling the mortar joints is also done very coarsely. Joints are either pointed with a small pointing trowel or rubbed out slightly with a rounded stick of wood or old broom handle. This rubbed joint, figure 25-4, reveals all the edges of the stone and is highly desirable. After the joint has been rubbed with the broom handle, it is brushed to remove all pieces of mortar. Random stonework can also be done by laying the stones dry (without mortar). This method is used for retaining and garden walls.

Mortar used for laying rubble stone should be mixed much stiffer than mortar used for brickwork. This is because of the hardness of the stone and its slow absorption rate. If a very hard stone is being laid, or if the weather is cold, it may be necessary to mix only enough water with the mix to blend the ingredients together. This is often the case when laying larger heavy rubblestone.

Fig. 25-4 Random stone wall with rubbed mortar joints. Notice how the edges of the stone stand out from the mortar joints.

Many attractive structures have been built of random rubble stonemasonry, figure 25-5. The work is slow compared to laying brick walls, but the stone walls provide maintenance-free masonry walls that last indefinitely.

Fig. 25-5 A new house being constructed of random rubble stonemasonry.

LAYING SQUARED RUBBLE STONE

Two Types Of Squared Rubble Stone

There are two types of squared rubble masonry: irregular squared rubble and coursed rubble. In both types the stone is squared on the edges. The mason does this on the job using stone axes and chisels.

The stone in irregular squared rubblework is not laid to a set course height that runs horizontally across the wall. Instead, the stones extend up and down in the wall in an irregular pattern. This is the most popular type of squared rubble stonework and is shown in figure 25-6.

In coursed rubble stonework the edges are also squared with a chisel and stone hammer, but the stones are laid to a specific course height, as in brick and concrete block work. This level course of stone usually occurs every 18 or 24 inches in height. Coursed squared rubble masonry costs more to build than irregular squared rubblework because the stone must be cut to the level course height.

Fig. 25-6 Squared rubble stone wall with raked joints.

No special pattern is used for irregular or coursed squared rubble stone. As the stones are fitted in the wall by cutting the edges and working one against the other, a pleasant, changing pattern results. It is not recommended, however, that all large or all small stones be laid together in the wall. Large and small stones should be divided equally for the best appearance. As mentioned previously, small chips or pieces of stones are not recommended for use; they are only used to fill in spaces where nothing else will fit.

Cutting Squared Rubble Stone

Stone that has recently been quarried or blasted from rock deposits contains moisture or what is called sap. It is much easier to work or cut stone when this sap is present. Once it is exposed to the air, the sap dries out and the stone becomes much harder. This is why stone should be cut as soon as possible after it has been quarried or blasted out.

The Stonecutting Bench

Stone can be cut more efficiently by laying it on a heavy wood bench, such as the one shown in figure 25-7, page 342. The wood absorbs the impact of the hammer and chisel, allowing the stone to give enough to prevent breakage in the wrong place. The stonecutting bench should be built high enough to be convenient for the mason doing the cutting. If the bench is too high, the mason must stretch to make the cut; and if the bench is too low, the mason must bend over uncomfortably.

A good general height for the top of the bench is about 28 inches. The bench should be built of heavy structural framing lumber so it will last a long time. The legs should be made of 2-inch x 6-inch lumber, and the top should be built of 2-inch x 8-inch lumber. The top of the bench should be approximately 32 inches x 32 inches. This dimension depends, however, on the needs of the particular job and the type of stone being cut. Remember that if a stonecutting bench is built too large, it is difficult to move from job to job. Enough braces should be nailed to the bench to make it solid and strong. The braces are stronger when they are run on a diagonal in the form of an "X".

> Note: Stone should never be laid on a concrete floor and cut, since there would be no "give" and the stone might break.

Cutting With The Pitching Chisel

The first step in cutting a stone for irregular or coursed rubble stonework is to chip or cut off the projecting knobs of stone. This is done with the stone axe (or stone hammer). The brick hammer can also be used if the knob of stone is not too large.

Once this has been completed, the edges of the stone must be squared. A pencil or soapstone line is drawn on the edges that are to be cut. This is usually done by holding a 2-foot framing square along one side of the stone and marking a line on the opposite side. The square does not have to lie on a perfectly flat place. It is used here to give an approximate place to start cutting.

Fig. 25-7 Stonecutting bench with assorted tools laid on it.

Fig. 25-8 A mason is squaring the edge of a stone using the pitching chisel. Notice the eye protection and glove worn by the mason to protect him from injury.

The mash hammer and pitching chisel are then worked along the line, cutting the stone. This type of cut face is referred to as a *rock cut face*. Figure 25-8 shows a mason cutting with the pitching chisel. Notice that the mason wears eye protection. It is also a good practice to wear a glove on the hand holding the chisel, as shown in the figure. A squared stone for use on corners is shown in figure 25-9. Figure 25-10 shows a mason using a brick hammer to chip off a small knob of stone to be used in a stone fireplace.

Fig. 25-9 This stone has been squared for use on a stone corner.

A point chisel is used to remove projecting knobs of stone or to square the final cut face of a stone. The pointed edge of the chisel only removes a small amount of stone, which decreases the chances of ruining the finished cut. A point chisel is also very helpful in chipping or removing any outcropping on the back of the stone.

Laying The Stone In The Mortar

The stone is laid in the mortar with the aid of a range line attached from one corner of the wall to the other corner. An ordinary mason's trowel is used for applying the mortar. Figure 25-11, page 344, shows a mason laying irregular squared rubble stone.

Fig. 25-10 Finishing the cut on a stone for a fireplace front. The brick hammer is very handy for trimming the edges of stones.

Random rubble stonemasonry is popular in stone fireplaces. Figure 25-12 shows a mason laying stone on a cantilevered fireplace. The rough brick in the front of the fireplace will be plastered to achieve a contrasting effect. The finished fireplace is shown in figure 25-13.

VARIOUS TYPES OF JOINT FINISHES FOR STONEWORK

There are several ways to tool and finish the mortar joints to establish character, prevent water from entering, and provide a sense of depth on the face of the stonework. Tooling the mortar joints also creates different textures and designs in the stonework.

Raised Flat Joint

Fig. 25-11 Mason laying squared rubble stone wall.

One of the most popular joint finishes on ashlar stonework is a slightly raised flat joint. The mortar joints are raked out to a depth of 3/4 inches when laid. At the completion of the job the joints are repointed with mortar and tooled with a slicker, creating a slightly raised flat joint, figure 25-14.

Fig. 25-12 Mason laying squared rubble stone on a cantilevered stone fireplace.

Fig. 25-13 Finished stone fireplace as it looks when the room is completed.

Raked Joints

Another popular joint finish for random or square rubble masonry work is the raked joint. This joint can be made in several ways. The mortar joints are raked out to a depth of approximately 3/4 inch. This is either done immediately after the joints are laid in the wall or after they have set enough to prevent the stones from moving out of position. Then the joints can be smoothed with a slicker or rubbed over with a rounded stick of wood, such as an old broom handle or brush handle. For random or squared rubble masonry the raked joint is very appealing. It highlights all of the edges of the stone and brings out the natural beauty of this type of stonework.

Figure 25-15 shows a portion of a squared rubble stone wall in which the mortar joints have been raked out and rubbed with a piece of broom handle. The joints are then brushed out to clean away any particles of mortar.

Fig. 25-14 Flat raised joint used in stonework.

Fig. 25-15 Squared rubble stonework with raked joints. Notice how the raked joints highlight the edges of the stones.

Rolling Bead Joints (Convex Joints)

Another unusual treatment of mortar joints in a stone wall is the rolling bead joint. When a rolling bead joint (also called a convex joint) is to be used, the stonework is laid up with mortar that matches the color of the stone in the wall. The mortar is then raked out to a depth of approximately 3/4 inch. The thickness of the mortar joints is not important to the appearance of the wall because these joints blend in with the stonework.

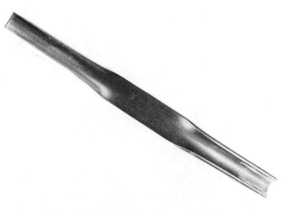

Fig. 25-16 Metal concave beading tool.

After the mortar has cured, it is repointed with regular mortar and finished off with a white portland cement mortar mixed with white sand and hydrated lime. The mortar areas are dampened with a brush and water. Then the mortar is pressed in and onto the face of the mortar joints with a concave beading tool, figure 25-16.

The advantage of a convex joint is that the rough mortar joints need not be a standard size. After they are pointed with the bead, the mortar joints appear the same size as the rolled edge of the concave jointer. Although the rolling bead joint is pointed at a later time, it will not crack or chip off if correctly formed. Figure 25-17 shows a random rubble stone wall that has been tooled with a concave beading tool. Notice the irregularities of the stone in the wall face.

Fig. 25-17 An example of stonework that has been tooled with a concave beading tool. Notice the size of the regular mortar joints and the size of the rolling bead joints. The white mortar makes the mortar joints stand out.

CLEANING OF RUBBLE AND SQUARED RUBBLE STONEMASONRY

Usually, rubble stonemasonry can be cleaned using a stiff brush, a hose with a spray nozzle, and water. Cleaning is not done until after the mortar has cured, which takes a week to ten days depending on weather conditions.

During construction of the stonework, be careful not to splash concrete, tar, or mortar on the wall. Anything splashed on the wall must be removed later. Also, at the end of the workday the stone wall should be covered with a sheet of plastic or canvas tarpaulin to protect it from the weather.

For a spot of mortar that is difficult to remove from a stone wall, a solution of muriatic acid with water can be used in a proportion of 1 part acid to 10 parts water. Rinse quickly and thoroughly with clear, cold water.

LAYING COBBLESTONES FOR WALKS OR PATIOS

The use of granite cobblestones has been decreasing in recent years. They are still, however, used for patios and walks, and for many public buildings such as museums and government buildings. Cobblestones are laid in a bed of portland cement and sand, which is slightly dampened so it blends together.

The mason cuts and fits the stone into position with the help of a line for leveling. The cobblestones are then tapped in place with the brick or stone hammer as shown in figure 25-18. This type of stonework lasts as long as any other type of paving. After the stone is laid, a mixture of portland cement and sand is swept into the mortar joints. When water, such as a spray from a hose, comes in contact with the mortar mix, it hardens into a solid joint.

FLAGSTONE

Flagstone is usually Vermont slate, New York flagstone, or fieldstone found in a local area, although it is produced in many states. The thickness of flagstone is usually no more than 1 1/2 inches; however, natural fieldstone varies in size.

Flagstone is popular for porches, patios, and walks. Vermont slate comes naturally in colors such as blues, greens, reds, and shades of gray and brown. Fieldstone varies in color depending on the mineral

Fig. 25-18 Mason laying granite cobblestones on a bed of portland cement and sand, with a small amount of water added to the mix. The stones are laid to the line and cut to fit with a brick hammer.

deposits in the stone. New York flagstone usually has a brownish-gray color. Ornamental stone such as marble can also be used for flagstone but must be specially cut for the job.

BONDS OR PATTERNS

One way to lay flagstone is in squared pieces. It can also be laid in a random pattern, where the irregular edges break into the pattern, creating a puzzle-type appearance. The random pattern is more popular because it readily fits into the design of outdoor structures.

MORTAR FOR FLAGSTONE WORK

Portland cement, lime, and sand are recommended for all exterior flagstone work. The most frequently specified mortar for flagstone is type M. The mortar is laid on a concrete base and the stone is laid in the mortar bed. If a porous type of flagstone is being laid, it is a good idea to mix some pure portland cement with water, to get a paste. The paste is then brushed on the back of the stone with an old paintbrush before the stone is laid in the mortar joint. This helps to keep the stone from shrinking away from the mortar joint. It is not necessary to do this with hard flagstone such as Vermont slate, because the mortar sets slowly due to the hardness of the stone.

Flagstone can also be laid mortarless. In mortarless work, stone dust or sand is used as the base. The stone is laid in the base materials. Joints are filled by using a broom or brush to sweep in sand, or portland cement and sand mix.

CUTTING FLAGSTONE

Flagstone can be cut with a chisel, as shown in figure 25-19, or with a masonry saw. To cut the stone with a flat-bladed chisel, first a line is marked on the stone at the point where it is to be cut. Then the stone is laid over a pipe or board so that the mark is in line with the edge of the pipe or board. The mason works the chisel along the edge of the line by tapping on it with a hammer. The stone should break along the line cleanly. Flagstone can also be cut by working the head of a brick hammer along the edges. The brittle, hard edges of the stone will chip or split off when the hammer is hit on a downward angle. This is a very useful method of cutting flagstone when only a small amount is to be removed from the stone. Hydraulic splitters can also be used for cutting flagstone.

Fig. 25-19 Cutting flagstone with a chisel and hammer, over a length of pipe, to produce a straight break.

LAYING AND POINTING FLAGSTONE

After the stone is cut, it should be laid in the mortar slightly higher than it will be when finished. A brick hammer handle or length of board is then tapped down on the stone to settle it into position,

Fig. 25-20 Mason laying the top of a flagstone porch. After all of the stones are laid, the joints are filled in with a mortar using a flat slicker tool. The stones are settled in place by tapping them with the brick hammer handle.

figure 25-20. It is important to settle all of the stones solidly into their bed or they can work loose later.

The stones should be cut as neatly as possible, using small points of stone sparingly. The mortar joints are either raked out approximately 3/4 inch deep and pointed later, or pointed in with the pointing trowel or slicker at the time they are laid. Use a rather stiff mortar mix for pointing because the mortar tends to bleed on the sides of the stone if it is too soft. If the raked joints are dry, dampen them with a brush and water before repointing.

CLEANING FLAGSTONE

Flagstone is cleaned with a solution of 1 part muriatic acid to 10 parts water. Immediately after scrubbing the flagstone with the acid solution, it should be rinsed well with running water from a hose. The true color of the stone comes out after the cleaning process. Stonework should never be cleaned, however, until it has thoroughly cured.

ACHIEVEMENT REVIEW

Part A

Select the best answer from the choices offered to complete each statement. List your choice by letter identification.

1. The old method of estimating stone that uses 24 3/4 cubic feet of stone is called a
   a. square yard.
   b. cubic square.
   c. perch.
   d. cord.

2. The roughest and simplest type of stonework is
   a. ashlar.
   b. rubble.
   c. random.
   d. roughly squared.

3. Stonemasonry requires the use of a portland cement mortar. The recommended type for general stonework is type
   a. O.
   b. M.
   c. N.
   d. S.

4. The hammer used for cutting with the chisel is sometimes called a striking hammer. It weighs from 2 to 4 pounds and is also known as a
   a. sledge hammer.
   b. pick.
   c. crandel.
   d. mash hammer.

5. The chisel that has a beveled blade and is used for cutting the rock face and squaring up the edges on a stone is known as a
   a. point chisel.
   b. pitching chisel.
   c. bush chisel.
   d. plugging chisel.

6. Mortar for random or rubble stonework should be mixed stiffer than mortar for brickwork because
   a. the stone has a fast absorption rate.
   b. the finished joint is harder.
   c. the stones are very hard.
   d. the work smears excessively.

7. Stone that has been blasted out of rock formations contains
   a. oil.
   b. sap.
   c. acid.
   d. film.

8. The tool used for dressing stonework is the
   a. plugging chisel.
   b. brick set chisel.
   c. point chisel.
   d. plumber's chisel.

9. The rolling bead joint used on stonemasonry is applied with a
   a. convex jointer.
   b. raked jointer.
   c. concave jointer.
   d. grapevine jointer.

10. The type of flagstone that has the most natural coloring is
    a. Vermont slate.
    b. New York.
    c. fieldstone.
    d. marble.

Part B

Describe briefly how the following tools are used in stonework.

    a. Mash hammer
    b. Pitching chisel
    c. Sledge hammer
    d. Point chisel
    e. Concave bead tool
    f. Stone axe

SUMMARY — SECTION 7

- Stonemasonry is the art of cutting and laying stone either dry or in mortar.
- The laying of stone is a craft that dates back thousands of years.
- The major advances in the laying of stonemasonry are the improved mortars used today and methods of quarrying the stone and handling, using power equipment.
- The outstanding properties of stone are its durability, uniformity, and strength.
- Stonemasonry can be grouped into two general classifications: rubblework and ashlar.
- Random rubble stonework is the laying of stone basically as it is taken from the field or rock outcropping with a minimum of cutting.
- Roughly squared rubble is stone that is squared on the job by the mason using chisels and hammers.
- Ashlar stone is precut at the mill or quarry and shipped to the job ready to be installed by the mason. The mortar joints are relatively small and uniform in comparison to rubblework.
- Stone that is too large to be lifted manually by the mason is lifted with the aid of power equipment.
- The most popular type of stone for the trim, sills, coping, and facings of buildings is Indiana limestone.
- Limestone is a fairly common sedimentary rock, and is composed of shell and shell fragments.
- Indiana is the location of the largest oolitic limestone deposits in the United States.
- Indiana limestone is highly desirable as a building stone because it is relatively easy to saw, cut, and adapt to the size needed for masonry construction.
- Limestone is carved at the mill for special designs specified by the architect.
- Indiana limestone is classified as fine, medium, and coarse and can be obtained in hand-tooled finishes.
- Indiana limestone should always be set in nonstaining mortar.
- Gin poles, derricks, and lewis pins are used on the job for setting limestone.
- Steel anchors and dowels are installed in the masonry work to help hold the limestone in position in the wall.
- The mason's standard tools are used for setting limestone. Special tooth chisels and hammers are used to dress the cut edges.
- Due to the weight of the stone, safe working practices should be observed when setting or handling limestone.

- If limestone must be tapped down in order to settle in the mortar bed, a 2-inch x 4-inch piece of wood should be used to protect the stone from damage.
- The head joints in limestone are best filled or pointed by blocking off the front of the joint and pouring with a grouted mortar.
- Limestone should be cleaned with a brush, using soap and water.
- Finished limestone work should be protected by building a wood box around any projected edges until the building is completed.
- Rubble stonework is estimated either by the square foot or the perch.
- Stone walls should be laid on a good footing below the frost line.
- Generally, larger stones should be laid on the lower courses of a rubble stone wall with the smaller stones towards the top of the wall.
- The use of large mortar joints in stonework is not recommended.
- When necessary, small filler pieces or points of stone should be used sparingly.
- Type N portland cement mortar is recommended for general stonework, with the exception of type M for stonework in contact with the earth.
- In addition to the mason's basic tools, a variety of hammers and chisels are needed to cut and lay rubble stone.
- The stiffness of the mortar used depends on the hardness of the stone and the weather conditions.
- Stonecutting chisels and hammers should be heated and sharpened by a blacksmith instead of having their edges ground.
- Stone can be cut more efficiently if placed on a wood bench made in comfortable proportions for the mason using the bench.
- Rubble stonework can be either of two types: irregular, in which the stone is not laid on specific course heights; and coursed rubble, in which the stone is laid to specific courses at selected heights.
- Mortar joints can be tooled in different ways. The most popular are the raked joint, raised flat joint, and rolling bead joint.
- The raked joint can be left with a smooth or rough-brushed finish. The raised joint is applied with the steel slicker pointing tool and the rolling bead joint with a concave striking tool.
- If rubble stonework is to be cleaned, usually only a stiff brush and a stream of water from a hose are needed to remove the surface dirt. For mortar stains, a solution of muriatic acid and water can be used.
- Flagstone is used for laying walks, porches, and patios.
- Flagstone is laid either in square pieces or irregular random patterns. Type M mortar is recommended for laying flagstone on a concrete base. Flagstone can also be laid mortarless on a base of sand or stone dust.

- Flat joint finishes are recommended for flagstone to prevent water from entering the joints.
- Flagstone is cleaned with a solution of muriatic acid and water for best results.
- Slate flagstone is available in various natural colors.
- Flagstone should always be tapped down solidly in the mortar bed.

SUMMARY ACHIEVEMENT REVIEW, SECTION 7

Complete each of the following statements referring to material found in Section 7.

1. The laying of stone in mortar has been done for _____ of years.
2. The most important advances in stonemasonry in recent years are _____.
3. The three most important properties of stone are _____ _____.
4. The two general types of stonemasonry are classified as _____ _____.
5. Large stones that are too heavy for the mason to handle by manual labor should be lifted by _____.
6. Limestone shells and shell fragments are cemented together by _____.
7. Most cut oolitic limestone used in the United States is found in the state of _____.
8. The select limestone that lies beneath an overburden in the quarry is called _____.
9. Limestone is cut at the mill by gang saws which have _____ _____ edges.
10. Limestone shipped to the job should be protected from moisture by stacking on _____.
11. The two types of limestone sills are _____ _____.
12. Limestone that is used as a cap for finishing off a wall is called _____.
13. A course of limestone that completely encircles a building for ornamental purposes is called a _____.
14. The drawings that accompany limestone when it is shipped from the mill are called _____.

15. Limestone should always be set in a portland cement mortar that is _____.

16. In order for the mason to adjust and move the limestone around in the mortar bed a metal _____ is used.

17. A heavy wood or metal pole that has a block and tackle attached to it for lifting stone to the wall is known as a _____.

18. Stone can be lifted with the aid of steel pins that are inserted in holes drilled in the stone. These pins are called _____.

19. A special chisel used for dressing limestone is known as a _____ chisel.

20. Limestone sills on the mortar bed are adjusted with the aid of a pinch bar and _____.

21. In stone trim, such as sills and coping, the head joints are filled with _____.

22. Limestone trim is cleaned with soap and water. Damage to the stonework can result if a _____ solution is used.

23. Limestone should be protected after installation by building a protective covering made of _____ around the stone.

24. Rubble stone should be laid on its broadest end. This is called its _____.

25. The recommended portland cement mortar mix for general stonework is _____.

26. The hammer used by the mason to strike the chisel when cutting stone is known as a _____.

27. The most common type of chisel used for cutting and dressing rubble stonework is called a _____.

28. Rubble stonework in which the stone is laid to level course heights is known as _____.

29. One of the most popular types of joint tooling in which the mortar joint protrudes outward in a rounded position is called a _____.

30. Flagstone can be laid in mortar on a base of _____ or laid mortarless on a base of _____.

Section 8
Alteration, Repair, and Maintenance of Masonry

Unit 26 CUTTING OUT, REPOINTING, AND REPAIRING MASONRY WORK

OBJECTIVES

After studying this unit the student will be able to
- list the procedure for removing old paint from brickwork.
- describe how old mortar is cut out and how repointing is done.
- explain working practices to follow when repairing or rebuilding fireplaces or chimneys.

INTRODUCTION

When masonry work is built using good materials, properly mixed mortars, and good workmanship, it lasts for a long time with little maintenance required. However, all masonry requires some repairs after a period of time. When old masonry buildings are repaired to their original condition, it is called renovation or restoration.

Conditions that make renovation or repairs on a building necessary are rain, snow, and changes in temperature. These eventually erode mortar joints. In the past, all mortar used for laying masonry units was of the lime and sand type. This mortar served the purpose, but the portland cement based mortars of today are much improved. The repointing of old brick or stone buildings with a portland cement, lime, and sand mortar mix prolongs the life of the masonry work.

The most important thing to remember in the restoration of old exterior masonry work is to prevent the entrance of water in the mortar joints. The elements can attack the masonry work and cause it to break down at a faster rate. Renovation and restoration includes not only the cutting out and repointing of the exterior walls but other tasks such as removing many coats of paint; replacing bricks that have softened; rebuilding interior partitions, fireplaces, and chimneys; and many other repairs.

Many masonry contractors do repairs and restoration as a part of their regular work, while others do only repair work. This type of work requires skill as well as creativity to restore a fireplace or exterior of a distinguished old home to its original condition, figure 26-1.

Before old chimneys and fireplaces can be used again, they must be made fireproof and safe. A mason who specializes in repair must know the correct mortars to use, the tool skills involved in cutting out and repointing work, and how to clean old masonry. It is also necessary to understand how an old building is constructed in order to be able to cut out and repair masonry without damaging the structure.

356 ■ Section 8 Alteration, Repair, and Maintenance of Masonry

Fig. 26-1 This old brick structure is being restored. The entire outside wall is being repointed. Deteriorated brick is cut and replaced.

In the past, this type of work was the job of only a few specialized masons. Today, repair work may be a necessary part of the job of masons employed full-time. Repair and renovation of masonry buildings supplements the mason's income when the building of new structures is slowed down or delayed.

This unit deals with the most common types of repair and restoration of masonry work. This involves cutting out and repointing exterior masonry and rebuilding interior partitions, fireplaces, and chimneys.

REMOVING OLD PAINT FROM BRICKWORK

When repointing old brickwork, one of the most frequent problems encountered is the removal of old paint from the face of the brickwork. A wire brush can sometimes remove the loose paint if it is not imbedded too deeply into the face of the brick. There are usually, however, many coats of paint to remove on an old brick building. Sandblasting using air pressure also removes old paint but in the process actually cuts away the face of the brick. This is not desirable in a building that has soft brick. (Many old brick buildings built before 1870 have a large amount of soft brick in the exterior walls.) The following procedure is a good method of removing old paint from old brickwork.

Procedure For Removing Old Paint

1. Obtain two 8-quart plastic pails and put 1 gallon of clean water in each pail.

   NOTE: Wear rubber gloves and eye protection for all of the following steps.

2. Pour one quart can of caustic soda (lye) into one of the buckets and stir.
3. Add 8 ounces of regular corn starch to the other bucket of water.
4. Stir and blend the corn starch with the water.
5. Mix the water containing the corn starch and the water containing the caustic soda together slowly while stirring the solution.

6. Apply the mix on the wall with a plastic fiber brush, being careful not to get any on the skin. If any of the solution does get on the hands or skin, flush it off immediately with plain water.
7. Leave the solution on the wall for 1 hour. The solution adheres to the wall like a gelatin. It works to "cook" or "eat" the paint off the wall.
8. Use a high-pressure washer with plain water on the wall to remove all cleaner.
9. Then rinse the wall with soap and water through the high-pressure washer.
10. Rinse the wall again with clear water through the high-pressure washer. This should remove all the paint from the surface. If all the paint is not removed, repeat the procedure.

REMOVING OLD MORTAR FROM THE JOINTS

Before attempting to repoint any mortar in the joints, the joints are cut out to a depth of at least 1/2 inch. This should be done very carefully with a plugging chisel, tuck-pointer's grinder, or with an air chisel, depending on the hardness of the mortar.

> Be sure to wear eye protection and a hard hat when cutting out mortar joints.

After the old mortar has been cut out, remove all loose materials with a brush or a high-pressure stream of water. Any loose or bad bricks should be cut out and removed as the joints are cut out.

Generally, it is more economical to tuck-point a given area at one time rather than at several times, due to the cost of moving workers and scaffolds back and forth. (To *tuck-point* a joint is to fill mortar in a joint with a tool such as a slicker, convex jointer, or pointing trowel.) The mortar joints should be cut out reasonably square to the 1/2 inch depth and not on an angle. If cut on an angle, the mortar being pointed back does not have the proper thickness to obtain a strong joint and a good bond. Also, if new mortar is pointed against loose sand, it does not adhere properly.

In figure 26-2 the small section of wall on the right side has been cut out for repointing. The wall above the hole and to the left has been pointed previously since the work is being done from the top to the bottom of the house.

Fig. 26-2 An old brick wall is in the process of being repaired, including repointing of the mortar joints.

MORTAR FOR REPOINTING WORK

Mortar for repointing should be carefully selected and proportioned. Although various colors can be added for special jobs, this is not discussed here since colors or additives are specified by the architect.

Mortar made of portland cement, hydrated lime, and sand is recommended for repointing work. Mortar that is too rich (high content of portland cement) is very hard when cured, and is subject to abnormal shrinkage. This is undesirable for pointing mortar because pointing mortar must bond in the joint. A high compressive strength is not necessary. The original mortar of the structure carries the load of the masonry work, not the 1/2 inch of pointed masonry. This is a common mistake in preparing mortar for repointing masonry work. Figure 26-3 shows how the mortar adheres to the brick.

Fig. 26-3 An example of how portland cement/lime mortar adheres to brick.

The recommended general mortar for repointing mortar joints is type N portland cement mortar consisting of 1 part portland cement to 1 part type S hydrated lime to 6 parts washed sand. The addition of lime to the mix prevents the mortar from drying out too fast; it also adds workability and provides a more waterproof mortar.

Even when lime is added to the mix there is still a certain amount of shrinkage involved when pointing mortar in old head and bed joints. Prehydration greatly reduces mortar shrinkage and also improves workability. To prehydrate mortar, first thoroughly mix all ingredients dry. Then mix again, adding only enough water to produce a damp, unworkable mix which retains its form when pressed into a ball. After keeping the mortar in this dampened state for one to two hours, add enough water to bring the mix to the proper consistency, somewhat drier than a regular masonry mortar mixed for laying bricks in the wall.

POINTING IN THE MORTAR

To ensure a good bond, the raked mortar joints should be dampened using a brush and water before filling them with mortar. The joints should not be soaked, but only dampened. Because the mortar joints should not be visibly wet, allow time for the water to soak into the wall.

The preferred method of tuck-pointing is to pack or tuck the mortar tightly into the joint, using a slicker or half-round (convex) pointing tool that matches the size of the mortar joints. Since mortar joints vary in size in an old building, several jointers of different sizes are needed. A mason's pointing trowel can also be used when pointing up around windows, doors, and any areas that may require a larger than normal joint. The mortar is placed on a *hawk* (a square platform, approximately 8 inches x 8 inches, with a vertical handle in the center). Hawks can be made of scraps of wood or metal obtained from a tool company.

When a large mortar joint or patch must be pointed, it is desirable to fill in less than half the amount needed and then wait until this is thumbprint hard. Then roughen the surface slightly and repoint in the rest of the mortar. This decreases the chance of cracking the joint. It is also helpful to remoisten the joint if it dries out before all of the pointing is done. This is especially important if a section of wall is being pointed in dry, hot weather. Spraying the wall lightly with water from a garden tank type sprayer is helpful.

When the pointing is done, the wall should be brushed lightly to remove all particles of dust or mortar. Figure 26-4 shows a very old brick wall that has been repointed with mortar made of portland cement, hydrated lime, and sand.

Fig. 26-4 A very old section of wall that has been repointed with hydrated lime mortar. Although the bricks shown are weatherworn from age, the mortar still bonds well to the edges.

POINTING STONEWORK

The raking out and pointing of stonework is done the same way as for brickwork. After the joints have been raked out, they are brushed clean and repointed, figure 26-5, with the same mortar mix used for brickwork.

REPAIRING AND REBUILDING OLD FIREPLACES

Fig. 26-5 Mason standing on a scaffold repointing a section of stone wall with new mortar.

The amount of repair needed for an old fireplace depends on its condition. If the bricks are badly burned out, it may be just as economical to tear out the entire fireplace and chimney, and rebuild. The permission of the owner should always be obtained before proceeding with a major repair such as this.

Since flue linings were not used years ago, it is a good practice to install them in an old chimney or fireplace when rebuilding. It is also a good practice to line the firebox with firebrick if the fireplace is going to be used. If the fireplace is not going to be used but only restored to its original condition, regular brick that matches the fireplace can be used.

GENERAL REBUILDING CONSIDERATIONS

Use well-filled mortar joints whenever rebuilding a fireplace or chimney. As when pointing, type N portland cement/lime mortar is good for rebuilding work. The lime in the mix creates a better bond with the old, softer bricks.

When covering over the top of the firebox, either a brick arch or an angle iron should be used to support the masonry wall above. Figure 26-6 shows masons building the brick wall over the angle iron that covers over the firebox. The rebuilt fireplace is constructed the same as when building a new one, except that all existing woodwork must be protected with at least 2 inches of bricks and mortar to prevent burning.

HEARTHS

The hearth can be relaid either in a sand base or laid in mortar. Usually a small bed of mortar is used to level the bricks in place. The finished head and bed joints are then either pointed in with mortar using the slicker or swept in with fine sand, depending on the desire of the builder or owner. Figure 26-7 shows a mason laying a hearth in an old fireplace. Notice the foam rubber cushion the mason is kneeling on to relieve the pressure on the knees.

CLEANING THE FIREPLACE

When the work is completed, it should be rubbed off with a burlap bag to remove excess mortar stains. If the mortar stains are still present, mix a solution of 1 part muriatic acid with 9 parts water in a nonmetallic container.

> Be sure to wear rubber gloves and protective eye covering when using this type of solution.

Fig. 26-6 Masons building a brick wall over the opening of a rebuilt fireplace. A jack arch is being built over an angle iron for extra support.

Fig. 26-7 Mason laying a hearth in an old, rebuilt fireplace.

First dampen the work with water and then scrub the affected areas with a brush and the acid solution. After the washing is completed, rinse the area with water. This procedure may be difficult if the floors are finished or if wood in a home is being restored. Therefore, good judgment must be used when cleaning the work. This is another reason to be as neat as possible and to keep excess mortar off the face brickwork. If water or acid solution cannot be used, then a stiff, dry brush will remove most of the stains.

REPAIRING CHIMNEY TOPS

As masonry structures age, the top of the chimney deteriorates (as shown in figure 26-8) because it is subject to moisture, gases, and extreme differences in temperature (due to the hot gases inside and the temperature outside). Before any work is started, a complete inspection of the chimney should be made to decide what work must be done. If, upon inspection, the only problem is that the mortar joints are coming out, they can be cut out and repointed as regular repointing work. However, if the chimney from the roof to the top is badly damaged or if the bricks are burned out, the best procedure is to tear down the brickwork to the roof line and rebuild. If this must be done, the property owner should be told beforehand since the expense will be greater.

Fig. 26-8 Chimney in need of repair on the roof of a house. The back of the chimney will probably have to be rebuilt.

One of the main problems in repairing chimney tops is getting to the work. Often building the scaffold and covering the roof is more work than the actual repairing of the chimney. This is especially true when the chimney is repointed.

First, a strong scaffold must be built and fastened to the building so it cannot pull away. Set the legs on a strong base and tie it to the structure with rope or wire attached to the scaffold frame.

Protection of any windows or parts of the house where the scaffold adjoins must also be provided by the mason. A sheet of plywood or covering can be installed to prevent damage. If the scaffolding is built on a busy street, a covered walkway must be constructed or the area must be roped off to protect passersby.

Ladders used to reach the scaffold must be tied to the scaffold at the top, middle, and bottom to prevent slipping. Any ladders laid on the roof should not be nailed to the roof because this will cause a leak. Tie ladders to make them steady and slipproof. A mason's helper should be on the ground at all times, ready to assist the mason.

The roof must be protected by covering with plywood or tar paper to prevent stains or damage from construction. Do not nail this covering into the roof for this can cause the roof to leak. Attach the covering with ropes or some type of fasteners that do not require holes to be punched in the roof. If a scaffold must be built on the roof, the *bucks* (scaffold frames, figure 26-9, page 362) should be made to fit the pitch of the roof so the top of the working platform is level. The scaffold is then tied to the base of the chimney so it cannot slip. If the roof is shingled, a sheet of plastic or a canvas tarpaulin should be laid over the roof in the work area to protect against mortar stains or dirt ruining the shingles. This covering is removed after the job is completed.

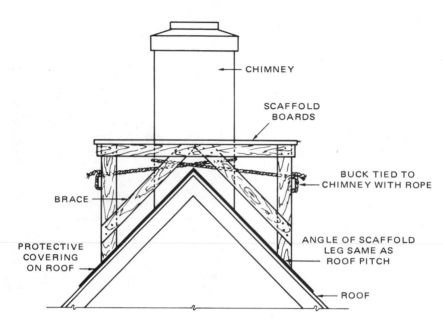

Fig. 26-9 Roof buck scaffolding made to fit the pitch of the roof. This provides a level working platform for the mason when repairing the chimney.

REBUILDING THE CHIMNEY TOP

Protecting The Heating System

Before tearing down any of the old brickwork, it is advisable to fill a burlap bag with straw and attach a rope to the end of the bag. Make the bag large enough so that it can be pushed snugly down inside the chimney without too much pressure. The bag should be pushed down just past the point where the work will be torn down for repairs. The bag of straw serves to catch any pieces of bricks or mortar that may drop in the chimney while repairs are being made. This is a very important practice to follow because any objects dropping down the chimney can cause damage to the heating system or cause soot to spread throughout the house on the ground floor. After the work is completed, the bag can be pulled up with the rope and the debris removed.

As the chimney is torn down, the old bricks and mortar should be loaded into buckets and lowered to the ground with a rope and pulley. The mason's helper can clean the old bricks for reuse or send up new materials by rope and pulley.

If possible, do not disturb the flashing at the roof line when tearing down the chimney. If the flashing is in poor condition, install new flashing being sure that it is walled in and waterproofed correctly so the chimney does not leak. All flashing should be counter-flashed into the roof. The flashing should be built at least 1 inch into the mortar joints for a good bond.

Build the chimney up to the height of the cap and install a mortar wash on top of the chimney. The wash should be sloped upwards from the cap toward the top of the flue to deflect air currents and to drain water from the cap. It is recommended to always use portland cement mortar for the cap regardless of the type of mortar used in the chimney.

Portland cement mortar lasts longer. A type M mortar is recommended as a wash. Figure 26-10 shows a mason rebuilding the top of a chimney.

When the job is completed and the roof coverings and scaffolding are being removed, care should be taken not to damage the roof or structure.

> The mason should always remember that working off the ground on scaffolding or on roofs is extremely hazardous and all safety precautions should be taken to prevent accidents.

Fig. 26-10 Mason repairing the top of a chimney on an old house.

Smoke Test

When rebuilding a chimney it may be advisable to conduct a smoke test. This should only be done after the mortar has thoroughly hardened and before any appliances are connected. A smoke test is recommended for each flue in the chimney.

This is done by building a smudge fire at the bottom of the flue and, while the smoke is moving freely from the flue, covering the top tightly with a damp rag. The escape of any smoke into other flues or through the chimney walls indicates openings that must be made tight.

> Smoke tests are *not* recommended when repairing a chimney top on a furnished house because smoke would saturate through the house. This test is only for new construction or when a new chimney has been built on an old building where a smoke test can be conducted without damage to the interior.

Cleaning Of An Old Flue

Chimney flues that have developed a large soot deposit can be cleaned using a weighted brush, bundle of rags, or burlap bags of straw, lowered and raised by means of a rope from the top of the chimney. In addition, the inside of the firebox and damper should be scraped free of soot.

Make sure that all openings (such as dampers) in a fireplace are closed until the cleaning process is completed because soot may come out of the opening into the house. After the cleaning process is completed, the soot can be removed by using an old vacuum cleaner. An excess deposit of soot can cause a chimney fire, so it is advisable to have the chimney cleaned periodically as the soot builds up.

Each repair job differs, based on the work needed to restore the structure to good condition. However, when any repair work is being done, consideration must be given to the safety of the occupants of the property, the safety of the mason, and the adherence to good workmanship practices. As a rule, it is more difficult to repair masonry work than it is

364 ■ Section 8 Alteration, Repair, and Maintenance of Masonry

Fig. 26-11 A beautiful old brick structure that has been repaired and repointed with mortar. Notice the old style of arches over all openings.

to build it originally. Figure 26-11 shows an old brick structure after being repaired and repointed with new mortar.

ACHIEVEMENT REVIEW

Select the best answer from the choices offered to complete each statement. List your choice by letter identification.

1. The best method for removing old paint from old brickwork is with
   a. muriatic acid.   c. caustic soda.
   b. soap and water.   d. wire brush.

2. Mortar should be cut out of the mortar joints of old brickwork for repointing to a depth of
   a. 1/4 inch.   c. 3/4 inch.
   b. 1/2 inch.   d. 1 inch.

3. The proper pointing tool to use if a flat joint is desired is a
   a. convex striker.   c. slicker striker.
   b. concave striker.   d. pointing trowel.

4. The mortar recommended for repointing work is type
   a. M.   c. O.
   b. N.   d. S.

5. The process to reduce shrinkage of pointing mortar is called
   a. tempering.   c. premixing.
   b. glazing.   d. prehydration.

6. The refilling in of mortar joints with mortar is known as
   a. slicking.
   b. packing.
   c. tuck-pointing.
   d. tooling.

7. Mortar used for pointing raked mortar joints must be elastic (workable), and must not shrink excessively. The material used in the mortar mix that adds this property to the mortar is
   a. portland cement.
   b. lime.
   c. sand.
   d. calcium chloride.

8. To protect mortar droppings or bricks from falling down the chimney during the repair process it is a good idea to
   a. block off the chimney with a piece of wood.
   b. close off the opening at the furnace.
   c. lower a bag of straw down the flue.
   d. stuff some old rags down the chimney.

9. The metal that is built into the chimney at the roof line for the purpose of deflecting water is known as
   a. bridging.
   b. wash.
   c. flashing.
   d. cap.

10. To prevent burning when rebuilding a fireplace in an old house, all woodwork should be filled around with brick and mortar a minimum of
    a. 1 inch.
    b. 2 inches.
    c. 3 inches.
    d. 4 inches.

PROJECT 17: REPOINTING A SECTION OF BRICK WALL

To enable the student to practice the techniques of raking out and repointing masonry work, a small, 4-inch brick panel wall can be built, 48 inches long and 32 inches high.

After the wall is built to the height stated, it should be allowed to set until thumbprint hard. The mortar joints should then be raked out to a depth of 1/2 inch with a skate wheel rake-out jointer or line pin.

The mortar can then be repointed back into the joints with a slicker or convex jointer. The wall should be brushed at the completion of the pointing.

Unit 27 REPAIRING CRACKS, REMOVING EFFLORESCENCE, AND DAMPPROOFING

OBJECTIVES

After studying this unit the student will be able to

- describe the various types of cracks that occur in masonry walls and understand how they can be repaired.
- explain what causes efflorescence and how it can be removed from masonry walls.
- describe the use of silicones and masonry cement paint to dampproof masonry walls.

INTRODUCTION

Repairing cracks, removing efflorescence, and dampproofing walls are the most frequent types of maintenance procedures necessary for masonry structures. These procedures are necessary primarily because of movements in the structure, moisture, and perhaps poor workmanship when the structure was built.

It is helpful to understand the cause of these problems before trying to determine a cure. Masonry structures are expected to stand with a minimum of repair for a considerable period of time. However, even the best construction requires occasional repair and maintenance if it is to stay in good condition and present an attractive appearance.

If a small crack or leak is not repaired, it will increase with the passage of time until it seriously affects the structural strength and stability of the building. This unit deals with some of the more common problems encountered in repairing and maintaining masonry structures.

CRACKING

There are four main factors that determine whether a masonry wall will function properly: the materials used, the design utilizing the materials, the construction methods used in carrying out the design, and the ability of the wall to keep out moisture. When any of these requirements are neglected the masonry wall probably will not perform satisfactorily.

Cracking is the failure that occurs most often. Cracks occur in many different ways, but there are typical signs that aid in determining their cause. The type and size of the crack often indicates the cause.

EXPANSION CRACKS

One of the most common types of cracks in masonry work is caused by expansion and contraction of the masonry wall. All ordinary building materials contract and expand with variations in temperature. If the material is not firm or restrained it gives without showing any crack or sign of movement. However, a masonry wall is hard and not likely to give without cracking unless some provisions are made for expansion when the wall is constructed.

Expansion joints greatly help to combat cracking from movement in the wall. As an example, a brick wall that is 100 feet long, if unrestrained, will increase in length (expand) approximately 3/8 inch when its temperature is increased 100 degrees Fahrenheit. These movements are due to temperature, and are called *thermal movements*. If prevented, these movements produce a force strong enough to cause cracking in the wall. For this reason, when the building is constructed, expansion joints (figure 27-1) should be built into the wall to relieve the pressure.

Fig. 27-1 Expansion joint built into a brick wall. The exterior of the joint is caulked with an elastic compound allowing the work to give.

REPAIRING AN EXPANSION CRACK

No crack should be repaired immediately after it occurs because the movement may still be taking place. It is recommended to wait at least one year after the crack appears before making any repairs. If the crack continues to move during that time (opening and closing), it probably will be necessary to saw a vertical slot in the wall at that point and install an expansion joint because the cracking reappears if the crack is patched.

Assuming that the crack is stable and does not move any further, the mortar joints at the affected spot should be cut out with a chisel to a depth of at least 3/4 inch. They should be brushed out clean to remove all dust. The crack is then dampened with water from a brush. A portland cement mortar, such as type N (1:1:6) should be prehydrated (the same as for tuck-pointing) and packed into the crack with a slicker. If the crack is deep, pack half of the mortar in the joint and let it get thumbprint hard before filling it in all the way. Fill the mortar joint until the mortar is 1/4 inch from the face of the wall.

There are various types of elastic materials that match the mortar joint finish. Using a caulking gun, fill the crack with caulking sealant until it is flush with the masonry wall. This elastic sealant stretches and blends in with the mortar joints, making the repairs less noticeable. Dark-type fillers such as tar should never be used on face work because they ruin the appearance of the finished wall. Figure 27-2, page 368, shows a staircase type of crack caused by expansion.

Fig. 27-2 Expansion crack in a brick gable. Notice how the masonry joints have opened up but the bricks are not cracked. The movement occurred along the mortar joints.

Fig. 27-3 Settlement crack in a concrete block wall.

SETTLEMENT CRACKS

Settlement cracks, figure 27-3, frequently occur when part of the foundation or footing settles unevenly with another part of the structure. They are more evident in small buildings, such as in a foundation of a home. Settlement cracks are usually larger at the top and get smaller toward the bottom, changing to a hairline crack at the very bottom. (This can be reversed, depending upon the direction and location of the settlement with respect to the wall length.) The tapering thickness of the crack, then, is the major distinction between an expansion crack and a settlement crack.

Settlement cracks are repaired by cutting out the affected area after the settling has stopped and filling in with mortar much the same as in an expansion crack. The force or shear of a settlement crack usually runs in a vertical position. Settlement cracks in a building are not necessarily a sign of poor construction but can be caused by different load-bearing strengths of soils under the structure. If the crack is in an outside foundation wall, the area that was repointed with mortar can be sealed with a tar compound.

PLACES WHERE OTHER TYPES OF CRACKS ARE FOUND IN MASONRY WALLS

Foundation Walls

Another common type of crack is the cracking of foundation walls in a horizontal position or plane. This is usually caused by earth being pushed in, up to and around the foundation, before the shell of the house is built on it. The wall cracks across the bed joint where the pressure is applied. This type of crack can be prevented by either bracing the inside of the foundation wall before backfilling with the equipment or waiting until the structure is built on the foundation so the weight of the structure can hold the walls in place.

The only correct way to repair this type of crack is to dig out the earth from around the foundation wall on the exterior side, where the crack is, and re-lay the concrete blocks in fresh mortar. This is an expensive, time-consuming procedure but if a crack is showing on the inside of the wall, then the wall must also be cracked on the exterior. Merely pointing mortar in the cracked joints does not solve the problem.

Corners

Cracks are also found at corners. This type of cracking is caused by thermal movement of the outside wall. It is particularly common in insulated cavity wall construction.

Offsets And Setbacks

Vertical cracks are quite common at wall setbacks or offsets. When one or more walls expand toward an offset, the movement produces bending in the offset, causing vertical cracks in the masonry.

Parapet Walls

Parapet walls (walls that extend above the roof) cause particular problems because there are two exposed surfaces which are subjected to moisture and extremes of temperature. Although through-wall flashing is often necessary, it creates horizontal planes of weakness. Expansion of roof systems can make the situation worse. Parapets also cannot resist expansion, for they have less weight than a solid wall. Expansion can cause the parapet to bow, crack horizontally at the roof line, and to hang over corners at the end of the building. Expansion joints spaced throughout the wall allow for expansion of the parapet wall.

Encased Columns

When structural elements such as steel columns are enclosed with masonry, they must have room to expand since the expansion rate of steel and masonry is different. Masonry work should never be built completely solid around steel columns or the wall can crack. These cracks will show up on the inside and outside of a building. Mortar should never be filled in solid around a steel column. Space must be left for expansion.

Curling Of Concrete

When concrete floor and roof slabs are placed directly on masonry walls, curling of the slab often occurs due to shrinkage, deflection, and plastic flow of the concrete. There are many different types of cracking but the kinds most frequently found are expansion and settlement cracks. It is difficult to repair any crack without it being noticeable. Figures 27-4 through 27-7, page 370, show various types of cracks in masonry walls.

REMOVAL OF EFFLORESCENCE STAIN

Efflorescence is the deposit of crystallized salts on the face of a masonry wall. This deposit is usually white but it can be other colors depending on the impurities in the materials. Efflorescence is caused by salts that originate in the interior of the wall in either the masonry

370 ■ Section 8 Alteration, Repair, and Maintenance of Masonry

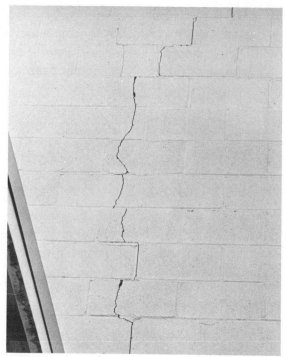

Fig. 27-4 Example of an expansion crack in a concrete masonry unit wall.

Fig. 27-5 An extremely bad crack. The lower corner of the building has actually moved out past the existing wall line. An attempt to repair this crack has been made by merely repointing with mortar. Eventually, the cracked portion will have to be completely cut out and replaced with new masonry.

Fig. 27-6 An example of cracking in a brick storefront. The crack is more horizontal than vertical, indicating pressure building up laterally (from side to side). This crack has been patched with a mortar that is too white, causing a poor appearance. When a crack is this bad, the wall may have to be torn down and rebuilt.

Fig. 27-7 Cracking of a section of retaining wall. Ice and snow have gotten behind the wall, resulting in severe cracking of the masonry work. This wall will have to be torn down and rebuilt.

unit, the mortar itself, or the backup. The entrance of water into the wall takes the salts into a solution. The solution moves to the face where the water evaporates, leaving the salts on the face of the wall. The following are the best ways to prevent efflorescence: keep the materials dry before laying them, keep the wall dry when in the process of building, and practice good workmanship. Generally, if there is no moisture problem, there is no efflorescence.

Efflorescence mars the appearance of masonry buildings because of the stain it produces. Figure 27-8 shows a typical example of efflorescence stain on a brick wall. When the efflorescence is not severe, it can be removed with a fiber brush and rinsed with clear water. If this fails, presoak the wall with water and use a wash of 1 part muriatic acid and 15 parts clean water in a nonmetallic container.

> Pour the acid into the water slowly. Never pour the water into the acid. The acid is harmful to the skin and eyes.
>
> It is also recommended to wear rubber clothes and gloves whenever using muriatic acid solution. If it does get on the skin wash it off immediately with clean running water.

It may be necessary to repeat the washing several times to remove all of the efflorescent stain. When finished washing, flush the wall with plenty of clean water from a hose with a nozzle.

Efflorescence is only removed from a wall temporarily. It recurs until all moisture is gone from the wall in the process of evaporation.

SILICONES FOR DAMPPROOFING

Although silicones are often called waterproofers, they are actually only *damp-proofers*. Instead of sealing the openings, silicones slow down water absorption by changing the contact angles between water and the walls of capillary pores in the wall. This creates a *negative capillary* which repels water rather than absorbs it. However, silicones are not 100 percent effective, so treated masonry still absorbs some water.

Masonry silicones are sold as a clear liquid that is available from many building supply dealers. It is used to repel water that has been entering masonry walls. The two major classifications are water-based and solvent-based silicones, each of which can be applied by spraying or brushing on the wall. In general, solvent-based silicones penetrate

Fig. 27-8 An example of efflorescence stain on a brick wall. This staining continues until all moisture has evaporated from inside the wall.

better due to their small molecular structures. However, water-based silicones are usually less expensive.

Silicones do not usually cause any color changes. However, when applying solvent-based silicones to walls having artificially colored mortars, apply the silicones to a small area of the wall to determine if any bleaching occurs. If no bleaching occurs after 2 weeks, it is likely that the silicones are not harmful to the wall appearance.

Silicones usually penetrate bricks to a depth of about 1/8 inch to 1/4 inch. Under wind pressure, water can flow through larger hairline cracks and pores in treated masonry. The effective use of silicones depends on the chemical formation, method of applying, physical characteristics of the masonry, length of time since application, and other factors. If the wall becomes dusty after the silicones are applied, it will be less water-repellent. Producers of silicones generally claim an effective life of five to ten years, after which the silicones must be reapplied. Silicones are available in 1 or 5 gallon sizes.

Since silicones do not seal the pores, water can evaporate through the wall face to the outside. However, the wall is resistant to penetration of water from the outside. This is called letting the wall *breathe*, and is highly desirable for a masonry wall. Before applying silicones to any brick wall, a careful study should be made of the correct application and precautions.

WATERPROOFING INTERIOR WALL FACES

The cause of a leaky basement is either water buildup against the exterior wall or groundwater from under and around the footings and wall edges. The best guard against a leaky basement is proper waterproofing of the basement when it is constructed, including the placement of drain tile around the perimeter of the foundation. This was discussed in Unit 5, Waterproofing the Foundation.

If the wall was correctly waterproofed when built, but the basement wall still leaks, there are several things that can be tried to stop the leaking. The possible solutions may or may not cure the problem, but they are the ones most often used before digging up all of the earth around the foundation and rewaterproofing the outside basement walls (which is very expensive).

Joint Waterproofing

If openings and cracks in the mortar joints are small, a grout coating can effectively seal them, greatly reducing water penetration. A typical recommended grout coating consists of 3/4 part portland cement, 1 part sand that passes a No. 30 sieve, and 1/4 part fine hydrated lime. Shortly before using, mix all of the ingredients with water to obtain a fluid consistency. Wet the joints thoroughly, but allow the masonry to absorb surface water before applying the grout. Then vigorously brush the grout into the joints with a fiber brush. If a neat appearance is desired, use a shield near the crack to prevent the grout from getting on the wall. Two coats of grout are usually required for a good job.

Applying Portland Cement Paints

Portland cement-based masonry paints can be used to reduce water penetration. These paints are sold by building supply and hardware dealers and are available in many different colors including standard white. They are produced in standard and heavy-duty types. These paints are mixed with water just before use. The standard type has a minimum of 65 percent portland cement by weight and is suitable for general use. The heavy-duty type contains 80 percent portland cement and is used where there is much moisture, such as in a swimming pool. Each type is available with a siliceous sand additive for use as a filler on a porous surface.

Portland cement paints set by hydration of the cement which bonds to the masonry wall. They are applied with a brush to a moist surface and are kept damp by a fine spray of water from a garden-type tank sprayed for 48 to 72 hours, until the paint cures. These paints are very successful when mixed properly and applied following the manufacturer's instructions.

Although portland cement paint can be used for brick, concrete, and concrete block, the most frequent use is on the inside of basement walls. Once the wall has been painted it requires periodical repainting to stay water-repellent.

Bituminous Coatings

While bituminous coatings work well on the exterior of foundation walls, a series of tests taken by the National Bureau of Standards indicates that they are not effective waterproofers when applied to the inside wall face. In general, the coating develops water-filled blisters, many of which eventually break. During the tests, the leakage rate through broken blisters was very high. Therefore, it is not recommended to use bituminous coats of waterproofing on the interior of foundation walls.

In conclusion, waterproofing applied to the interior face of walls is not as satisfactory as that applied to the exterior face. While the interior coating may stop water from passing completely through the wall, it does not prevent water from entering the inside of the wall. Pressure does, however, build up, so that eventually the water breaks through. As stated previously, the correct remedy is to prevent large quantities of moisture or water from entering the wall at the start. This can be done by following good construction practices and waterproofing before the earth is filled in around the foundations.

The problems discussed in this unit are those most often encountered in the repair of masonry work. They are, of course, not all of the problems that can occur, but rather the ones that the mason will come in contact with most often.

ACHIEVEMENT REVIEW

Select the best answer from the choices offered to complete each statement. List your choice by letter identification.

1. A step-type crack is usually a sign of
   a. settlement.
   b. expansion.
   c. curling of concrete.
   d. pressure exerted against the outside of a masonry wall.

2. Cracks in masonry walls should not be repaired until after a waiting period of

   a. 3 months.
   b. 6 months.
   c. 9 months.
   c. 12 months.

3. A patched crack in an exposed brick wall can be filled with an elastic filler that stretches. This filler is a type of

   a. paint.
   b. tar compound.
   c. oakum.
   d. caulking compound.

4. The whitish stain that appears on the surface of masonry walls, caused by salts inside the wall leaching to the surface, is known as

   a. bleaching.
   b. efflorescence.
   c. prehydrating.
   d. crazing.

5. The best way to prevent efflorescent stain on a wall is to

   a. plaster the wall.
   b. keep the wall dry while building.
   c. treat the wall with chemicals to prevent water from entering.
   d. build with a very high-strength mortar.

6. Silicones are used on masonry for the purpose of

   a. waterproofing.
   b. insulating.
   c. dampproofing.
   d. improving the appearance of the masonry work.

7. Silicones penetrate bricks to a depth of

   a. 1/8 inch to 1/4 inch.
   b. 1/4 inch to 1/2 inch.
   c. 1/2 inch to 3/4 inch.
   d. 3/4 inch to 1 inch.

8. An interior basement wall that has small cracks can be waterproofed by using

   a. a grout coating.
   b. tar compound.
   c. plaster.
   d. an epoxy compound.

9. Paint applied to foundation walls to stop the entrance of water has a base of

   a. lime.
   b. epoxy.
   c. portland cement.
   d. latex.

10. If a liquid cleaner is used to remove efflorescence from the face of masonry work, the best choice is

    a. sulfuric acid.
    b. soap and water.
    c. a caustic soda solution.
    d. a muriatic acid solution.

Unit 28 BRICK VENEERING EXISTING STRUCTURES

OBJECTIVES

After studying this unit the student will be able to

- describe different methods for installing a footing for brick veneer on an existing structure.
- describe the details of construction required when brick veneering an existing structure.
- explain how the caulking is done on an existing structure that is being brick veneered.

INTRODUCTION

Brick veneer construction against wood framing, figure 28-1, is the most popular type of masonry construction used for building single-family houses in the United States. For satisfactory performance of brick veneer, the following four points must be followed: the veneer must have a good foundation to rest upon; the frame backing must be sufficiently strong and well braced; the brick veneer must be anchored to the frame backing with wall ties nailed into the studding; and work must be done using good materials and good workmanship.

Brick veneer is not designed to support any weight other than its own. It serves as an attractive facing material that needs little or no upkeep, strengthens the house, is durable, and provides some amount of insulation.

Fig. 28-1 A new brick veneer home under construction. Notice the use of corner poles as guides for building the walls.

376 ■ Section 8 Alteration, Repair, and Maintenance of Masonry

Old buildings, particularly of frame construction, can be made to look like new by veneering with brick. Brick veneering an old house greatly increases its value and adds many years to the life of the structure.

Since the structure has been standing for a period of years, the installation of brick veneering may offer problems that would not be present on a new building. Some of these problems involve fitting to and around windows and doors, waterproofing around and over openings, and building around bulges or bumps in the old building without it showing in the finished brickwork.

Although there is a considerable amount of labor involved in brick veneering an old house or building, it is less expensive than building a new one. Brick veneering old buildings is becoming a good source of work for the mason and continues to grow in popularity because of the cost of constructing a new building.

This unit deals with some of the problems encountered when brick veneering an old building. It also describes the process of brick veneering from the beginning footing to the completed job.

FOOTING AND FOUNDATION

When the original footing around an existing structure is wide enough to carry the new facing work, the work should be started at that point. This extra weight usually does not create any problems because the weight is not enough to affect the footing. This method requires that the earth around the building be excavated to allow working room for the mason. It is also very important that the footing be cleaned free of any earth or clumps of mortar remaining from the original construction so the new brickwork can be started level, with a good bond. The footing should be below the frost line, as for any building. The frost line is determined from the building code in the particular area where the work is being done.

If the footing of an existing structure is not wide enough for the new brick veneer, then an additional footing should be placed that is wide enough to carry the new work. The new footing must be tied to the original foundation. This can be done by cutting into the masonry foundation, just above the original footing, and inserting short lengths of steel reinforcing rods. The steel rods should be grouted in solid with concrete or mortar. The new footing is then placed approximately 4 inches higher than the original footing, resting on the original footing, extending outward so it carries the new work.

Placing a new footing by merely butting against the old one may result in separation of the new footing concrete from the old footing. Figure 28-2 shows a better method of placing a new footing by tying the new footing to the old one.

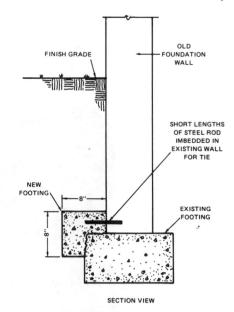

Fig. 28-2 Method of placing a new footing against an old one. Short steel rods are added to tie the new footing to the old one. The new footing should be 8 inches wide and 8 inches deep.

The addition of steel reinforcement rods inserted in the footing before placing the concrete greatly strengthens the new footing, in addition to the short tie rods walled into the old foundation wall. This method also requires that the earth be firm and packed in around the existing foundation since any filled earth can cause future settling, resulting in a crack in the new veneer above. Figure 28-3, page 378, shows various methods of constructing a new brick veneer over an old building.

Another method of providing support for the veneer is to bolt continuous noncorrodible steel angle irons to the existing foundation wall at or slightly below the finish grade line. If angle irons are used they are not bolted into the wood plate or any of the wood framing members. They should be bolted to the masonry foundation itself. This method must be used with caution because the load being placed on the irons should be calculated and the mason must be sure that the angle irons can carry the weight.

The size and spacing of the bolts anchoring the angle irons to the wall must be carefully considered, not only for the loads to be carried but also for the bearing value of the foundation wall itself. In general, this method of support for masonry veneer should be used only for one-story structures that have a total wall height (to the top plate) of not more than 14 feet. A regular steel angle iron placed near or below the grade line is subject to severe corrosion or rusting, therefore an angle iron made from noncorrosive metal or coated with bituminous materials should be used.

Another method of installing a strong base for the new brickwork is to place a continuous grade beam of concrete around the entire structure. This beam is reinforced with steel rods and is built below the frost line. The grade beam must be on firm ground if it is going to support the load. Refer to the alternate foundation detail in figure 28-3 to see how the grade beam appears in a section view.

The type of footing or foundation to be used depends upon the particular job conditions. It should only be decided after the site is inspected.

TYPE OF MORTAR TO USE

The part of the foundation wall that is below the grade line can be built of concrete block or rough brick, whichever is available and within the cost of the job. Regardless of which is used, any of the work that is below grade or in contact with the earth should be built with type M portland cement mortar. The section of the wall that is above the grade line is built of type N portland cement mortar. If masonry cement mortar is used, it should be equal to the mortars recommended here.

LAYING OUT THE VENEER WORK

Before any of the masonry units are laid out on the footing or base, a plumb bob should be dropped down, as shown in figure 28-4, page 379, from the point where the top of the wall will finish against the framework. The usual allowance for brick veneer on a new building when the framing is fairly plumb and straight is approximately 4 1/2 inches. If the framing on an existing house is bowed out and not plumb, an allowance may be required in order to lay the brick past this point without too much cutting. If most of the wall is plumb, then the bricks can be cut at the spots that bulge out. This decision is based on exactly how far the old framing work extends out.

378 ■ Section 8 Alteration, Repair, and Maintenance of Masonry

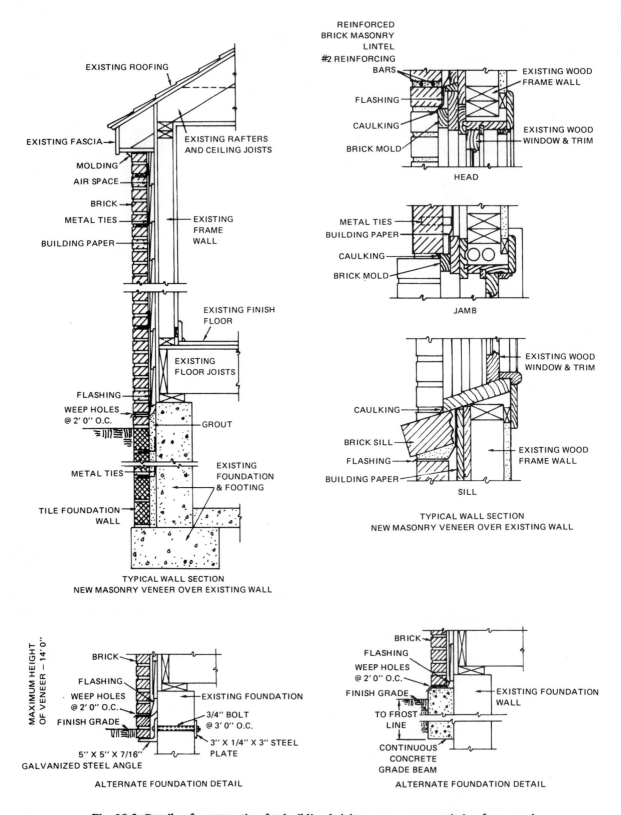

Fig. 28-3 Details of construction for building brick veneer over an existing framework.

It is also a good practice to attach a line at each corner of the house, from the footing to the soffit, before laying any brick. The brick corner can be laid to the line and the wall built from a line attached to the corners.

BUILDING THE WALL

Good workmanship should be followed to ensure that the wall is resistant to moisture penetration. Mortar head and bed joints must be full, with good bonding of the mortar to the masonry units. The mortar should be tooled as soon as it is thumbprint hard.

While the wall is being built, do not drop or fill in any mortar on the back of the bricks being laid. Mortar that is dropped behind the bricks will build up and may push out the wall. Do not attempt to cut the mortar off the back of the bricks with a trowel. Never fill in behind the wall with leftover mortar. Figure 28-5 shows an existing house that is being brick veneered.

FLASHING

The head of all wall openings such as windows or doors (except where there is a large overhang), as well as the sills of all windows and the juncture of any porch roofs, should be properly flashed. It is recommended that building paper such as 30-pound felt be applied to the framework before laying brick, to prevent

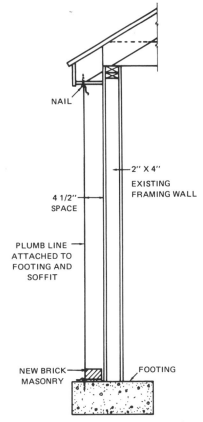

Fig. 28-4 Establishing a plumb line for a brick veneer wall. A plumb bob is dropped from the overhang to the footing. Notice the 4 1/2-inch space required for building the brickwork from the plumb line to the framework.

Fig. 28-5 Starting the brick veneer work to enclose a frame house. Notice the new foundation of concrete blocks on the end wall of the house, used for supporting the new brickwork. A chimney is also being added to the house.

dampness or moisture from seeping through the wall. Nail the building paper with a large-head nail (roofing nail) or rounded metal roof clip so it holds the paper in place without pulling out. The paper should also be nailed where the studs of the framing wall occur. This should be 16 inches on center, except in very old homes where this may differ.

In areas of heavy rainfall, it is recommended that continuous flashing be installed at the base of the wall, just above the finish grade line, and that weep holes (small holes left in the mortar head joints for seepage of water from within the wall) be provided to allow any water to drain from the inside of the wall. A short length of standard cloth line rope can be used to form the hole and then removed from the wall just before the mortar hardens. This is another reason to be careful not to get any mortar down the back of the wall or it will close up the weep holes. It is recommended to space weep holes 2 feet on center for maximum drainage. Refer back to figure 28-3 (alternate foundation detail) to see how this appears. Flashing should also be installed at the heads and sills of all windows and doors.

ANCHORING THE VENEER TO THE FRAMING

The brickwork should be tied or anchored to the framing using noncorrodible metal veneer wall ties every 16 inches in height and 16 inches apart. Ties should be nailed to the studding with an 8d common nail. The wall ties stabilize the work and prevent it from shifting. Make sure that the wall ties are well mortared in as the bricks are laid over the ties.

FRAMING AROUND OPENINGS

Steel angle irons should be used to support bricks over openings. The angle iron should bear at least 4 inches into the masonry on both sides of the opening. Angle irons can also be used over frame porches. They should follow the slope of the porch roof and be firmly fastened to the studs with lag bolts. Refer back to figure 28-3 to see how the veneering of an existing house progresses from the footing to the roof.

CAULKING

Caulking is an essential part of the construction process. Its purpose is to fill up any cracks around windows, doors, and other openings, and to keep moisture out.

Outside joints at the edges of doors and window frames should be 1/4 inch to 3/8 inch wide, and should be cleaned to a uniform depth of at least 3/4 inch before caulking. These joints are filled with an elastic caulking compound that is forced into place with a pressure caulking gun.

If the joints to be filled are larger than stated, a piece of wood molding is nailed to the frame and the correct spacing is left for the caulking material. It is difficult, if not impossible, to fill a joint of less than 1/4 inch to a depth that will stay. In a large joint (1/2 inch and over) the caulking compound sags and pulls away from the edges of the joint. In cold weather, caulking tubes must also be warmed up to at least room temperature before they are used.

Joints between masonry, door and window frames, and expansion joints are the most common places where rain penetration occurs. If the joints are deeper than stated, oakum, fiber filler, or styrofoam rope can be used as backing to press the caulking material against.

Fig. 28-6 Brick veneer being applied over siding on an existing frame house. Notice how the old steps are walled in with brickwork to form a brick porch. The addition of brick veneer helps insulate the house, and requires very little maintenance.

Caulking is one of the important jobs in finishing the brick veneering of an existing building. Figure 28-6 shows a frame building in the process of being veneered. Notice how the mason has built in the existing porch with bricks.

PROTECTION OF MASONRY WHILE UNDER CONSTRUCTION

Partially completed masonry walls that are exposed to rain may become so saturated with water that they require months to dry out. This can cause efflorescence or water soaking into the framing.

While the brickwork is being built, it is the responsibility of the mason to be sure that the walls are covered at all times (when not being worked on) to prevent this. The covering can be of plastic, canvas, or some other suitable material that not only covers the top of the wall but hangs over at least 2 feet on the face. It should also be weighted down to prevent the wind from getting under it and damaging the wall. The common practice of laying a heavy board on top of the wall at the end of the workday does not keep the work protected and can cause the masonry underneath to sag or bow out of position.

INSULATION

The addition of brick to an existing wall lowers the cost of heating a building. The air space that is trapped between the brick and the framing acts as an insulator. There are also other insulating materials that can be built against the framing of the wall to help insulate the structure.

Brick veneering an existing structure is a practical and economical way to rehabilitate an old structure, while adding many years to its expected life. The wide range of colors, sizes, and designs of brick allow the builder and homeowner an opportunity to create a new house out of an old one. Brick veneering of old buildings also opens up another source of work for the mason.

Figure 28-7, page 382, is an example of how a house would look before and after brick veneering. One would have to look closely to realize it is the same house.

382 ■ Section 8 Alteration, Repair, and Maintenance of Masonry

(A) Before brick veneering

(B) After brick veneering

Fig. 28-7 These two illustrations show the same house, before and after brick veneering.

ACHIEVEMENT REVIEW

Select the best answer from the choices offered to complete each statement. List your choice by letter identification.

1. The recommended spacing for brick veneer from the exterior of the framing to the face of the brickwork is
   - a. 4 inches.
   - b. 4 1/2 inches.
   - c. 5 inches.
   - d. 6 inches.

2. For a brick veneered job on an existing building, the minimum depth for the footing (below the frost line) should be
   - a. 6 inches.
   - b. 8 inches.
   - c. 10 inches.
   - d. 12 inches.

3. If masonry work for brick veneer is in contact with the earth, the correct mortar to use is type
   - a. N.
   - b. O.
   - c. M.
   - d. S.

4. The correct mortar to use above grade for the veneer work is type
   - a. N.
   - b. M.
   - c. O.
   - d. S.

5. The best method for determining the plumb line of the wall when veneering a wall on an existing building is to use a
   - a. transit.
   - b. level or plumb rule.
   - c. plumb bob.
   - d. straightedge.

6. Before the brickwork is veneered around an old building, the exterior of the framed wall should be protected with
   - a. copper sheathing.
   - b. plastic covering.
   - c. building paper.
   - d. plaster on wire lath.

7. Wall ties should be fastened to the framing work to tie the veneer masonry to the building. The correct spacing in height is every
   - a. 12 inches.
   - b. 16 inches.
   - c. 18 inches.
   - d. 24 inches.

8. Weep holes to drain the moisture from a veneered wall should be spaced in the head joints just above the grade line. The correct spacing is
   - a. 16 inches apart.
   - b. 24 inches apart.
   - c. 32 inches apart.
   - d. 48 inches apart.

9. Wall ties should be fastened to the studding with
   - a. 4d nails.
   - b. 6d nails.
   - c. 8d nails.
   - d. 12d nails.

10. The angle irons that are laid over openings should bear on the brickwork a minimum of
    a. 4 inches.
    b. 8 inches.
    c. 12 inches.
    d. 16 inches.

11. The opening around windows and doors should be filled with
    a. mortar.
    b. insulation.
    c. caulking compound.
    d. wood strips.

12. When the brickwork is covered at the end of the workday, the covering should extend down over the face a minimum of
    a. 12 inches.
    b. 24 inches.
    c. 36 inches.
    d. 48 inches.

SUMMARY – SECTION 8

- The biggest enemies of masonry work are penetration of moisture and changes of temperature.
- The portland cement mortars used today are a great improvement over the lime and sand mortars of the past.
- The repair and renovation of old masonry work provides the mason with a necessary source of income.
- Old paint can be removed from brickwork with a caustic soda (lye) solution.
- Safety precautions should be taken whenever using caustic soda for removing paint from brickwork.
- All old, loose mortar should be removed from the joints before repointing.
- Mortar joints that are to be repointed should be cut to a minimum depth of 1/2 inch before repointing.
- The mortar joints should be dampened with water before pointing in with fresh mortar.
- Soft bricks that are flaking or cracking should be replaced with new bricks when repointing an old brick structure.
- It is important to use a portland cement/lime based mortar that is not too rich for the best results when repointing. The new mortar does not have to support a load but it must bond well to the bricks.
- Type N portland prehydrated cement mortar is recommended for repointing work.
- The most practical method of repointing mortar in old masonry work is to dampen the joint and pack or tuck the mortar in the joints with a flat metal tool called a slicker.
- Fireplaces that are old and have not been used for a long time should be inspected before use to make sure they are safe.
- Flue linings are always installed when rebuilding a fireplace or chimney.

- Well-filled mortar joints and good workmanship should be used when rebuilding old fireplaces or chimneys.
- When the fireplace in a home is rebuilt, the masonry work should be kept as clean as possible.
- A solution of 1 part muriatic acid and 9 parts water can be used to clean the fireplace, providing it does not ruin the interior of the house.
- If a chimney top is too badly damaged to be repointed, it should be torn down and rebuilt.
- Great care should be taken when erecting scaffolding and preparing to repair a chimney top. The protection of the property and the safety of any people who pass by on the ground level must be considered.
- A smoke test should be performed on a rebuilt fireplace and chimney to see if any of the flues leak.
- Excessive deposits of soot should be cleaned from old chimneys to prevent chimney fires.
- Repairs of cracks, removal of efflorescence, and dampproofing are the most frequent repairs needed for masonry structures.
- Expansion cracks in masonry walls are caused by differences in temperatures and changes in seasons.
- Cracks in masonry work should not be repaired for a period of one year because the crack may still be moving.
- When repairing cracks, the cut out place should be dampened with water to promote better bonding between the masonry units and the mortar.
- Expansion cracks are usually in a step fashion; settlement cracks are in a vertical position.
- Cracks in masonry walls are also caused by other parts of the structure that have a different expansion rate, such as masonry units walled in solid around steel columns.
- Efflorescence can be removed by a stiff brush or with a solution of muriatic acid and water.
- Safety should be exercised at all times when handling acid solutions.
- Silicones are dampproofers, not waterproofers.
- Silicones are obtained as a clear chemical liquid that repels water by changing the contact angles between water and the walls of capillary pores in the wall.
- Silicones are applied to a wall either by spraying or brushing.
- The most important advantage of silicone is that it keeps water from entering the wall while allowing moisture inside the wall to evaporate.
- Portland cement paints help to dampproof basement interior walls and also can add color to the walls.
- Problems with wet, leaky basements can be greatly reduced if the foundations are properly waterproofed when constructed.

- Brick veneer construction is the most popular method of building a masonry house in the United States.
- Brick veneering consists of building brick walls around an existing frame structure.
- When veneering an existing structure, it is important that the brickwork is started on a good footing or base.
- Footings for brick veneer over an existing structure can be done in three different ways: placing on the old footing if there is room; placing a new footing next to the old one; or bolting angle irons to the old foundation wall and laying the masonry units on the irons.
- Type M portland cement mortar should be used below grade and type N portland cement mortar is used above grade for brick veneer over an existing structure.
- A plumb line is established for the brickwork from the footing to the top plate before any bricks are laid. This is to ensure that there are no obstructions.
- Flashing is installed over all openings, windowsills, and under masonry courses that contain weep holes.
- Building paper should be nailed to the existing frame wall before brick veneering.
- The brick veneered walls are anchored to the framing by wall ties nailed to the studding.
- All open cracks or spaces around doors, windows, and other openings are filled with a caulking compound.
- The addition of brick veneering not only improves the appearance of the structure but also helps to insulate it.

SUMMARY ACHIEVEMENT REVIEW, SECTION 8

Complete each of the following statements referring to material found in Section 8.

1. Paint can be removed from old brickwork by applying a solution of _____.
2. The correct chisel to use when cutting out mortar joints for repointing is the _____.
3. The mortar joints of an old wall that are being cut out for repointing should be cut out to a minimum depth of _____.
4. The reason for prehydrating mortar for pointing work is to decrease the chance of _____.
5. The mortar joints are dampened with water and a brush to slow the _____ of the mortar.
6. To reduce the risk of the chimney burning out it should be lined with _____.

7. All woodwork surrounding a chimney must be protected from heat or fire by walling around with brick and mortar to a thickness of _____.

8. The major cost involved in preparing to repair an old chimney is the _____.

9. A smoke test is performed on a rebuilt chimney to determine if _____.

10. Old flues of existing chimneys can be cleaned of soot by using _____.

11. The most popular type of brick construction for a home in the United States is _____.

12. Footings for brick veneering an existing house must be below the _____.

13. When a footing is not used for brick veneering an old structure the alternate practice is to use _____.

14. The recommended mortar to use for brick veneer work below grade is _____. The mortar for brick veneer work above grade is _____.

15. A plumb line is attached from the footing to the soffit of the building as a guide for the corner. This line is found by _____.

16. _____ should be installed over the head of all windows and doors for protection from moisture.

17. Brick veneer walls are tied to the framing with _____.

18. Brickwork is supported over window and door openings with _____.

19. Caulking material is filled in around openings with a _____.

20. The movement of masonry work caused by changing temperatures is called _____ and _____.

21. When a crack appears in a masonry wall it should not be repaired for one year. This is because _____.

22. Walls that extend above the roof of a building are called _____.

23. Settlement cracks almost always appear in a _____ position.

24. When concrete is placed on a masonry wall and then shrinks causing a crack, this is known as _____.

25. A clear dampproofing solution that is brushed or sprayed on masonry walls and allows the wall to breathe is called _____.

26. If small hairline cracks are present in the interior of a basement wall, before the masonry portland cement paint is applied, the cracks should be sealed with a _____.

Section 9
Trends and Developments in Masonry Construction

Unit 29 CONSTRUCTING MASONRY IN COLD WEATHER

OBJECTIVES

After studying this unit the student will be able to
- explain proper cold weather storage of masonry materials.
- describe the procedures for preparing mortar and masonry units before they are laid in the wall in cold weather.
- describe the various types of protective shelters used to protect masonry work.
- list the most important factors to consider when building masonry work in cold weather.

INTRODUCTION

When masonry construction is done in temperatures below 40 degrees Fahrenheit, the problems are different from those found at other times in the building year. The coming of winter, which brings cold and freezing temperatures, must be considered when building masonry walls. Not only can the work be damaged from freezing weather but the mason is also affected. The mason must dress differently and adjust to the cold temperatures. A mason's productivity decreases considerably with the coming of cold weather unless the job is winterized.

For years building materials and labor have been in short supply in the warm months of the year. During the cold months, however, there is an abundance of materials and labor. This imbalance causes winter layoffs, so that many people in the construction industry are out of work in the colder months. This is a drain on the nation's economy and a loss to the mason.

The matter of combatting cold weather has been a major concern to the government, labor, and builders for many years. Since all of the trades must cooperate with each other to create a structure, anyone who is not working eventually affects the others.

In many countries that have long winters and prolonged freezing temperatures, programs have been developed so that building can continue in the cold months without sacrificing good construction principles. European programs have been developed that consider factors such as planning and scheduling the work; excavations; heating the masonry materials; storing the materials on the job in a way that frost, ice, and snow cannot penetrate; heating the structure; and protecting the masonry during and after construction.

The mason must learn to work in colder temperatures to achieve full employment. Construction work does not have to stop due to cold weather. However, the same types of planning and methods that work in the warm months do not work in the cold months. Construction methods are different depending on the size of the structure, its design, how close it is located to protected areas or windbreaks, and the weather conditions in a particular area. Therefore, no one solution serves all masonry structures.

The single most important factor for the mason to understand in cold weather is how to prevent the mortar and materials from freezing or being damaged. All other points of information revolve around this major concern. Also, in order to work, a worker must be relatively comfortable. In winter weather, this means being warm enough to perform the tasks associated with the laying of masonry units.

The information in this unit deals with these problems. This unit describes how planning ahead and learning to live with the weather can make the masonry trade a full, year-round job.

STORAGE OF MATERIALS

When stored on the job, all masonry units and mortar materials should be completely covered with tarpaulins or plastic. They must also be stored off the ground so moisture cannot be absorbed into the materials. Laying loose board planks or plywood on top of the brick or block piles does not adequately protect them. Masonry materials should not be left unprotected to be coated with ice or snow. Careless storage of materials increases the cost of laying the masonry because it is necessary to remove ice and snow and completely thaw the masonry units before construction can be started. Figure 29-1 shows an example

Fig. 29-1 An example of masonry materials that have been poorly covered on the job in winter. Notice how the snow is laying on the concrete block and lintels. No provisions were made to keep the work going during the cold weather.

of materials that have not been covered properly on a winter job. Notice the snow on the concrete masonry units.

Masonry materials should also be located close to the area where the work is being done. When board runways must be built and materials moved a long distance, the mason's productivity is lessened. This is especially true if there is a heavy snowfall.

If the site of the job allows, it is a good practice to build an overhead shelter for the storage of masonry materials. The sand should be covered with a canvas or plastic covering to prevent the entrance of moisture from rain or snow. Once the sand is frozen, it is very difficult to remove from the pile and must be thawed before using.

PREPARATION OF MORTAR

As the temperature falls below freezing, the different ingredients in mortar become colder. Mixing the mortar thus becomes more difficult. The heat-liberating reaction between the portland cement and water is slowed or stopped when the cement paste is subjected to temperatures below 40 degrees. Hydration and the strength of the mortar depend on the temperature being above freezing and enough water being available. However, masonry work can continue in cold weather in below freezing temperatures. The mortar ingredients must be heated and, as the surrounding temperature decreases, the masonry units and the building under construction must be kept above freezing temperature during the critical hours after the masonry has been laid and set in the mortar.

When mortar is mixed from materials that are cold, but not frozen, it possesses plastic properties that are different from mortar mixed when the temperature is warm. Mortar mixed at low temperatures has a lower water content, longer setting and hardening times, higher air content, and lower early-age strength. When materials are heated for the mortar mix, the mortar has the same characteristics as mortar in the normal temperature range. Therefore, it is recommended to heat the sand and water that will be used for mortar in cold weather.

HEATING SAND

Masonry contractors should provide heated sand for all masonry work done at temperatures below 32 degrees whether it is specified or not. Even if it is covered, sand always has some moisture in it if stored outside on the job. The moisture turns to ice when the temperature drops below freezing. Therefore, the sand should be heated before using it to remove the ice. Sand should be heated slowly and evenly to prevent scorching. Scorched sand has a reddish color and should not be used in mortar.

Proper heating techniques can be accomplished by piling the sand around a horizontal metal pipe, barrel, or smokestack section. (Be sure the drum or barrel is clean and free of any foreign materials that can affect the mortar.) After piling the sand around one of the above items, a slow-burning fire is built in it, using wood scraps found on the job. As the sand thaws, all of the ice is removed, and the sand is heated, making it suitable for mixing.

Manufacturers have developed and marketed devices for heating sand and water, such as the one shown in figure 29-2. A gun-type oil burner is controlled by an aquastat that regulates the heat and prevents overheating. Sand is dumped over the heated tube and both the sand and water are heated and kept warm automatically. This device eliminates the need to start fires in the morning on the job. The mason has warm mortar to start off the day.

HEATING WATER

An easy method of increasing the temperature of mortar is to heat the mixing water. There are several methods of doing this, in addition to building a fire around the water barrel. One of the fastest methods is the injection of steam (where available). Regardless of the type of heat supplied to warm the water, it should not add any materials to the water, such as wood ash, that can cause problems in the setting and hardening of the mortar. One of the most important points to remember when heating water is that the mixing water should never be above 160 degrees Fahrenheit (F), because of the danger of a *flash set* (rapid setting of cement caused by adding water that is too hot) when the portland cement is added.

After combining all of the ingredients, the temperature of the mortar should be between 70 degrees F and 120 degrees F. If the mortar temperature is over 120 degrees F, excessive hardening may occur, resulting in lowered compressive strength and reduced bonding to the masonry units.

It is also a good idea, at the end of the workday, to drop a block of wood in the water barrel. This is a great help the next morning when removing any ice that has formed.

Fig. 29-2 Oil-fired device with a metal tube used for heating sand and water.

MIXING THE MORTAR

Once the mortar is mixed in a mechanical mixer, it should be removed and transported to the mason before it cools. Dumping the mortar in a mortar box or wheelbarrow, letting it sit until chilled, and then taking it to the mason defeats the purpose of heating the mortar.

On small jobs in which the steel mortar box is used, the box should be raised off the ground and leveled on bricks or concrete blocks. Small fires, built of wood from waste scraps, can be started under the box. This keeps the mortar warm until it is delivered to the mason. Mortar should not be mixed far ahead of time and left in the box. Usually a good rule to follow in cold weather is to mix only enough mortar beforehand that can be used in a one-hour period.

CEMENTITIOUS MATERIALS USED IN THE MIX

Cold weather does not require any drastic changing of the mortar mix. The mortar should conform to the requirements of the Standard Specifications for Mortars and Unit Masonry (ASTM C270) during both normal and below normal temperature construction. If a portland cement mortar is to be used, it may be advisable to substitute type III, high early strength portland cement for type I, normal portland cement. Type III provides greater internal protection for the mortar because of its more rapid early setting. This setting occurs from the time it is mixed through 72 hours or before freezing. It is recommended for use in all types of cement-lime mortars and in type S and M portland cement, masonry cement mortars. It is important, however, not to let type III mortar freeze before the cement sets.

LIME

Lime in the dry form, called *hydrated lime,* is used in the mix. The older type of slaked lime causes problems due to the moisture it contains, and the slaking process makes it undesirable for cold weather construction.

ACCELERATORS AND ANTIFREEZES

Admixtures of antifreezes added to lower the freezing point of mortars should not be used. The amount of such materials needed to lower the freezing point of mortar is so great that it would have harmful effects. The mortar strength may be lessened. Too much salt added as antifreeze can contribute to efflorescence and may cause spalling through recrystallization. The reason that most commercial antifreeze compounds are effective is because they act as accelerators (speeding up the setting action), which in most cases results from the calcium chloride they contain.

In the past, calcium chloride was used, causing the mortar to set faster during the protected periods. Recent studies indicate that metals imbedded in mortar corrode faster by adding calcium chloride. Therefore, when metal ties for bonding masonry are used or when other metal objects are in the mortar joints, the addition of calcium chloride to the mortar is not recommended.

When calcium chloride is used in mortar it should never be added in amounts greater than two percent (by weight) of the portland cement. It is added to the mixing water and then put in the mix. If the water is too hot and calcium chloride is added, this causes the mortar to set too fast.

It should be remembered that the adding of calcium chloride to mortar does not replace other protective methods. Its real value lies in the rapid set and strength gain in the mortar. The chemical action keeps the mortar warmer during this process.

PREPARATION OF THE MASONRY UNITS

To prevent the sudden cooling of warm mortar in contact with cold masonry units, it is recommended that all masonry units be heated when the temperature is below 20 degrees Fahrenheit. The masonry units should be heated to about 40 degrees Fahrenheit. If the weather is damp and cold, heating the masonry units increases the mason's work output.

Wet, frozen masonry units should be thawed without overheating. If the masonry unit is too hot, the mortar sets too fast, resulting in a loss of bond strength.

One of the most effective methods of heating masonry units uses oil-fired blowers that are directed against the pile of masonry units. This movement of air warms the units quickly. The major problem with this treatment now is the high cost of oil and fuel. Enclosing the masonry units in a plastic covering that is subject to the sun's rays uses solar energy heat to prevent the units from getting too cold. However, solar heat cannot be used alone to heat frozen masonry units, while the oil-fired blower is effective alone.

During cold weather, bricks having a high rate of absorption should be sprinkled with warm or hot water just before being laid. In this way they do not absorb the water from the mortar too fast. Those units with low absorption rates can be laid dry, with no prewetting.

BUILDING ON A SOUND BASE

Masonry work should never be built on a snow- or ice-covered base because no cementitious bond can be established. When the base thaws out there is the danger of movement.

If the walls are properly covered when the work is stopped at the end of the workday, there will be no accumulation of ice on the base. If there is ice on a footing, or if the covering has blown off and ice or snow is present, the ice must be removed before any masonry can be laid. A portable blowtorch or live steam can be used to thaw and remove the frozen material. The heat should be concentrated on the area long enough to dry out the base or masonry before attempting to lay any masonry. It is also important to replace any frozen or damaged masonry units before building on top of the existing work.

It is recommended to cover masonry work at the end of the workday to prevent snow, sleet, or rain from entering. The protection should cover the top and extend a minimum of 2 feet down all sides of the masonry. Weight down the covering or tie so it cannot blow off.

PROTECTING THE MASON AND MASONRY WORK

Chill Factor

One of the most difficult problems when constructing masonry work in cold weather is the chill factor. The *chill factor* is a combination of the temperature plus the wind speed, so that the air feels colder than the thermometer indicates.

Driving winds do not allow the body to retain its normal heat. This causes discomfort to the masons, affecting their ability to work, and thus production is lessened.

The development and use of lighter, warmer, insulated clothing has relieved this problem in recent years. The insulated clothing allows greater freedom of movement while still providing warmth. However, when the weather is so bad that insulated clothing does not help, some type of windbreak or enclosure must be constructed so the work can continue.

Protecting The Wall With A Tarpaulin

An enclosed site for masonry, if maintained at a temperature of 40 degrees F or greater, is ideal for building masonry work in cold weather. The important considerations are to protect the masonry from freezing, and to allow the masonry to cure properly without any damaging side effects. In addition, masons must be comfortable enough so that the cold weather does not affect their productivity. The work must be done on schedule in order for the other trades to begin their work.

The simplest type of protection is a tarpaulin covering the masonry wall. This is used when the mortar is warmed, and the daily temperature is above 25 degrees F. The tarpaulin protects the wall from the cold, allowing the mortar to set properly.

When the temperature is between 20 degrees F and 25 degrees F, an insulating blanket can be used for the completed walls. Metal stack heaters, called *salamanders,* burn oil and may be used to decrease the chilling effect of the cold weather on the work and the mason.

Windbreak

When winds are above 15 miles per hour (mph), and temperatures are below 25 degrees F, windbreaks should be provided. A windbreak helps to prevent a rapid loss of surface heat from the masonry wall being built.

Windbreaks can be built of sheets of plywood braced around the work, sheets of polyethylene (plastic), canvas, or similar materials. Any of these can effectively reduce the chill factor.

> When erecting any type of windbreak, care should be taken to properly construct and brace it. If improperly constructed, a windbreak could collapse or blow over, possibly injuring the mason or damaging the masonry work.

Enclosing The Structure

The entire structure and area where the mason is working can be protected by enclosing it with a variety of materials. This is the most efficient method of winterproofing a building. The cost of the enclosure must be considered beforehand to determine if it is worthwhile for the particular job. Some of the important points to consider when enclosing a masonry structure are the severity of the weather, type of masonry work being built, type of scaffolding being used (if any), type of building and design, and force of winds expected against the covering.

The types of materials used for enclosures are canvas, synthetic coverings (reinforced polyethylene and vinyl), and lumber (plywood or insulated sheathing). Stretched sheets of plastic, figure 29-3, usually cannot withstand the wind unless in frames and braced into position. A technique that has gained wide acceptance is the use of prefabricated plastic panels, figure 29-4, page 396, which are covered with polyethylene reinforced for greater strength. The carpenters on the job build the panels and these panels are braced into position. The transparency of the plastic permits solar radiation to help heat the enclosed space. It also allows light to filter into the work area, reducing the amount of temporary lighting needed.

When selecting enclosure materials, consider the following features: strength, durability, transparency, fire resistance, flexibility, and ease of installation. Often, an unheated enclosure alone reduces the chill factor and is all that is needed to protect the masonry work if the materials are heated.

Enclosing Scaffolds

Scaffolding can also be enclosed in cold weather to protect the walls and workers. Enclosures for scaffolding are constructed and secured around the scaffolding in the same

Fig. 29-3 Sheets of unreinforced plastic were used to cover the windows in this building under construction. Notice how the wind has torn the plastic loose from the wood frames.

Fig. 29-4 A building under construction that is enclosed in panels of plastic, allowing the continuation of the work in cold weather.

way as an enclosure for a building. The safety of the workers must always be considered when working in enclosed areas, with regard to the dangers of high winds, fires, and air contamination. The scaffolding should be properly built and braced, and anchored to the structure so that it is safe for loads such as snow, materials, and workers. Swing scaffolds are completely enclosed for weatherproofing, with an opening or flap to allow the flow of materials to the mason. Figure 29-5 shows a scaffolding enclosed with plastic panels.

When The Exterior Masonry Shell Has Been Built

If the exterior masonry walls have already been constructed, and the interior partitions are being built, then the windows, doors, and all openings can be covered, and a few

Fig. 29-5 The scaffolding of this building under construction is enclosed in plastic panels. This greatly reduces the chill factor and prevents the masonry from freezing.

carefully placed heaters can be used to protect the masonry. Oil-fired blowers (heaters) provide the most effective method of heating because they move the hot air quickly throughout the structure. Remember that walls should not be heated only on one side with no protection on the opposite side. The warm air must be circulated on both sides of the wall.

SUMMARY OF RECOMMENDED PRACTICES FOR BUILDING MASONRY IN COLD WEATHER

The following points are considered to be the most important factors in laying masonry work in cold weather.

- Planning and scheduling of the work must be done beforehand if masonry work is to be built in cold temperatures.
- Take advantage of warm days by working on the outside of the structure, saving the inside work for the colder days.
- Store all masonry units close to the structure. Be sure units are covered and off the ground to prevent moisture or frost from penetrating.
- Build a mortar mixing area covered with a roof. Keep the sand pile covered to prevent moisture, ice, or snow from penetrating.
- Preheat the sand and water before mixing the mortar.
- Use type III, high early strength portland cement for a quicker set.
- Do not use antifreeze in the mortar. Calcium chloride is considered to be an accelerator, not an antifreeze. Never use calcium chloride if there is metal in the mortar joints.
- Preheat the masonry units before laying them if the temperature is very cold or if frost or ice is present.
- Build protective shelters such as windbreaks and enclosures to protect the mason and the masonry work.
- Observe good safety practices when building shelters to prevent them from collapsing or blowing over, causing damage and injury.
- Take protective measures at the end of the workday to protect the work and to ensure that work is started on time the next day. Some protective measures are covering the work and piles of materials, draining the hoses, cleaning out the mortar pans, and placing a block of wood in the water barrel so the water does not freeze.
- Think ahead to the next day, and even the next week, about things that need to be done to keep the job running smoothly. Keeping the mason working all winter helps to build morale and keeps the mason on the payroll the entire year.

ACHIEVEMENT REVIEW

Select the best answer from the choices offered to complete each statement. List your choice by letter identification.

1. The heat-liberating reaction between portland cement and water is either slowed down or stopped when the temperature drops below
    a. 50 degrees F.
    b. 46 degrees F.
    c. 42 degrees F.
    d. 40 degrees F.

2. The mason should heat all sand to be used in mortar when the temperature drops below
    a. 45 degrees F.
    b. 40 degrees F.
    c. 36 degrees F.
    d. 32 degrees F.

3. Sand that is overheated becomes scorched. Scorched sand is usually
    a. blue.
    b. reddish.
    c. brown.
    d. orange.

4. When heating water for mortar it is very important that the mixing water is not heated to more than
    a. 95 degrees F.
    b. 110 degrees F.
    c. 125 degrees F.
    d. 160 degrees F.

5. If an earlier set is desired for portland cement mortar used for laying masonry units in cold weather, use type
    a. I.
    b. II.
    c. III.
    d. IV.

6. Calcium chloride is used in mortar because it
    a. prevents the mortar from freezing.
    b. delays the set of the mortar.
    c. speeds up the setting time of the mortar.
    d. prevents the mortar from changing color and makes the mix stronger.

7. The objection to using calcium chloride in a mortar mix is that it
    a. bleaches the mortar joints white.
    b. weakens the mortar strength.
    c. causes corrosion of metals in the mortar joint.
    d. causes efflorescence in the masonry.

8. Metal stack heaters that burn oil are used to heat the working area for masonry work. This type of heater is known as a
    a. convector.
    b. salamander.
    c. injector-type heater.
    d. blower.

9. The most efficient method of reducing the effect of cold temperatures is by

   a. building windbreaks of plastic.
   b. framing in with plywood.
   c. enclosing with plastic panels.
   d. installing heaters around the wall.

10. It is recommended that all masonry units be heated when the temperature is below

   a. 40 degrees F.
   b. 30 degrees F.
   c. 32 degrees F.
   d. 20 degrees F.

Unit 30 ENGINEERED BRICK MASONRY

OBJECTIVES

After studying this unit the student will be able to

- explain the advantages of brick bearing wall construction in comparison to steel and concrete frame construction.
- describe how brick bearing walls are constructed for multistory buildings.
- list the important factors to be considered when building engineered brick masonry.

INTRODUCTION

There is a growing trend in the construction industry to return to load-bearing brick masonry walls for buildings. The concept behind brick bearing masonry walls is that the brick wall supports the floors, contents, roof, and structural weight of the building. This type of construction is becoming increasingly popular because of the discovery that brick bearing wall buildings can be built economically, with relatively thin walls and at a rapid pace.

The first step in building brick bearing walls is to build the brick wall to the height of the first story or where the concrete floor is to be installed. After it cures enough to support the weight of the concrete floor, the floor is placed, or concrete floor planks are laid so that they rest on the back half of the wall. The brickwork is then continued, working from the floor until the second floor level is reached. Then the process is repeated. The brick bearing wall supports the building without the need for steel or reinforced concrete columns.

Before 1965, the design of a building was based primarily on past experiences and the building codes for minimum wall thickness and maximum height. As a result, bearing wall construction for buildings higher than three to five stories was considered uneconomical, and other methods of support (steel or concrete skeleton frame) were used.

Since 1965, the design of modern masonry wall construction has been based more on reasonable structural analysis rather than on outdated absolute requirements. This type of brick bearing wall construction is known as *engineered brick masonry*. Much of the interest shown was focused on European construction practices of building many load-bearing brick buildings higher than ten stories.

One of the tallest brick bearing wall buildings was constructed in Zurich, Switzerland. This 18-story building has 6 to 10 inch thick interior load-bearing brick walls. The exterior walls are 15 1/4 inches thick; the thickness in this case was determined by the requirements for thermal insulation rather than structural requirements. In Grenchen, Switzerland, there is a 16-story apartment building which utilizes cavity wall construction, figure 30-1. The exterior walls of this building consist of a combination of a 6-inch wythe brick inner bearing wall, a 1 1/2-inch cavity, and a 5-inch wythe exterior brick wall. The floor loads are carried by the 6-inch inner wythe and the 6-inch interior brick bearing partitions. The 5-inch outer wythe of wall is self-supporting on the foundation. This outer wythe of wall is tied only at each floor level with 1/4-inch stainless steel anchors embedded in the edge of the concrete slab approximately 20 inches on center. Another example of a high-rise brick bearing wall building is a 14-story building in Lucerne, Switzerland, in which the outside bearing walls are only 7 1/4 inches thick. Figure 30-2 shows the construction of this building at the floor line. Regardless of the thickness of the wall or the height of the building, the floors must be supported by the bearing walls.

An example of high-rise engineered brick masonry in the United States is in Denver, Colorado, where a 17-story high-rise building was constructed. There have also been other brick bearing wall buildings built. The information in this unit deals with the construction techniques and practices in engineered brick masonry.

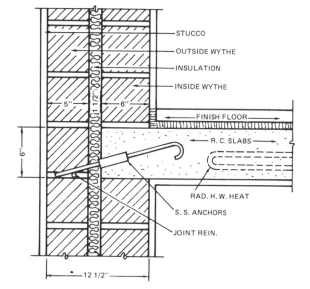

Fig. 30-1 A brick bearing cavity wall shown at the floor line. Notice the use of insulation in the center of the wall.

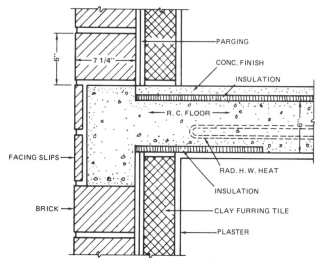

Fig. 30-2 Section view of a 7 1/4-inch exterior bearing wall. This building is 14 stories high. Notice how the concrete floor ties into the masonry wall.

BUILDING CODES

In 1966, the Structural Clay Products Institute (now called the Brick Institute of America) published the first standard for the rational design of brick masonry. The standard was widely adopted by building code authorities and was made a part of each of the model codes, and many state and local codes. The Institute has continued its research and publishes new reports on the subject of brick bearing walls as new findings are made.

COMPARISON OF BRICK BEARING WALLS WITH STEEL AND CONCRETE FRAME CONSTRUCTION

With the return of bearing wall masonry to multistory construction, new masonry techniques and methods were necessary. It is also necessary to have total cooperation between the mason and general contractor if the job is to move smoothly.

To examine brick bearing wall construction techniques, comparison can be made with two other main types of construction practiced in the United States. These types are steel frame construction and reinforced concrete frame construction. In steel frame construction, figure 30-3, the steel is erected story by story. In many cases the floor is set as the framework goes up. Work always slows down as the concrete floor is placed on a steel framework building. Then the exterior walls are built before other building trades can complete their work.

In concrete frame construction, figure 30-4, more effort must be expended by the carpenters building the forms. A maze of lumber and reinforcement steel is needed to build a reinforced concrete frame structure. After the concrete forms are in place, the concrete is placed and left to cure. It is usually one or two weeks or longer (depending

Fig. 30-3 Steel framework building in the process of construction.

on the weather), before the forms can be removed and the exterior walls constructed.

In brick bearing wall construction, figure 30-5, the masons set and regulate the pace for the building. They build the walls for each story or portion of a story in one procedure. Then the floor is installed either by placing concrete or setting precast concrete floor planks resting on the bearing wall. After the floor is completed, the brickwork is continued. The mason works off the floor and lays the brick overhand. This greatly reduces the amount of scaffolding needed. Scaffolding is built on the concrete floor, and the brickwork is built up to the next floor level.

IMPORTANCE OF SCHEDULING THE MASONRY WORK WITH OTHER TRADES ON THE JOB

Building masonry walls and installing concrete floors requires careful scheduling by the contractors involved, if the job is to be finished on time. As soon as the floor is in place and has cured, the other trades can start working on this floor. This is a great advantage because the bottom floors of the structure can be finished sooner resulting in early use or rental of the structure.

TYPES OF BRICK UNITS

Although different types of masonry materials can be used, this unit deals primarily with brick walls. Standard brick may be used; usually, larger masonry units reduce the cost of building the wall up to a limit. However, brick that is too heavy or too large for one mason to easily lift and lay is not the most economical to use. For single-wythe brick bearing wall construction, a very popular size brick unit is the 6-inch through-the-wall unit. This

Fig. 30-4 Constructing a building using concrete form construction.

Fig. 30-5 Mason building a single-thickness masonry bearing wall.

unit is also available in a 12-inch length, and in either a standard height or a 4-inch height. Figure 30-6 shows a typical load-bearing brick used in brick engineered masonry.

MORTAR

Most of the mortar tests that have been performed for engineered brick masonry are based on portland cement/lime mortars. The standards for this type of work provide for three different types of mortars: M, S, and N, as described in ASTM C270. However, in this type of work, the mortar must consist of portland cement (type I, II, or III), hydrated lime (type S, non-air-entrained), and aggregate (sand) when the allowable stresses included in the standards are used. The standards also provide that "other mortars may be used when approved by the building official, providing strengths for such masonry construction are established by tests." Specific formulas for mortars are usually contained in the job specifications.

Fig. 30-6 The man in this figure is holding a typical single-wythe, load-bearing brick unit used in the construction of a load-bearing wall.

When designing a multistory bearing wall building, it may be necessary to specify a high-strength mortar to develop the required wall compressive strengths, for at least the lower stories of the structure. This decision is the responsibility of the designer and architect. After the mortar is specified, it becomes the responsibility of the mason to make sure the mortar is mixed correctly according to the specifications. In engineered brick masonry, proper supervision and control of the batching and mixing is of great importance. No revisions or changes should be made by the mason on the job. If the design of the structure requires a specific strength of masonry units and mortar, any deviations from those requirements without the approval of the designer could result in a structural failure.

Remember that there is no single mortar that is best for all purposes. For a particular project, the basic rule for the correct selection of mortar is the following: do not use a mortar that is stronger (in compression) than is required by the structural requirements of the project. This general rule must be used with good judgment, since it would be impractical to change mortar types for various pieces or parts of a structure. However, the basic idea should be followed.

High-bond mortar with special additives for greater strength can also be used, if needed, in brick bearing construction. This requires an even greater control in the batching and mixing procedures. Careful supervision is also very important in this part of the masonry building process.

USE OF PREFABRICATED METAL STAIRS IN BRICK BEARING WALL CONSTRUCTION

Sometimes prefabricated steel stairs are installed ahead of the masonry work. These stairs provide guides for story heights, wall line guides, and coursing of the masonry work. The use of prefabricated metal stairways is also economical in the overall construction of the building. It is essential, however, before walling in with masonry work that these stairs are set in place, level and plumb, by the ironworkers. The mason foreman should use a plumb bob to see if the stairs are true before the walling in process begins. Figure 30-7 shows a metal stairway being installed in a brick bearing wall building.

SCAFFOLDING

The scaffolding is usually built on the concrete floor, and the brickwork is laid overhand. This greatly reduces the cost because usually only one section of scaffolding (in height) is needed to construct the wall to the story height, figure 30-8. (If the scaffolding was built on the exterior of the wall, it would have to be constructed all the way from the ground up to the top of the structure.) This is a big savings for the contractor and builder, resulting in a savings for the owner. That is one of the main reasons that brick bearing wall construction is becoming increasingly popular.

MORTAR JOINT APPLICATION

Bricks are laid in the usual manner, with the mortar joints well filled. The use of corner poles instead of the traditional prebuilt masonry corner helps to reduce the cost of the masonry work.

Fig. 30-7 Prefabricated metal stairway that is installed before the masonry work. The stairway must be set in position, level and plumb, before walling in with masonry.

Fig. 30-8 Masons building a load-bearing wall off the scaffold. Usually one section of scaffold is high enough to build to story height.

The *shove joint* technique of applying mortar to the wall is recommended for constructing engineered brick masonry walls. This is done by spreading the mortar on the wall with a trowel and flattening out the mortar with the trowel blade instead of furrowing with the point of the trowel. The brick is then shoved into place. The shoving movement positions the brick, and when properly done, fills the head and bed joints completely. On-the-job experience has proven that increased production and well-filled mortar joints result from this method. It is important to use solid mortar joints on all engineered brick masonry to obtain total wall bearing strength, and ensure watertight walls.

MATERIAL HANDLING

Two main advantages of multistory bearing wall brick construction are speed and economy. Efficient material handling is therefore necessary if the job is to be profitable.

Much of the multistory bearing wall construction involves the building of office buildings, apartments, hotels, and motels. Because of the floor design and the placement of the concrete (in conjunction with the building of the masonry for the next story), it is very important that the masonry materials are not overloaded on the floors. Wood runways should always be built on the floors to protect them from damage as wheelbarrows or powered equipment are operated.

A good method of providing a runway for the movement of materials is to lay 4-foot x 8-foot sheets of 3/4-inch construction grade plywood over the area where the materials are to be transported. In addition to protecting the floor, this helps to distribute the weight of the materials over a large area. It is very important to set scaffold frames or bucks on 2-inch planks.

Store all masonry materials on ground level near the hoist (if one is being used), and handy to the loading platform. The materials can then be transported easily to the waiting elevator.

Small gasoline-powered lifting machines, called *walking buggies* (figure 30-9), can be used to speed the movement of materials from the main stockpile to the elevator. With this machine, one person can move a large pallet of masonry materials to the elevator. The materials can then be transported to the masons who are working on the wall without anyone handling the units a second time. This has been a breakthrough in the handling of materials for brick bearing wall construction. Mortar also can be moved in specially powered wheelbarrows.

It is very important to keep all materials covered to prevent moisture from penetrating. This should be done in warm as well as in wet and cold weather. Since

Fig. 30-9 Mason using a walking buggie to transport a pallet of concrete blocks to the waiting elevator.

the building of engineered brick masonry demands the close coordination and scheduling of the concrete work and the brick masonry, the building schedule can only be met if materials are properly handled and protected. The materials must be dry and available to the mason at all times during the workday.

WORKMANSHIP

Quality workmanship is very important in building bearing wall masonry for several reasons. First, the required transverse (lateral) and compressive (downward) strengths must be obtained in the walls. Also, in many cases, engineered brickwork involves the use of thinner clay masonry walls than those used in regular load-bearing work. Good workmanship greatly helps to prevent the entrance or absorption of moisture from forces such as wind-driven rain or snow.

As previously mentioned, full mortar joints are essential to obtain the maximum load-bearing strength of the masonry wall. Mortar must be carefully mixed and tempered; only water lost through evaporation should be added back when tempering the mortar.

As in all masonry, the walls should be level and plumb. The more plumb the walls, the more evenly the load is transferred to the footing. Wall ties, if specified, should be installed and fully mortared.

Mortar joints are struck or tooled as soon as they are thumbprint hard. It is also recommended to use a sled runner striking tool if the joint finish is either concave or V joint. All pin or nail holes should be filled when striking. Any mortar that has been forced out around the edge of the brick in the striking process should be cut off with the blade of a trowel before brushing. Brush the wall after it has set enough so that it does not smear or show brush marks.

Do not use chipped or broken bricks in the face work. The bricks should be dry and free from mud or earth. Take care when cutting the bricks to produce a straight clean edge.

At the end of the workday, the scaffold board nearest the wall should be turned away from the wall. This is done so that if it rains, the rain cannot beat against the scaffold and stain the wall.

SUMMARY

Through the use of engineered brick masonry techniques, many types of buildings can be built at a lower cost. This method makes it possible to complete the structure faster than can be done with steel or concrete frame construction.

Structural clay masonry units such as bricks, having high compressive strength, are well suited for use in engineered bearing wall design. Because of their great strength, properly engineered brick walls built entirely with clay units can be made thinner and built higher or with more stories, without bad results.

Clay units are available as standard production items in a variety of designs, sizes, and shapes. Likewise, improved mortars, developed through a greater knowledge of what constitutes a better mortar and how it will react in a bearing wall, will add to the success of the job.

Providing adequate protection for the masonry work and for the mason in cold weather is especially important in bearing wall construction. The enclosing of scaffolding can be very helpful when building work in cold weather.

Fig. 30-10 A completed section of bearing wall construction. Notice the thinness of the masonry wall.

For bearing walls that are higher than two stories, the mason and general contractor must fully coordinate their operations. More efficient methods of bricklaying, scaffolding, and job scheduling must be used if the work is to proceed on schedule.

The main objectives of engineered brick masonry include good job supervision by the mason foreman, careful attention to the proper design and mixing of mortar, advance ordering of materials, and good workmanship practices. The return to the use of brick bearing walls as the method of supporting the load of a structure provides the mason with another source of work which is not available in concrete and steel frame construction. Figure 30-10 shows a completed section of bearing wall masonry.

ACHIEVEMENT REVIEW

Select the best answer from the choices offered to complete each statement. List your choice by letter identification.

1. The basic concept behind engineered brick masonry is that

    a. the brickwork is reinforced with steel rods.
    b. the concrete is placed, and the brickwork is built under it.
    c. the brick wall acts as the support for the floors and roof.
    d. a steel framework is veneered with brickwork.

2. The first standard for engineered brick masonry was published in the United States in

    a. 1950. c. 1966.
    b. 1960. d. 1970.

3. The return to the use of brick bearing walls (instead of steel or concrete) for multistory buildings has become popular because

   a. steel is scarce.
   b. they are more economical and faster to construct.
   c. brick walls last longer than steel or concrete.
   d. the designs and construction practices for brick walls are unlimited.

4. To obtain full mortar joints when laying bearing wall masonry, it is recommended to use a

   a. double joint.   c. buttered joint.
   b. shove mortar joint.   d. grouted joint.

5. The most economical method of erecting scaffolding for load-bearing masonry is to

   a. build swinging scaffolds attached to the framework of the structure.
   b. use tower scaffolding erected on the outside of the building.
   c. build scaffolding on the concrete floor, and lay brick overhand.
   d. use steel tubular scaffolding erected on the outside of the wall and built as the work proceeds to the top of the structure.

6. The main objection to using concrete form construction is that

   a. it is not as strong as brick bearing masonry.
   b. materials are difficult to obtain.
   c. there is a long waiting time while the concrete is setting and curing.
   d. an excessive amount of scaffolding is needed.

7. Brick bearing wall construction is especially interesting to the owner of a structure because

   a. brick bearing walls are more fireproof.
   b. the brick structure is more pleasing in appearance.
   c. the brick structure is easier to insulate.
   d. this type of construction may allow early rental and occupation.

Unit 31 PREFABRICATED MASONRY PANELS

OBJECTIVES

After studying this unit the student will be able to

- describe how prefabricated masonry is built by using hand tools or conventional methods.
- explain how prefabricated masonry is done in a factory setting.
- describe how prefabricated masonry panels are erected and attached to the structure.
- list the advantages and disadvantages of using prefabricated masonry panels.

INTRODUCTION

The desire of the construction industry to decrease on-site labor has brought about the use of prefabricated building components. Prefabrication methods have been developed by several segments of the masonry industry: mason contractors, brick manufacturers, equipment manufacturers, and others closely associated with the industry.

There are several recent developments which make the prefabrication of masonry panels possible. The most important is the development and acceptance of a rational design method for masonry based on using engineering knowledge. Other factors such as new and improved brick units and mortars, and research have been responsible for the fast progress in the prefabrication of masonry panels.

The development of high-bond mortars for use in prefabricated masonry has been a breakthrough. Several advanced techniques are responsible for this new trend in the masonry business. These techniques include the building of masonry panels in plants under ideal conditions; tested methods of installing metal ties and anchors to the panels, and installed on the job site with cranes and heavy equipment; and architects' and designers' acceptance of prefabricated masonry work in place of conventional masonry or precast concrete.

A few years ago, the building and installing of prefabricated masonry panels was thought to be impossible. Now, however, it is possible to store the masonry in a protected building, transport it to the job, and install it in place. This greatly influences the architect's decision in selecting prefabricated masonry over other materials. Often the brickwork for a prefabricated building can be finished and stored at a separate site awaiting the completion of the substructure. Work scheduling can also be done very efficiently because weather is not a factor. This unit discusses how prefabricated masonry work developed and how prefabricated panels are constructed.

HISTORY OF PREFABRICATED MASONRY PANELS

Prefabrication of brick masonry began during the 1950s in France, Switzerland, and Denmark. Shortly after this, the Engineering and Research Division of the Structural Clay Products Research Foundation, (now part of the Brick Institute of America) developed a prefabricated brick masonry system. This system was known throughout the industry as the SCR® building panel, and was used in the construction of several structures in Chicago which are still in service today.

Most of the early methods of building prefabricated masonry panels were attempts to mechanize the bricklaying process to produce standard panels, using unskilled labor. Later trends, however, have been to use skilled masons and standard masonry construction practices. The emphasis is on developing different ways to increase the mason's work output.

There are several systems of prefabrication now being used in the United States. Some of these are available on a local franchise. Other systems involve mechanized equipment which can be obtained from suppliers, while still others are not patented but are merely methods of prefabrication developed by individual manufacturers or mason contractors.

Brick and glazed structural facing tile can both be used in the building of prefabricated masonry panels. As a rule, however, panels of masonry built with high-bond mortar or grout are one story high with varying lengths up to a maximum of 25 feet. Longer panels are made in special cases. Building masonry panels allows the designer to create complex shapes with returns, soffits, arches, and openings in a plant-type operation with no loss of working time.

MORTAR AND GROUT

The mortar or grout can be either conventional mortar or mortar with one of the high-bond additives developed in recent years. Prefabricated panels are usually subjected to greater stress when transported from the plant to the job site than regular masonry walls built in place. The high-bond mortars give the needed higher tensile bond strength which permits earlier handling of the finished panels. It is important to determine the compatibility (how well the brick and mortar bond together) of the brick with the high-bond mortar. Some bricks or masonry units do not perform as well with high-bond mortars as others. Tests with the mortar and masonry units should be performed in advance to determine if the combination does produce the required strengths for the projects.

BRICK FOR PREFABRICATED MASONRY PANELS

There are specially made bricks available for prefabrication work but this unit deals with bricks of standard size and shape. Both solid and hollow bricks have been used in prefabrication work. Solid brick masonry units are those which have coring of less than 25 percent of the bedding area. Hollow bricks are those cored more than 25 percent, but less than 40 percent of the bedding area. The hollow bricks are suitable for and used in applications where vertical reinforcement may be required.

Units with large face sizes have been used both in solid and hollow unit coring patterns. These bricks have a larger finished face than the standard modular unit of 2 1/4 inches x 7 5/8 inches and provide some advantages. The main advantage of large face units is the savings in cost produced by the increased work output of the mason. In a single-wythe wall, a 4-inch x 12-inch face unit gives an equivalent face area of 2.25 times that of the standard modular unit, and an 8-inch x 8-inch face unit provides 3 times the face area of a standard modular unit. Single- and multiple-wythe units are used in prefabricated masonry.

GLAZED STRUCTURAL FACING TILE FOR PREFABRICATED MASONRY

Glazed structural facing tile can also be used for building prefabricated masonry panels. A regular load-bearing glazed structural facing tile is recommended. The tile is laid in a high-bond portland cement mortar with special additives that make the prefabricated masonry panel stronger than if laid with conventional mortars.

VARIOUS METHODS OF BUILDING PREFABRICATED MASONRY PANELS

Several manufacturing methods are presently used for panel construction. There are six general factors which affect the manufacturing process and resulting properties of prefabricated panel masonry. These factors are as follows: hand method, casting, equipment, masonry units, mortar, and grout. The traditional hand-laying method is the simplest process of prefabricating masonry panels.

HAND METHOD

In the hand method of building prefabricated masonry panels, the standard masonry tools are used. The panels can be built inside a building away from the job site or on the job in an enclosed area. If they are constructed in a plant or building away from the actual job site, the trucking of the masonry panels to the site presents an added expense. If built on the job, the cost is less, but there is seldom enough space on the job site to allow this type of work. Most mason contractors either rent or lease a large building nearby when there is a need for panel masonry. The work can then be built and transported at a reasonable cost.

It is important to have ample floor space. The panels are bulky and power equipment must have room to pick up and move the panels from the inside of the building to the storage yard when finished. In addition to the standard tools, corner poles, jigs, and templates are also used for special shapes and to increase the mason's production.

Construction Practices Using Hand Method

Although there are other ways of building prefabricated panels by hand, the following procedure is based on a typical operation. Before any masonry units are laid (whether they are brick or glazed structural facing tile), the plans should be studied carefully to make sure that the mason fully understands how the wall is to be built. Once the panel is laid and has cured, it cannot be changed due to the high-strength mortar used.

The wall is laid out on the floor or base with a chalk line. Then the bond is laid out dry to make sure it works. Next, the first course is laid out to the chalk, figure 31-1. It is recommended to use a solid, flat mortar shove joint with no furrowing of the bed joints. This produces the strongest mortar joint.

Fig. 31-1 Mason laying out the first course of bricks using a nonfurrowed mortar joint.

Fig. 31-2 Mason raising the line for another course of brick by sliding up on the corner pole.

After the first course is laid out, the line is raised on the gauged corner pole, figure 31-2, and the brickwork is continued until scaffold height is reached. Metal angle iron clips are built into the back of the brick panel so the panel can be lifted and transported.

The scaffolding is built and the panel is laid to story height, figure 31-3, with window or door openings built in. The mortar joints should be struck as soon as they are thumbprint hard, using the specified striking tool.

Fig. 31-3 A prefabricated brick masonry panel in the process of being built up to story height.

After the panel has cured, it is cleaned with a chemical solution. It is then washed down with water from a pressurized spray to remove all dirt and mortar. Figure 31-4 shows a completed panel drying in the plant.

After the masonry panel is removed from the plant, it is stood upright on a wood board and braced into position so that it cannot fall over. A coded number and letter are marked on the back of the panel to indicate specifically where the panel will be located when installed in the structure on the job. Figure 31-5 shows the finished panels resting on boards at the plant before being trucked to the job site.

Fig. 31-4 A completed prefabricated brick masonry panel drying inside the plant. This section of wall is cut on a rake as it will be placed on the end of a building.

CASTING

Another method of prefabrication is called casting. *Casting* involves the combining of grout into a prefabricated panel similar to precast concrete. The casting method is performed while the panel is either in a vertical or horizontal position. This method usually lends itself to automated equipment and requires a form, some method of placing the units, and a system for grouting. Casting generally takes place in an off-site plant.

The usual practice is to place the units (either by hand or machine) in a form, either horizontally or vertically, and to fill the joints with grout. The casting method uses jigs and forms which provide the spacing for the mortar joints. The face of the unit must be protected from smearing with the grout. Pressure forces the grout into the mortar joints.

Fig. 31-5 Finished prefabricated masonry panels situated on boards in the storage yard outside the plant, awaiting shipment to the job.

AUTOMATED PLANT PREFABRICATION OF MASONRY PANELS

Certain plants are used specifically for building masonry panels. Specially designed pneumatic equipment and motorized scaffolds are operated. Masons working in the plant can produce one and one-half times the usual on-site production using this automated approach. This system offers fast delivery, highly uniform workmanship, good working

Unit 31 Prefabricated Masonry Panels ■ 415

Fig. 31-6 Mason laying a prefabricated glazed structural facing tile panel working next to the wall on a motorized scaffold.

conditions, moving scaffolds, and precise working practices. It can be used for brick or tile panels. Figure 31-6 shows masons building a prefabricated glazed structural tile wall panel.

CONSTRUCTING A GLAZED STRUCTURAL FACING TILE PANEL

Glazed structural tiles are laid on a bed of high-bond portland cement mortar with an additive for extra strength. The mortar is applied to the wall with specially designed pneumatic equipment. This method is a faster and more efficient way of applying mortar to the bed joint and creates a very strong bond to the tile. Figure 31-7 shows a mason applying mortar to the joint using a pneumatic tube.

Fig. 31-7 Mason applying a high-bond mortar (portland cement mortar with a special additive), using specially designed pneumatic equipment.

Fig. 31-8 Glazed structural facing tile prefabricated panel being removed from the plant with the aid of handling clamps. The use of special high-bond mortar makes it possible to use single-wythe walls.

After the panel has cured in the plant, special handling equipment, figure 31-8, is used to move the panel to the area where it will be placed on a truck and transported to the job site. The high tensile strength of high-bond portland cement mortar (with additives) makes it

Fig. 31-9 A lowboy tractor trailer transporting the prefabricated panels to the job site.

possible to use a single-wythe panel. Because it is stronger, the panel can be picked up and moved without damage. Figure 31-9 shows a "lowboy" truck transporting the panel to the job site. Special racks hold eight or ten prefabricated panels. The panels rest on a rubber pad braced by a crossbar. The crossbar also acts as a separator.

When the panels arrive at the job, the 8-foot x 11-foot prefabricated glazed structural tile panels are set into place with a crane. This crane is specially adapted for the job with a rotating head mounted on a telescoping boom. Each glazed tile panel weighs slightly over 20 pounds per square foot, for a total weight of about one ton per panel. Figure 31-10 shows the panels being set into place in a tunnel.

Fig. 31-10 A mason is setting glazed structural tile panels into position in a tunnel job, using a special device mounted on a truck bed. Notice that the tiles are being erected in a soldier position for design. Each panel weighs about one ton.

ERECTING PREFABRICATED MASONRY PANELS

Prefabricated masonry panel walls can be either bearing walls or curtain walls. Bearing walls support the weight or load of the structure, such as floors, contents, and roof. Curtain walls are non-load-bearing enclosure walls on the exterior of the building that act as a shell for the building. The curtain wall is attached to and supported by the previously erected steel or concrete structural framework.

The following discussion of how a prefabricated brick panel is installed on a building is concerned more with the use of curtain walls than bearing walls. To better understand how this is done, the progress of a typical prefabricated brick panel wall is followed from the time it is built in the plant until it is in place on the structure. Figure 31-11 shows the completed brick panel in the plant ready to be shipped to the job site. The bricks are laid approximately the same way as the glazed structural facing tile panels.

Unit 31 Prefabricated Masonry Panels ■ 417

Fig. 31-11 A completed brick panel in an automated-type prefabrication plant process.

Fig. 31-12 Angle iron clip plate attached to the back of a prefabricated brick panel. This angle iron will be fastened to a corresponding angle iron in the structural frame of the building.

Method Of Connecting The Prefabricated Panel To The Structure

The masonry panel is attached to the structural frame by either bolting or welding angle iron clips (that have been imbedded in the back of the panel when constructed) to the angle iron clips built in the structural frame. The angle iron clips are as secure as the bricks, themselves, since they were fully mortared in with high-bond portland cement mortar (with special additives). Figure 31-12 shows an angle iron clip that has been built into a prefabricated brick panel.

Lifting The Panel Into Place On The Structure

When the panel arrives on the job site, it is removed from the truck with a crane. The crane has either a large sling or hooks that attach to the panel. The hooks are set

Fig. 31-13 Prefabricated brick panel being lifted with a crane to the position it will be located on the structure. The panel is either bolted or welded to anchor it in the structural framework of the building.

far enough apart so that the masonry panel cannot swing or turn when being lifted. Figure 31-13 shows a section of prefabricated brick masonry being lifted to be installed on the concrete frame of a high-rise building.

The panel is usually set in place on the structure with temporary shims or wedges until it is finally leveled and plumbed into position. After the mortar has cured, the wedges are removed and the joints pointed or caulked. The angle iron clips in the masonry panel are

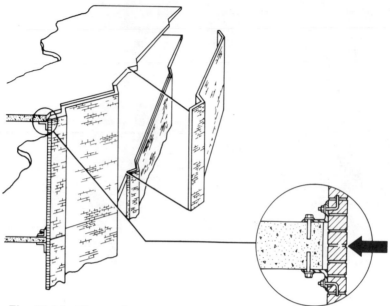

Fig. 31-14 The prefabricated brick panel is bolted in place to the structure of the building.

then bolted or welded into the angle iron clips in the structural frame. This can be clearly seen in the cutaway section view in figure 31-14, showing the installation of a masonry panel in the concrete framework. Cranes that are mounted on top of the structure can also be used to lift the panels into position, as shown in figure 31-15.

Decorative reinforced brick panels with overhanging masonry soffits can be prebuilt in the plant, transported to the job, and then set in place to create an ornamental appearance. The masonry panel being set into place in figure 31-16 would be very difficult to construct

Fig. 31-15 A crane mounted on the roof of the structure is lifting a prefabricated brick panel with a sling-type arrangement.

Fig. 31-16 Prefabricated brick masonry panel with a soffit that was preformed at the plant. This is an example of the holding strength of high-bond mortar.

using conventional bricklaying methods because special forms and scaffolding would have to be constructed on the outside of the building. This is one of the more valuable advantages of prefabricated masonry panels.

Sealing The Joints Between The Panels

As previously mentioned in curtain wall construction, every panel is supported by the structural framework, separate from all other panels. Therefore, there is a joint between each panel, called a *soft joint,* that must be made waterproof.

This joint can be sealed by using a good elastic sealant the same color as the mortar joints in the panel so it is not noticeable. An alternate method is to point the joints with mortar to match the panels. This is not recommended because there is a chance of shrinkage in the pointed mortar to the edge of the masonry. A sealant installed behind the joint before pointing helps to provide a watertight joint. It is also recommended that flashing be installed for horizontal joints. If any moisture does penetrate inside the panel it is collected by the flashing and drained to the outside of the wall through weep holes.

ADVANTAGES OF PREFABRICATED MASONRY PANELS

There are a number of advantages of using prefabricated masonry panels which are not found with conventional masonry.

- The most important advantage is that the work is done inside an ideal working environment that is maintained at a temperature which is perfect for the setting and curing of masonry work. This also allows for steady, year-round work for the mason, resulting in a more stable job since bad or cold weather is not a factor. Prefabricated panels also greatly decrease the amount of winterizing needed on a structure in cold weather.
- By using panel construction, exterior scaffolding is eliminated for laying masonry walls.
- If an off-site plant is used, the work and storage area for masonry materials is kept to a minimum.
- The proper scheduling for delivery is maintained. The panels can be erected as they are delivered, eliminating the need for large areas of on-site storage.
- The use of panels allows the fabrication of complex shapes. These shapes can be accomplished without expensive forms and supports which are necessary for regular masonry construction. Complicated shapes such as arches, soffits, and returns can be built using jigs and forms that are reusable. The more the jigs and forms are reused, the lower the cost of the panels.
- The requirements of very strict quality control can better be met in a factory-controlled condition than on the job. Mortar batching systems can be tightly controlled by automation of sophisticated equipment. The curing of the mortar is more consistent, since there are no weather changes to contend with.
- Masonry work can be ready when needed, with no delay, saving valuable construction time for the completion of the project. On some projects, the masonry panels can be built starting as early as ground breaking, thus keeping the work far enough ahead of schedule so that the work of other trades is not delayed.

The job superintendent can schedule the job more efficiently, resulting in a more exact time of completion because bad weather is not a factor in delaying the masonry work. When building masonry walls with other methods, the job is often delayed for months or for the entire winter, if the weather is not suitable. Using panels gives a great advantage to the owner and to the contractor building the structure. It results in cost savings for the owner since any time saved reduces the length of time for high-interest financing and allows earlier rental of the structure. Both of these factors are important considerations for the building owner.

DISADVANTAGES OF PREFABRICATED MASONRY PANELS

As with any type of new construction method, prefabrication has some disadvantages as well as advantages.

- Prefabrication of masonry has not been as inexpensive as desired. In construction today, costs are approximately the same or a little higher than conventional laid-in-place masonry based on a square-foot comparison.

- Its use is limited to certain types of construction. The size of prefabricated masonry panels is limited mainly due to transportation, erection methods, and capabilities. The use of prefabricated masonry is also limited to some degree by its basic materials: the masonry unit and the mortar. The panels must be designed to withstand the loads that occur in a structure.

- In-place, conventionally laid masonry allows the bricklayer to build the walls to fit or accommodate other parts of the structure by adjusting the mortar joint thickness over a large area so that the joints are not noticeable. This is not possible with prefabricated masonry panels. When prefabricated masonry panels are used on the job, other trades must construct their work with greater accuracy, beyond the standard construction practices usually used, to make sure all the work fits together.

SUMMARY

The design and construction of prefabricated masonry panels has increased in recent years and may become a very important part of the mason's work in the future. It is being accepted gradually because it does require methods and practices that are new to the mason.

The design is based on rational engineering criteria resulting from tests and research. One big advantage is that the masonry panel can be laid in a protected environment, transported to the job site, and installed in sections on schedule with other phases of the construction process. This is a very appealing factor for architects, builders, and owners.

The more prefabricated masonry that is done, the less expensive it should become. The future looks bright for this new, advanced type of masonry work.

ACHIEVEMENT REVIEW

Part A

Select the best answer from the choices offered to complete each statement. List your choice by letter identification.

1. The prefabrication of masonry panels has become a reality in recent years. The most important reason for this is the use of

    a. stronger mortars.
    b. a rational design based on engineering principles.
    c. better masonry units for the wall.
    d. improved work methods.

2. Prefabrication of brick masonry panels had its earliest beginning in

    a. South America.
    b. the United States.
    c. Europe.
    d. Australia.

3. Prefabricated masonry panels made with high-bond mortar or grout should not be built more than

    a. 10 feet long.
    b. 20 feet long.
    c. 25 feet long.
    d. 40 feet long.

4. High-bond mortars with special additives make the prefabricated masonry panel much stronger than conventional mortars. In order to handle the panels sooner after construction a certain strength must be developed called

    a. tensile.
    b. compressive.
    c. lateral.
    d. flexible.

5. A method of building prefabricated masonry by placing the masonry units in a jig or form and filling the joints with grout is called

    a. casting.
    b. veneering.
    c. facing.
    d. pressurizing.

6. Prefabricated brick walls that are non-load-bearing and are attached to the frame of a structure by means of angle iron clips imbedded in the back of the panel are known as

    a. parapet walls.
    b. curtain walls.
    c. bearing walls.
    d. engineered brick masonry.

7. Mortar should not be filled in between prefabricated masonry panels once they are erected on the structure because

    a. there is discoloration.
    b. there is shrinkage.
    c. the mortar is not strong enough to carry the weight of the panel.
    d. it is difficult to get to the work to point the mortar in the joint.

8. For best results, it is recommended to fill in the mortar joints on prefabricated masonry panels with

   a. a metal strip.
   b. sealant.
   c. mortar.
   d. wood strips.

9. Prefabricated masonry panel curtain walls are tied to the frame structure by

   a. metal veneer wall ties.
   b. concrete beams interlocked from the structure into the masonry wall.
   c. angle iron clips bolted or welded to the frame.
   d. masonry ties built onto the floor levels.

10. The single most important advantage of prefabricated masonry work is

    a. stronger masonry walls for high-rise buildings.
    b. more intricate designs and patterns for architectural effects.
    c. it allows for building the walls under near perfect conditions in a factory or plant and shipping them to the job in a completed condition.
    d. cost savings over conventional laying of masonry walls by the mason.

Part B

List 4 advantages and disadvantages of prefabricated masonry panel construction stated in this unit.

Unit 32 ADVANCES IN MASONRY EQUIPMENT

OBJECTIVES

After studying this unit the student will be able to

- explain the advantages of using mechanization in handling masonry materials.
- describe some of the types of equipment used in handling masonry materials.
- explain why it is important for the mason to accept new equipment and methods on the job.

INTRODUCTION

Competition exists between masonry building materials and other products such as plastic, glass, steel, and aluminum. Therefore, the efficient handling of masonry materials on the job, at a lower cost, is necessary if the masonry contractor is to remain in business.

New equipment has been developed in recent years to reduce the amount of labor needed on the job. Mechanization is one of the most important developments in the masonry industry. Many types of machines and equipment that were used in the past have been greatly improved to meet the demands of new methods of construction, such as the prefabrication of masonry.

Years ago, brute strength was needed to handle the materials used to build masonry structures. Today, many types of equipment allow one person to do the work that once required many people.

With the introduction of this new equipment comes the need for more skilled workers, both masons and laborers, to operate and maintain the equipment. Everyone benefits from this because the quality of the work improves with better equipment and methods. Even more important, new jobs are created since the equipment must be serviced and kept in good condition. The use of improved and mechanized equipment also decreases the cost of the masonry work, resulting in a savings to the contractor and owner. It also enables the mason to compete with other types of construction. The mason regains some of the building market that was lost when new materials were developed that replaced masonry work for the exterior of a building.

One of the important advances in the area of equipment development is material handling with the use of such machines as tractors, loaders, elevators, lifts, masonry saws, masonry hydraulic cutting and splitting machines, cranes, mortar spreading devices, and mortar pumps. The constant improvements in steel scaffolding have allowed faster erection of the scaffolding and greater safety for the mason.

The Federal Occupational Safety and Health Act, passed by Congress in 1970 (also known as OSHA), establishes the standards for safe working conditions in all places of employment in the United States. It is the responsibility of both employer and worker to see that its requirements are met. If these requirements are not met, the employer can be fined. This strict law has resulted in the development of new and better types of equipment in the construction business to make each job safer.

This unit deals with some of the more important types of equipment and machinery that are used in the masonry business. While some of these mentioned may not be completely new, they are being constantly improved and have greatly increased the productivity of the mason.

CRANES

Huge steel cranes are used on large high-rise jobs to lift materials to where the work is being done. They can be built from the ground up and anchored into position, or built in the center of the building with the structure constructed around them. When the job is completed, the crane is taken apart and removed from the structure.

Cranes such as the one shown in figure 32-1 have a high tower and a long boom. A hook and cable arrangement is attached to the boom to lift the materials from the ground to any place in the building. The operator sits in a cab high on the tower and operates the crane to move materials where needed. Giant cranes of this type are usually rented by

Fig. 32-1 Large crane used in the construction of tall buildings. Cranes of this type are for general construction work.

Unit 32 Advances in Masonry Equipment ■ 425

Fig. 32-2 Crane mounted on wheels. This is a common type of construction crane used by the mason for lifting and setting cut stone, such as Indiana limestone trim.

Fig. 32-3 A small type of material elevator used on masonry jobs. The elevator is operated by a gasoline engine and stops at floor levels by means of weights on the cable.

the general contractor because of the cost. They provide a method of lifting and placing concrete on high-rise buildings.

Smaller cranes built on the back of a truck body are used for buildings of medium height. The mason utilizes this type of crane especially in the setting of large stones. Figure 32-2 shows a motorized crane mounted on a truck frame. This type of crane can be driven to various parts of the job without requiring any special setting up procedures.

ELEVATORS

Different types of material elevators are used depending on the height of the building. The large cage type is used for tall buildings. An elevator (or hoist, as they are more commonly called), works by means of a cable attached to a drum and a gasoline or diesel engine for power.

One of the most popular types, used for lifting masonry materials on a small to medium height building, is the self-operated elevator, figure 32-3. This elevator is mounted on a metal frame with wheels. It is pulled up to where it is needed next to the structure and secured into position.

A wood runway is built level with the floor of the elevator, both at ground level and at each floor where it will stop to unload materials. The mason's laborer loads the elevator with materials at ground level, then raises it up to the height needed, and unloads the materials there. The advantage of this type of elevator is that it can be obtained with a set of weights that automatically stop the cage at the correct floor level.

> **Material elevators are only for the lifting of materials. No workers are ever allowed to ride on the elevator. This rule is strictly enforced by safety officials.**

HYDRAULIC-EQUIPPED MATERIAL TRUCKS

It is a very strenuous task to unload masonry materials by hand from a truck. There is also an added cost because of the long hours of labor needed in the unloading process.

Special truck bodies equipped with a hydraulic arm, figure 32-4, are now used to pick up and set materials on the truck at the plant, and unload them at the job site. In the past, one-half day to a full day was required to *hack* (unload bricks with brick tongs) bricks from a truck. The same was true when unloading concrete masonry units. Now, the truck driver simply drives on the job, lowers hydraulic stabilizer pads to prevent the truck from tipping, and unloads the masonry material by pushing a button on a hand-held control box. There are holes in the pallets of materials or in the cube of bricks in which the metal forks are inserted. The hydraulic lifting arm then places the materials exactly where needed. The arm can pick up a cube of bricks (500 standard brick) and set them next to the mason.

Fig. 32-4 Truck driver unloading a load of bricks with the hydraulic unloading arm, mounted on the truck body.

No one should ever be under a load of materials while it is being lifted or moved to its unloading position. There is always the possibility of the materials slipping or falling.

FORKLIFT TRACTORS AND HYDRAULIC-OPERATED MACHINES

Some tractors are equipped with hydraulic lifting devices, figure 32-5, that allow the operator to pick up masonry materials with forklifts and deliver the materials to the masons working on the ground or scaffold. This type of tractor lifts up to only one story in height and is particularly useful for low masonry construction. Notice that a roll bar is built over the operator's head to protect the driver.

When higher lifting is needed, hydraulic-operated tractors that have large traction-type rubber tires and an adjustable hydraulic boom are used. This boom enables the operator to reach up and out to place materials on a scaffold or higher story of a building, figure 32-6.

Fig. 32-5 Forklift tractor equipped with a hydraulic lift that picks up mortar or masonry materials.

Fig. 32-6 Tractor equipped with a long extending boom that works on a hydraulic system. It enables the operator to place materials near the mason even when the mason is working up several stories.

CONVEYORS

Belt-type conveyors are very popular for moving mortar, masonry materials, or general building materials from the ground level to the floor of a building. They can be operated by hydraulic pressure or a chain that is powered by a gasoline engine. The moving belt or cleats operate continuously, making a loop from the bottom to the top of the floor level. They are used for buildings of moderate height.

Figure 32-7 shows a conveyor transporting concrete to the third story of a building. Bricks, blocks, and mortar can be moved the same way.

CONCRETE SAWS

Unit 6 discussed the use of the masonry saw to cut masonry units for the wall. There are also large concrete masonry saws that can cut through concrete, blacktop, stone, or other similar material. Figure 32-8 shows a large

Fig. 32-7 Conveyor that can be used to transport masonry materials from the ground up to the floor levels of a structure. Mortar, masonry units, or concrete can all be moved on this type of conveyor.

Fig. 32-8 Large diamond-bladed saw that can be used to cut through concrete, blacktop, stone, etc.

diamond-bladed saw, lubricated by water, cutting out a section of concrete.

HYDRAULIC MASONRY CUTTING MACHINES

Hydraulic masonry cutting machines have been developed and improved in recent years. They are available in a wide range of sizes depending on the job needs.

A standard hydraulic cutting machine (also called a splitter) operates by applying hydraulic pressure through a foot pump. This pressure forces the cutting blades to compress against the masonry unit being cut.

This type of unit is very valuable when solid concrete blocks and small stones must be cut. The smaller cutting unit can be set up on the scaffold, and the cutting is done on the spot where it is needed. Figure 32-9 shows a hydraulic cutter mounted on a set of wheels being used to cut a building stone.

WHEELBARROWS POWERED BY A GASOLINE ENGINE

Metal wheelbarrows powered by a small gasoline engine can be used to transport a large amount of mortar from the mixer

Fig. 32-9 Masonry cutting machine used for cutting masonry materials by hydraulic pressure. This machine is pumped by a foot control until enough hydraulic pressure builds up to fracture the unit.

to the mason. They may be driven up a runway and into the area where the mason is working.

Powered wheelbarrows, figure 32-10, can also be driven onto elevators and lifted to the floor level where needed. Some have a dump body that is very useful when handling concrete.

BRICK BUGGIES

Brick, concrete block, and masonry units can be handled and transported to the area where needed by using either gasoline-powered engines or manually operated carts. These are available in full or half pallet models with hydraulic forklift operation. Figure 32-11 shows a brick buggy that is used to move masonry materials to the work area. This machine can easily load a first section of scaffolding. It is powered by a small gasoline engine. Figure 32-12 shows another brick buggy that is made for a lower lift. This is also equipped with a gasoline engine. Machines of this type require a fairly good, flat running surface.

MORTAR SPREADERS

In recent years, electrically operated mortar spreaders have been developed that can apply mortar to the wall more efficiently and with less waste than by manual application. A mortar spreader consists of a rugged, all-metal hopper with a sturdy base assembly that fits over the masonry unit previously laid. Twin mortar metal gates are adjusted to deposit the correct amount of mortar on the bed joint. The gates can be adjusted to spread a full-width mortar bed joint for bricks.

For concrete blocks, the machine is operated in the following manner. The mason places mortar in the hopper until it is full. The button is then pushed and the mason pulls the spreader along the course. Vibrations move the mortar out of the hopper through the gates on both sides of

Fig. 32-10 Mason tender shoveling mortar from a gasoline-engine powered wheelbarrow.

Fig. 32-11 Brick buggy used for picking up masonry materials and placing them on scaffolds.

Fig. 32-12 Brick buggy made for a lower lift. Notice how the load is carried between the wheels.

Fig. 32-13 Applying mortar to a concrete block wall with a mortar spreader.

the course. Smooth, even ribbons of mortar are deposited as the mason rolls the spreader along the course.

The advantage of this type of machine is that the correct width of mortar is spread on the bed joints with none dropping down inside the core of the blocks. The machine can spread up to 30 feet of mortar without being refilled.

After spreading the mortar on the wall, the concrete blocks are laid to the line in the conventional manner. Figure 32-13 shows how the mortar is spread on the wall with the machine.

SUMMARY

As stated previously, it would be difficult to mention all the types of equipment and machinery that have been developed for use in the construction business. The effect of mechanization in the masonry industry is resulting in the swifter, more efficient movement of materials to the mason working on the wall. Delivering masonry materials to the job in packages and metal bands also improves equipment handling. The new equipment is constructed with regard to safety and allows faster handling of materials.

More and more advances are being made in masonry equipment methods. The end result will be a better masonry job at a lower cost. Masonry students should understand the importance of efficient handling of masonry materials in relation to job profit.

ACHIEVEMENT REVIEW

Part A

Select the best answer from the choices offered to complete each statement. List your choice by letter identification.

1. A cube of standard size brick contains

   a. 250 bricks.
   b. 500 bricks.
   c. 750 bricks.
   d. 1,000 bricks.

2. The main advantage of the hydraulic arm mounted on masonry material trucks is that
   a. there is less chipping of materials.
   b. less time is needed to unload the truck.
   c. materials can be lifted up high and placed on the different floor levels.
   d. they are safer to operate.
3. The term *hack* refers to
   a. the cutting out of mortar joints.
   b. reducing the time it takes to unload materials from a truck.
   c. the number of bricks on a pallet.
   d. the process of unloading bricks with brick tongs from a truck.
4. The most powerful type of lifting devices used to handle materials operates on
   a. electrical current.
   b. battery power.
   c. hydraulic pressure.
   d. high-cranked gear ratios.
5. The most important advantage of the automated mortar spreader is that
   a. it spreads more mortar than is possible by hand methods.
   b. it uses a higher strength mortar.
   c. the mortar can be spread to a smaller thickness than by hand.
   d. there is very little waste as it spreads an almost perfect joint.

Part B

List 8 individual developments discussed in this unit that deal with the handling of materials.

SUMMARY, SECTION 9

- It is necessary to develop different work methods when building masonry in cold weather.
- In many places where cold weather is a problem for much of the year, progress has been made in cold weather construction techniques.
- Masonry materials that are stored on the job in cold weather should be laid on boards, off of the ground, and covered with tarpaulins to keep out moisture.
- Sand and water should be heated when mixing mortar in cold weather.
- Sand should not be heated so much that it scorches. Scorched sand has a reddish color and should not be used in mortar.
- Water used for mixing mortar should not be heated above 160 degrees F because of the danger of a flash set when the portland cement is added.
- Once mixed, mortar should be kept warm until used.

- Type III, high early strength portland cement can be used in mortar in cold weather. It causes the mortar to set more quickly than the regular type I portland cement.
- Calcium chloride is not recommended as an accelerator for mortar in which any metal ties are being used because it causes corrosion. Even when metal ties are not used, the calcium chloride should not exceed 2 percent by volume.
- Masonry units should be heated when the temperature is below 20 degrees F.
- Masonry work should never be laid on a base that is ice or snow covered because there will be no bonding.
- Maintaining the temperature at or above 40 degrees F is necessary for proper curing of the mortar.
- Using windbreaks and enclosing the structure with a covering helps to provide a temperature that is beneficial for the laying of masonry walls.
- Canvas, regular and reinforced plastic, and vinyl coverings are all effective as windbreaks or for enclosing a building under construction.
- Oil-forced air heaters and salamanders that burn oil are devices that raise the temperature inside a structure so that masonry work can be built without freezing.
- Safety should be practiced when working around heating devices to prevent fires or accidents.
- All exterior masonry should be covered at the end of the workday to protect it from moisture and cold temperatures.
- There is a growing trend in the construction business to return to the building of load-bearing masonry walls.
- Brick bearing walls that are designed by engineers using a rational structural criteria are called engineered brick masonry.
- Tall buildings built of brick bearing walls can be constructed with relatively thin walls using high-strength mortar.
- The construction of brick bearing walls has an advantage over steel or concrete frame construction because it supports the floors and load of the building and does not require an inner skeleton frame.
- Close cooperation among the construction trades and efficient scheduling of work is critical in the construction of load-bearing masonry buildings.
- The correct mortar mix for load-bearing walls is very important if the load of the structure is to be supported properly.
- The shove mortar joint, which prevents the furrowing of mortar, is recommended for load-bearing masonry construction.

- Because the laying of masonry work by the mason is done off the interior floors working overhand, it is important to protect the floors during construction.
- Well-filled mortar joints and good workmanship are very important in the construction of load-bearing masonry walls.
- Although masonry bearing walls can be built of different types of masonry units, bricks are very popular because of their high compressive strength.
- Good supervision of the masonry work is necessary if construction schedules and high standards are to be met.
- The desire of the construction industry to decrease the amount of on-site labor has created a demand for prefabricated masonry work.
- The prefabrication of masonry panels has been made possible by the development and acceptance of rational structural design, based on engineering knowledge and practices.
- High-strength portland cement mortars with special additives create a very strong mortar that bonds well to the masonry unit.
- Prefabricated masonry panels can be laid in a protected, heated plant; trucked to the job; and erected into position on the structure without weather being a major factor.
- The building of prefabricated masonry panels inside a plant can provide year-round work for the mason.
- Prefabricated masonry construction started in Europe and has now become an important part of masonry work in the United States.
- Prefabricated masonry panels are usually no longer than 25 feet because of the problem of shipping and erecting on the job.
- Standard brick and glazed structural facing tile can be used for the construction of prefabricated masonry panels.
- Prefabricated masonry can be built either by conventional hand methods of building walls or by using automated methods in which the mason rides on a motorized scaffold. Both of these methods can be used inside a plant-type operation.
- Another method of constructing prefabricated panels is by casting, in which the masonry units are set in a form and the mortar is grouted under pressure into the mortar joints.
- After the panels have cured, they are trucked to the job and set into place by large cranes.
- Prefabricated masonry panels are connected to the structure by either welding or bolting metal anchors in the back of the panel to corresponding anchors built in the structure.

- Two of the main advantages of prefabricated masonry panels are that complex shapes can be made in the plant and extensive scaffolding is not necessary on the job.
- The joints between the prefabricated panels can be filled with mortar or sealant or a combination of both.
- Builders like the prefabricated masonry panel because it allows the masonry work to be built ahead of the rest of the structure. Also, the panels can be installed with no loss of time due to bad weather. It is possible to schedule the total job more accurately, and sometimes allows earlier occupancy of the structure.
- Two disadvantages of prefabricated masonry work are that the individual courses of masonry cannot be adjusted on the job, and the panels must fit without any major alteration.
- The structural frame of the building must be constructed very accurately in order for the prefabricated masonry panels to fit.
- The future of prefabricated masonry looks very bright. As its use increases, its cost should decrease.
- The development and use of mechanized equipment in the masonry business is essential if masonry is to stay competitive with other building materials in the market.
- The Federal Occupational Safety and Health Act (OSHA), passed in 1970, requires that every employer maintain a safe place of employment. The machines used in handling masonry materials must be built to meet these safety standards.
- The main goal in developing equipment and machines to handle masonry materials is decreased cost and improved productivity of the mason.

SUMMARY ACHIEVEMENT REVIEW, SECTION 9

Complete the following statements referring to the material found in Section 9.

1. Masonry units should be stored off the ground on _____.
2. The heat-liberating action between portland cement and water is slowed or stopped when the temperature falls below _____.
3. The sand used in mortar should be heated when the temperature falls below _____.
4. If the water used in mixing mortar is heated above 160 degrees Fahrenheit there is the danger of a _____ when the portland cement is added.
5. Type III portland cement is used for mortar in cold weather because it will _____.

6. When calcium chloride is added to mortar to speed setting time, it is very important that the calcium chloride is not added in amounts greater than _____ by volume.

7. Masonry work should never be built on a base that has ice or snow on it, because there will be no _____.

8. Polyethylene (plastic) is used for enclosing masonry work in cold weather so that _____ and _____ will pass through.

9. When load-bearing brick masonry walls are constructed using rational structural analysis and engineering data, this is known as _____.

10. The three main types of construction methods used in the United States are masonry load-bearing walls, steel skeleton frame, and _____ _____ construction.

11. The principal objections to concrete reinforced structures are the amount of forming needed and the _____.

12. One of the major advantages of load-bearing masonry walls is that the scaffolding can be built _____.

13. Brick load-bearing walls are built by laying the bricks in a _____ position.

14. To finish masonry bearing wall buildings on time, all of the trades must work together efficiently on the job. This demands careful _____.

15. The mortar recommended for building load-bearing masonry walls is _____ type.

16. For the best results when building brick load-bearing walls, it is recommended to use a mortar bed joint that is not furrowed with a trowel. This type of bed joint is known as a _____ joint.

17. To protect the floors of brick bearing wall buildings, the scaffolding should always be set on _____.

18. Bricks are highly recommended for bearing wall construction because of their size and high _____ strength.

19. The building of masonry work in a plant under ideal conditions, trucking the work to the job, and installing it as a finished product is known as _____.

20. Mortar used for building prefabricated masonry work is composed of portland cement, lime, sand, and special _____.

21. The type of mortar described in question 20 is classified as a _____ mortar.

22. The prefabrication of masonry work began in _____.

23. Solid bricks are those which have coring (holes) of less than _____ percent of the bedding area.

24. The main advantage of using large masonry units in a load-bearing wall is the _____.

25. Prefabricated brick panels are bonded to the structural frame of a building by attaching _____, either by welding or bolting.

26. The prefabrication of masonry panels by casting consists of placing the masonry units in a form, either horizontally or vertically, and filling the joints with _____.

27. Non-load-bearing masonry walls that enclose the exterior of a structure and act as a shell for a building are called _____.

28. When the joints between prefabricated panels have been set into place on the structure, they are known as _____.

29. The most important advantage of prefabricated masonry panels to the builder is _____.

30. The most important benefit of prefabricated masonry walls for the mason is _____.

31. The two most popular types of masonry units used for building prefabricated masonry panels are _____ and _____.

32. The automated system of building prefabricated masonry panels in a plant is highly efficient because the mason rides on a motorized _____.

33. Handling of masonry units and mortar can be performed more efficiently by using _____ equipment.

34. The Federal Occupational Safety and Health Act, which deals with the safety of a worker on the job, is also known as _____.

35. The handling and lifting of masonry materials is usually done with machines that are equipped with _____ systems.

Section 10
Planning and Managing The Job

Unit 33 JOB PLANNING AND PRACTICES TO IMPROVE THE MASON'S PRODUCTIVITY

OBJECTIVES

After studying this unit the student will be able to

- explain labor versus material cost and proper mason-to-tender ratio.
- state why a good relationship between the mason and laborer is important in improving the productivity of the mason.
- describe job practices that can improve the mason's efficiency and productivity.

INTRODUCTION

Efficiency is the key word in masonry construction today if the masonry contractor is to make a profit and stay in business. A knowledgeable contractor knows that it takes more than just good materials and labor to make a successful job. The best masons and materials are wasted if the job is mismanaged and poorly planned.

New and improved equipment and tools, and efficient setting up of the job help the mason perform the maximum amount of work with the least amount of effort. The same is true in relation to the helper or laborer. The basis of a well-run, profitable job is good planning before the work starts and careful management as it progresses.

LABOR VERSUS COST OF MATERIALS

It is important for the student mason and apprentice to understand the division of cost for labor, materials, and other expenses of masonry construction. According to the Brick Institute of America, the average in-place masonry wall cost breaks down in the following manner: labor = 50.4 percent; materials = 23.1 percent; and overhead, profit, contingency, scaffolding, and cleaning = 26.5 percent. These are only average figures and may vary with different localities. However, they point out the fact that labor (mason and tenders) represents more than half the cost of the total job. In order to reduce costs, or stay even in the face of inflation, efforts to increase the mason's work output must be accomplished.

MASON – TENDER RATIO

There should always be enough tender labor to service the masons. The ratio of tender labor to mason varies from 1:1 to 1:2 1/2, depending on the job conditions and where the mason is working on the job.

If the masons are working on the first floor level of a building and materials can be handled efficiently with various types of equipment, the amount of labor needed can be decreased. However, if the masons are working high up on scaffolding, and all of the materials must be lifted and transferred to the work area, more labor is required. An example of this is when masons are working in or around a maze of pipes or mechanical equipment such as found on a commercial building. The task is not only to get the materials to the immediate area, but to move the materials to the location where the mason is working.

The type of work should always be taken in account when the job is estimated and the proper adjustment made for labor.

Fig. 33-1 Mason laborers (tenders) stocking a scaffolding with bricks for the mason

This is one of the problem areas when estimating a masonry job. Regardless of the type of work, it is essential to have enough tender labor to keep the mason busy at all times, since the progress of the work is based on the mason's productivity. Figure 33-1 shows two mason tenders working together to stock the scaffold in preparation for the mason.

THE TENDER'S JOB

A large part of the mason's productivity depends directly on the tender's work. To be efficient, the laborer should be properly trained and equipped, and have the right attitude toward the job. With a good attitude and sense of contributing to the job, the tender will be primarily concerned with keeping the masons busy. Therefore, good public relations among the job foreman, laborer, and mason are necessary if the job is to progress smoothly and efficiently.

For a successful masonry job, the mortar must be the correct mix for the work being done and must suit the individual mason's preference. On one job, the mason may prefer to temper the mortar, while on another job the tender does the tempering. If the tempering is not done correctly, the mason's work output is greatly affected. Usually at the start of the job, the masons and laborers discuss how they like their mortar tempered.

JOB PRACTICES TO INCREASE PRODUCTIVITY

As previously mentioned, the adoption of good practices helps to make the mason more productive. The following practices are suggested in different areas of the masonry trade.

Placement Of Materials

On the job site, the masonry units and mortar mixing materials should be located as close as possible to where they will be used, so they do not have to be moved a long distance.

Fig. 33-2 A typical masonry job in progress.

The manufacturers of most masonry materials deliver them to the job either banded together or on wooden pallets which allow easier handling with power equipment.

All materials should be covered with a tarpaulin to keep out moisture when they arrive on the job. This helps to prevent the delay of work due to rain or snow. If the masonry units become soaked with water, they do not set in the mortar and the wall cannot be built in a level or plumb position.

Main traffic lanes should be established for the movement of materials to the mason's work area. Keep these lanes free from objects or obstructions. The runway or area in which the laborer is transporting the materials should be the shortest route possible.

If the masonry units are stocked on earth at the work area, a wood plank should be laid down first and the materials should then be stacked on the plank. Otherwise, stains and mud from the earth can absorb into the units. This is especially important if the materials will not be used that day. Figure 33-2 shows a typical masonry job in progress. The materials are stacked in piles, allowing the mason tenders free movement when bringing materials from the stockpiles.

Leveling The Earth In The Work Area

The earth around a wall should be filled in and leveled off before stocking the materials or building the wall. If this is not possible, wood planks or scraps of plywood are sometimes laid down to make a level, sound, work footing. Walking up and down over humps of earth and rocks will tire the masons and cause their work output to drop noticeably.

Federal and state laws also require that safe working conditions be maintained for all workers. Wood runways and slippery plywood bases can be made safer by sprinkling with sand.

Laying Out The First Course Ahead Of The Crew

One of the most important items in increasing job production is proper layout of the masonry work before the crew starts building the wall. The masonry units should be laid out with the proper size mortar joint as specified on the plans. Avoid any cutting if possible. Large mortar joints should not be laid out because they require more time to form; also, excessive shrinkage of large mortar joints can cause hairline cracks and create a poor appearance.

The height of the first course should be correct according to the bench marks posted near the corners. Care should be taken that the walls are laid out true to the batter boards which establish the wall lines.

Once the first course is laid out, the laborer can stack the needed materials. The layout course should be given ample time to set before a crew starts on the wall. In addition, cuts that are made with the masonry saw can be stocked next to the wall and allowed to dry. This saves much time when building the wall. Mechanical or electrical outlet cuts of masonry units can be made and placed in the work area. The saw operator's time is used more efficiently when cuts are made beforehand, because the operator will be very busy most of the time as the job progresses. Figure 33-3 shows a mason laying out the first course of brick around a door buck.

Fig. 33-3 Mason laying out the first course of bricks around a steel door buck frame.

Making A Story Pole And Gauge Rod

A story pole and gauge rod should be made by the foreman or mason before any of the walls are constructed. The making of a story pole and gauge rod was described in Unit 3 and can be reviewed.

Building The Corner Ahead Of The Line

After the story pole and gauge rod have been laid out, the corner, figure 33-4, is built. The mason should have an advance start, laying at least 6 courses of bricks or 3 courses of concrete blocks before the crew begins to fill into the line. If the mason building the corner is overly rushed, the corner may be pulled loose

Fig. 33-4 The brick corner has been built ahead of the wall to the scaffold height. Notice that the scaffolding is loaded with materials ready for the mason.

when the line is tightened by other masons working on the wall. Starting all of the masons (including the corner mason) on the wall at the same time usually results in confusion, making it difficult for the corner mason to work correctly or efficiently.

Setting Door Bucks Ahead Of The Wall

When metal door buck frames, figure 33-5, page 442, are installed, they should be set into position by the carpenter before any masons start building the wall. The frames should be braced in such a manner that they do not interfere with the construction of the wall.

Although it is the job of the carpenter to set the door bucks plumb and level, it is the responsibility of the mason to recheck to be sure they are correct before building around. It is inefficient and costly if door bucks must be torn out and rebuilt.

Use Of Corner Poles

In recent years, manufactured metal corner poles, figure 33-6 and 33-7, have become available for use in place of the conventional masonry corner. There are two types of metal corner poles; one can be attached to the framework on a masonry veneer building, the other is erected using bracing on the exterior of a regular solid masonry wall. The corner pole can be set up for outside and inside corners. Coursing of the brick or concrete blocks is marked on a steel tape on the pole and can be adjusted as needed. The line holder is slid up the pole for each additional course. Using corner poles eliminates the need for building masonry leads or corners, as was traditionally done.

Fig. 33-5 Metal door buck frames set into position ready for the mason to wall in place. Notice how the bracing is nailed back out of the mason's way.

Field studies have shown that a mason can lay six or seven bricks to the line in the same amount of time required to lay one brick in a typical corner lead. In addition, job

Fig. 33-6 Corner pole setup on a brick veneer house job. The corner pole eliminates the need for building the regular brick corner.

Fig. 33-7 Corner pole setup for a concrete block foundation for a house

Tools should be set back under the mortar pan or hung up on a nail close by so they are convenient but out of the way. Stock only enough masonry units along the work area to build the wall. Stocking too many units in the work area results in extra cost to the contractor at the end of the job when the extra material must be removed.

Efficiency In Handling The Mortar

It is critical to have good mortar available at all times if the work is to progress on schedule. In the morning before the mason goes to work, the mortar mixer operator and a few of the key laborers should begin fifteen minutes to one-half hour before the mason. The actual time will depend on the size of the job and where the mortar has to be set up.

The goal is to prepare the materials and work area so that the mason is ready to go to work immediately. All mortar pans should be dampened with water and emptied again before filling with mortar. This helps keep the mortar from drying out. It is especially important in hot weather when the drying rate is faster. It is a good practice to have either a small bucket or can of water at every mortar board so the mason can temper the mortar, if desired. As mentioned before, many masons would rather temper their own mortar than have the laborer do it because they want it to be a certain consistency. The mason can also dampen the edges of the pan or board periodically to prevent overdrying. This greatly reduces the amount of tempering needed. Figure 33-9 shows how mortar is tempered with water and a shovel.

When the masons and laborers leave for lunch, provisions should be made for several laborers to return 15 minutes early to temper the mortar before the masons return. (To compensate these laborers, they are usually allowed to leave 15 minutes earlier at the end of the day.) In this way, the mason does not have to wait while the mortar is tempered and can get a faster start on the second half of the workday. A competent foreman always sees that this is done. The mason tender should check the mason's mortar pan and stockpile materials to know when additional materials are needed. On large jobs a labor foreman is hired for this purpose.

Near the end of the workday, the foreman should check the work area to make sure that all of the masons have enough materials to last until quitting time, and that they have enough to get started the next day. Checking with the mortar mixer operator about 1 hour before quitting time is a good safeguard to make sure the masons do not run out of mortar. In order for the mason to work right up to quitting time, there must be a small amount of mortar left to be discarded. The theory behind this practice is that it is better to have some mortar to discard than to have the masons run out before the end of the workday. However, there should not be too much mortar left. Careful planning by the foreman in determining

Fig. 33-9 Mortar being tempered with a shovel to replace the water it has lost through evaporation and drying out on the wood mortar board.

the amount of mortar needed results in the least amount of waste.

Paying the mortar mixer operator and scaffold builders slightly more than the other laborers results in workers who will look after the required work without constant supervision. This is a common practice on masonry jobs since the person who runs the mortar mixer and the scaffold builders have more responsibilities than the average mason tender. Figure 33-10 shows the mortar mixer operator emptying the mixer at the end of the day. This person is one of the most important workers on the job, because the key to a productive masonry job is good mortar.

FACTORS AFFECTING THE MASON'S PRODUCTIVITY

Fig. 33-10 Mortar preparer cleaning out the mixer at the end of the day.

Masonry Unit Size And Weight

The productivity of a mason is directly related to the size of the unit being laid. As the face area of the unit increases, the total wall area laid in place for the day increases.

This statement is true but has limits if a very large, heavy masonry unit is laid. As a mason grows tired, productivity also drops off. An example of this is when large masonry units are stacked too high on a scaffold in which extenders are being used. The mason has to reach up and drag a unit off the top of the stack, creating stress on the arm. The continued jerking motion and the heavy weight of the unit reduce the amount of masonry units the mason can lay in the wall.

Although a larger face unit increases productivity, size is not the only determining factor. Other factors that affect productivity include handling of the units, stocking the scaffolding, mason's fatigue, cutting the unit with the masonry saw, and chipping of the unit.

Thickness Of Mortar Joints

Studies conducted at the University of Texas sponsored by HUD (Department of Housing and Urban Development), the Brick Institute of America, and 20 other masonry groups indicate that mortar joint thickness also affects productivity. It was found that 1/2-inch joints decrease productivity about 15 percent as compared to the more frequently used 3/8-inch joints. The majority of masonry units manufactured are made to be laid with a 3/8-inch joint. Since the mason lays the 3/8-inch joint more often than the 1/2-inch joint, this is probably the reason the 3/8-inch joint is more efficient.

Of Scaffolding On Productivity

The mason can be more productive and efficient when working at the proper work on a scaffold. The proper height is that which does not require the mason to bend too much. This is the main reason for mortar stands.

The most popular type of scaffolding used is known as *tubular scaffolding*. It is made of tubular steel frames with adjustable supports, such as extenders. Extenders allow the mason to stand on a bracket-supported board near the wall. The bracket can be raised as the wall is built to keep the mason at a comfortable work height.

Other adjustable scaffolds are also used. These include swinging scaffolds where the working platform is suspended from steel cables connected to beams projecting from the top of the structure. This type of scaffolding is frequently used on tall buildings where it would not be economical to construct scaffolding from the ground up. As the work progresses, the scaffolding is cranked or winched up.

There is also a tower type of scaffolding that can be moved up as the work progresses. The purpose of any type of scaffolding is to keep the mason at a productive working level.

Regardless of the type of scaffolding used, it should be kept a distance of 3 to 4 inches away from the wall at all times to prevent bumping the wall and splashing it with dirt and mortar. Scaffolding, of course, should meet safety regulations and should not be overloaded with materials.

It is important that scaffolding always be built before the masons begin working so they have somewhere to move. This requires planning ahead by the foreman on the job. A job should never be built entirely to scaffold height before some scaffolding is erected and some of the walls are topped out. This would interfere with the other trades and create problems with the job schedule. A good mason foreman must be able to think ahead three to five days to know what materials will be needed and where the crew will be working. Figure 33-11 shows scaffold builders erecting scaffolding for the masons.

Working Overhand Versus Facing The Work

Working overhand when laying masonry units means working from one side of a masonry wall only, even though both wall faces may be finished. The mason does this, as shown in figure 33-12, by leaning out and building one side of a wall on scaffolding off of a floor and backing up the opposite side level with both. In terms of building the scaffolding, this is economical because only one set of scaffolding is needed. However, the strain on the mason's back and the increased difficulty in building the wall true make overhand work less efficient than if scaffolding is built on both sides of the wall.

Fig. 33-11 Scaffold builders erecting the scaffolding for the masons.

experience has shown that the corner pole saves from 20 to 37 percent on labor costs when laying bricks. The corner pole can therefore be used to improve and increase the mason's productivity.

SPACING OF MASONS AND MATERIALS

The proper spacing of masons and materials along a masonry wall is very important. For the best results, one mason should not lay bricks alone on a long length of wall. The recommended length of wall per mason is 8 feet to 12 feet. This depends, of course, on the size of the job and the number of workers. In hot weather, this is especially efficient because there is less tempering of the mortar due to excessive drying out. Also, the mason will not tire as when laying bricks on a long length of wall. Figure 33-8 shows the proper spacing of masons on a brick wall to divide the work evenly and cause less fatigue.

Fig. 33-8 Masons spaced properly at work on a decorative brick wall. The work is divided up evenly so each individual mason experiences less fatigue.

Mortar boards or pans should be spaced back from the wall about 24 inches and placed so that the mason can pick up and spread mortar by merely pivoting around from the board to the wall in one motion. The mortar boards should also be set up along the wall so that mortar boards are placed into position, followed by a stack of the masonry units. This stocking procedure should continue along the entire length of the wall. A mortar board or pan is placed on the point of the corner so that masons working on both walls can use the same board, especially if corner poles are being used.

Elevate the mortar boards or pans to a height of approximately 20 inches, on a metal stand or on concrete blocks. The mortar should be of the correct consistency for the type of unit being laid, and the mortar pan or board should not be overloaded.

> Always check the mortar boards or pans for cracks, splinters, nails, and protruding or rough edges before setting them up along the wall. Metal mortar pans sometimes become split along the edges as a result of being thrown off scaffolds and other poor treatment. A sharp edge can cause injury to a worker on the job. If the edge cannot be welded and repaired, the pan should be discarded and replaced.

The masonry units are stacked back from the wall approximately the same distance as the mortar pans. The units should be placed so that the mason does not have to turn them over before use.

Keep the stacks of masonry units approximately 12 inches away from the edge of the mortar pans on both sides. This will prevent the units from being splashed with mortar. It is also a good practice to interlock the masonry units on the pile every other course to prevent them from falling over.

Fig. 33-12 Masons working overhand laying up a brick wall on an apartment building.

The exception to this occurs when building brick bearing walls where the floors of the structure can support the scaffolding. The productivity of the mason can be increased significantly by building scaffolding on both sides of the wall and being able to work facing the finished wall. On tall buildings special scaffolding, such as swinging or adjustable tower scaffolding, can be used for this purpose. Figure 33-13 shows an example of facing the finished masonry wall while laying the bricks. It is evident that the mason can work more efficiently and obtain better results when facing toward the finished wall. It is also less tiring because the mason does not have to bend over and reach out as much as when working overhand.

Providing Protection From The Weather

Protecting the mason and the work from cold weather is an important element in productivity. Unit 29 deals with the techniques and practices involved in working in cold weather.

There are several important points that the mason should remember in relation to working in different weather conditions. When the weather is suitable for working outside, the masons should be working there. If inside work is necessary on the job, it should be saved for cold days when windbreaks or heat are available.

The study previously mentioned (University of Texas) showed weather to be a definite factor in productivity. The important fact that the study established is that maximum productivity or best working conditions occur at 75 degrees F and 60 percent relative humidity. Information such as this should be taken into

Fig. 33-13 A mason working facing toward the brick wall. The work can be done much more efficiently when facing the work than when working overhand.

consideration when estimating the job. Contractors who have been in business for many years can benefit from past experiences, and thus find it easier to predict the mason's productivity as it relates to changing weather conditions.

Keeping masonry units dry and having covered materials available at the end of the workday also help increase the amount of work that can be done.

SUMMARY

It is impossible to cover all of the work practices that are used in the masonry business to increase productivity in one unit. However, careful study of current work practices and constant attention to details are important considerations.

The use of new, improved equipment and methods of handling materials is an area where much progress is being made. If the masons are kept supplied with materials, they will generally produce.

Have enough well-trained laborers on the job to tend the masons. Try to promote a good attitude among all workers, since satisfied, happy people usually work more efficiently than those who are unhappy. Make the working conditions as good as possible. Be sure the mortar (which is the key to doing a productive day's work) is mixed consistently the same, is readily available at all times, and suits the materials being laid. Promote safety on the job; everyone loses when there is an accident.

Laying masonry units is a skill involving the correct movements performed efficiently. If a mason has one incorrect movement in spreading mortar, and this movement is repeated many times throughout the spreading process, the amount of time spent in these false moves greatly reduces the mason's productivity. The same type of motion study applies to everything the workers do in the course of a day's work. Efficiency and productivity are a blend of preplanning, good working practices, and the most effective use of materials and equipment.

ACHIEVEMENT REVIEW

Part A

Select the best answer from the choices offered to complete each statement. List your choice by letter identification.

1. When estimating a masonry job, the percentage allowed for labor is
    a. 25 percent.
    b. 40 percent.
    c. 50.4 percent.
    d. 60.5 percent.

2. Runways are built on the job to allow the movement of materials. They are usually constructed of
    a. metal sheets.
    b. plastic forms.
    c. wood boards.
    d. bricks or blocks laid on a firm base.

3. For maximum productivity, masonry work should be laid out before the crew begins a wall. The top of the first course of masonry units should be laid level with the
    a. footing.
    b. bench mark.
    c. mason's rule setting on the footing.
    d. thickness of a masonry unit plus a standard mortar joint.

4. The term *stocking the wall* means
   a. to strike the mortar joints.
   b. plumbing the wall.
   c. building the scaffolding.
   d. stacking and supplying materials.

5. The mason building the corner should be given a head start before the rest of the masons are placed at the wall. For a brick wall, it is recommended to allow
   a. 3 courses of bricks.
   b. 6 courses of bricks.
   c. 9 courses of bricks.
   d. 12 courses of bricks.

6. The laying of masonry walls can be done more efficiently by
   a. one mason building the corner and other masons laying the wall to the line.
   b. building wood forms at the corners on which to fasten a line.
   c. using metal corner poles that can be fastened to the structure.
   d. spotting a brick on the corner, attaching the line, and running the course through.

7. To allow for working room, mortar pans should be spaced back from the face of the wall at least
   a. 12 inches.
   b. 18 inches.
   c. 24 inches.
   d. 36 inches.

8. To relieve strain on the mason's back, mortar pans or boards should be set up on stands or blocks to the height of
   a. 12 inches.
   b. 20 inches.
   c. 24 inches.
   d. 36 inches.

9. The majority of masonry units are made to be laid with a mortar joint that is
   a. 1/4 inch thick.
   b. 3/8 inch thick.
   c. 1/2 inch thick.
   d. 3/4 inch thick.

10. It is a poor practice to build an entire masonry job up to scaffold height before building any scaffolding. This is true because
    a. it will require too much scaffolding.
    b. the masons will have no place to work.
    c. it is difficult for the laborers to stock the scaffolds.
    d. it will take too long to top out the walls.

Part B

Using the information from this unit, write a report for class discussion on how to make a masonry job more efficient. Discuss various areas such as the mixing and handling of mortar, stocking of materials, mason-to-labor ratio, laying out work ahead of the crew, and use of corner poles. The report should be for a job that would be starting soon.

Make the report at least two pages long. Discuss this assignment with the instructor before beginning, to get some suggestions or special comments.

Unit 34 LEADERSHIP ON THE JOB — THE MASON FOREMAN

OBJECTIVES

After studying this unit the student will be able to

- describe the qualifications of a good mason foreman.
- explain the importance of a foreman having a good attitude when dealing with workers on the job.
- describe some of the problems that a mason foreman must deal with on the job.
- explain the mason foreman's responsibility to run a safe job.

INTRODUCTION

The mason foreman has a very important role in the construction of any masonry job. Successful jobs result from good leadership. The mason foreman should be a good craftsman and must also meet certain responsibilities on the job, such as ensuring the job is done on time, keeping all workers satisfied to keep production up, and making sure the contractor receives a profit.

A mason foreman with good leadership qualities is an asset to the trade and to the contractor and plays an important role in management-labor relations. It is the foreman who usually hires the workers, and to whom the workers report. Therefore, the foreman must also be able to judge whether a person would be a good worker.

Workers often feel that the actions and attitude of the foreman represent the feelings of the company they work for. If the foreman runs the job smoothly and efficiently, the masons and tenders assume that the company is well organized and a good one to work for. If the workers do not like the way the job is run, they assume that the company, in general, is to blame. This causes a problem in management-labor relations from the start.

When there is confusion and signs that the mason foreman cannot get along with the workers, it is evident that the construction job is not being run well. Other indications of incompetent leadership include the following: workers standing around without knowing what they are supposed to be doing, a lack of materials, lack of bench marks, no walls laid out in advance, the laborers arguing with the masons. On the other hand, a visit to another job may reveal that everything is moving along smoothly, work is progressing in an efficient manner, the foreman is supervising the work in an orderly way, the masons have plenty of materials to work with, and everything is well coordinated. These are sure signs of good leadership.

An incompetent foreman will not have the job for long. The foreman must be able to get the workers to efficiently produce enough work, and work that is of high-quality, if the contractor is to make a profit on the job. Therefore, it is important that in addition to being an expert in the masonry trade, the mason foreman must also have the qualities of leadership necessary to inspire and maintain good relations with and among all of the workers.

QUALIFICATIONS OF A GOOD MASON FOREMAN

What qualifications make a foreman a good leader? To be a good leader, a foreman must be a self-starter and not need constant supervision; provide leadership, not merely give orders and watch from a distance; get the most production from the masons and laborers with the least amount of effort; be concerned with providing a safe work area; provide an atmosphere where the workers enjoy their work; and maintain a good morale so the workers are happier and work together well to make the job progress smoothly. A foreman is considered a good leader when the workers do a good job not because they have to, but because they want to.

An apprentice who nears the end of training will naturally start thinking about future positions of responsibility. From a journeyman, the next step up in the masonry business is to job foreman. The information in this unit deals with different methods of managing a job and the techniques involved in handling people efficiently.

MAINTAINING A GOOD ATTITUDE

There is an old saying on the job, "I would not ask you to do anything I have not already done myself." This is a good way for the foreman to begin when asking workers to perform a task that may not be pleasant. It really works. A good foreman acts on this by assisting on the job if necessary. Often the foreman does emergency repairs on an elevator, repairs a mixer, helps to change a tire on a piece of equipment, or does some other repair job. These are all part of the foreman's job and performance of these tasks is sometimes necessary before a qualified repairer arrives.

A foreman should present a positive approach. After making a mistake, the foreman should admit it, and not try to shift the blame. No one is perfect and thus the foreman, too, is subject to mistakes. Masons and laborers should be treated as the foreman's equals, and not be taken advantage of. Taking away a person's dignity can only result in bad feelings.

Workers must be shown that they have the foreman's confidence. Job expectations and requirements should be explained, but workers should be made to feel that they can do the job. The foreman should, however, follow up to be sure the workers are performing to their best capabilities. Workers who produce good work should be complimented and encouraged. People doing a job need to feel that they are contributing and that their job is important. Many masons and laborers have quit a job not because they were dissatisfied with the pay, but because they felt they were not appreciated.

When faced with a problem on the job, the foreman should face it with a positive attitude. If a foreman seems lost and shows weakness when confronted with a problem, the workers will soon lose confidence. Also, a foreman should not make a problem appear larger than it is.

All of the areas mentioned deal with attitude. If the workers are expected to have a good attitude about their work, then the foreman must also display a good attitude.

BECOMING FAMILIAR WITH THE JOB BEFORE STARTING TO WORK

If a foreman is going to be running a job very soon, it is beneficial to both foreman and contractor for the foreman to obtain a set of plans and specifications beforehand. The plans and specifications should be studied where there is no pressure from everyday job conditions. They can be taken home and reviewed in a quiet place so the foreman can become acquainted with the upcoming job. Many contractors provide the plans and specifications to the foreman several weeks before the job is to start; few contractors, however, allow any length of time on the job for studying the plans. Studying the plans beforehand usually results in a smoother running job and less problems for everyone.

A good foreman never starts studying the plans the same day the masons begin to work. This causes confusion and mistakes. Many companies pay bonuses based on production and efficiency, so the time spent studying the plans beforehand pays off. If the plans are studied at home, the time spent checking details of construction on the job, figure 34-1, will be kept to a minimum.

Fig. 34-1 Mason foreman checking the plans for a detail of construction on the job. If plans are studied at home or off the job, only a minimum of studying is needed on the job.

STARTING THE JOB ON TIME

A job that is not started on time usually causes bad feelings and a lack of incentive among the masons and laborers. The same applies at quitting time if there is frequently a lot of mortar to use up. When this situation exists, the workers know they are being taken advantage of and will only do their work when the foreman is watching over them. This type of tactic usually results in the masons seeking another job.

Since almost all masons and tenders work by the hour rather than by salary, any time lost results in a smaller paycheck. Workers become very discouraged if much time is lost. The same is true if the foreman has the workers leave immediately because of a light rain shower. Instead, if the foreman has the workers wait a short time, the rain may stop, and work may resume. There are, of course, exceptions to this; an agreement may have been established that requires the mason contractor to pay waiting time until the workers go back to work. This is done on some union jobs. The foreman's good judgment must be relied upon for each case.

An efficient foreman always starts the laborers 15 to 30 minutes earlier in the morning to make sure the mason has enough mortar and materials to begin work. This is also done at lunchtime when a laborer starts working 15 minutes early to temper the mortar that has dried out over the lunch break. This means that the laborer is allowed to leave the job 15 minutes earlier and have an earlier lunch break.

DEALING WITH EVERYDAY PROBLEMS

When building a masonry job that is more than one story high, figure 34-2, it is a good practice to find out beforehand if any of the workers are afraid of heights. Upon finding a mason who is afraid of heights, the foreman should try to assign the individual to a place on the job that encourages the worker's productivity where there could be no fear of falling. Many good masons who are afraid of heights are very productive if placed in a work station that is not near the edge of a building.

WORKING TO ONE'S HAND

Productivity is increased if a left-handed mason is paired with a right-handed mason when working on a job. This is known as *working to one's hand.* Insisting that a left-handed mason work out from a corner which is in a right-hand position will

Fig. 34-2 Masons working high up on the gable of a building. A good foreman finds out in advance if heights bother a mason before placing the mason high on a wall.

not only decrease work output, but results in poor workmanship from working in an unnatural position. The alert foreman should be aware of this situation and be able to handle it properly.

LIFTING HEAVY MATERIALS OR LINTELS

When there are heavy concrete blocks such as semisolid or solid units, the stress on the individual mason can be reduced if two masons lift each unit together. This method can also be followed when heavy concrete or steel lintels must be lifted in place on the wall over openings. (The word *heavy* refers to materials that a person or persons can lift without suffering any injuries.) Machines should be used to lift lintels and beams that are heavier than usual.

DRINKING WATER

Drinking water should always be available on a job. In hot weather it is the responsibility of the mason foreman to have ice added to the water. Sanitary disposable paper cups should also be available.

Drinking water should never be run through a garden hose into a can, because the hose can cause a bad taste in the water. Laborers and masons have quit jobs where the mason foreman has shown a lack of concern by not furnishing suitable drinking water.

COFFEE BREAKS

On many construction jobs it is common to have a short coffee or soda break in the middle of the morning. This is a good incentive to the workers if it is not abused.

Company policy usually dictates whether the workers will receive coffee breaks. A quick break in the middle of the morning usually increases the productivity of the workers and promotes good relations between employees and their employer.

GENERAL RESPONSIBILITIES OF THE FOREMAN

Working With The Other Trades On The Job

A competent mason foreman establishes good relations with all of the other trades on the job, such as plumbers, electricians, pipe fitters, and carpenters, since they must all work together to successfully complete the job. Obviously it is antiproductive to have all of the trades on a job working against each other.

An example of a job on which trades work together is the installation of electric boxes and fixtures in a masonry wall. The electrician installs the conduit and electrical boxes in the masonry wall before the walls are built so that the face of the box will be level and plumb with the masonry units after they are laid by the mason. The mason foreman should spot-check the wall before starting masons there to make sure that the boxes are installed correctly. If there is a problem, the foreman should immediately contact the electrician to correct the situation, figure 34-3. The electrician will appreciate this, since it will save time and effort later. When temporary lighting is needed inside a darkened area of the building, the electrician can install it. By creating good relations with the other tradespeople and contractors working on the job, the foreman can effectively speed the masonry work and eliminate unnecessary delays.

A good mason foreman always consults the foremen of the other trades a day ahead of time if it is known that the masonry workers will be moving into a certain area. This gives the other foremen the opportunity to be sure that their part of the work is installed properly before the masons begin. This spirit of cooperation helps the job progress smoothly.

Making A Story Pole And Establishing Bench Marks

Another very important part of the mason foreman's work is to make sure that a story pole, with all coursing and heights indicated, is made before the masons start building any of the walls. This, as previously explained, is determined by studying the plans.

Bench marks and level points should be established close to where the wall will be built. Although it is not necessarily the responsibility of the mason foreman to actually use the transit level, the building engineer must be contacted to perform the task.

Keeping Good Records On The Job

One of the most important tasks of the foreman is to keep an accurate record of the number of hours worked by the masons and laborers. This must be done daily and turned in to the bookkeeper at the end of the workweek to ensure that the payroll is made up on time.

The amount of masonry units laid per day by the mason should also be calculated daily or weekly (whichever is the policy of the particular company). This practice is known as *making the count*, figure 34-4.

Fig. 34-3 Mason foreman inspecting an electrical box with the electrician to determine if there are any problems.

Fig. 34-4 Mason foreman making the daily count of materials laid by the masons during the day. This practice is essential to determine if a profit is being made on the job.

Generally, a standard form simplifying the listing of materials used and number of units laid is supplied by the contractor's office. There are two main reasons why this practice is followed. First, it is possible to keep track of the amount of masonry units actually being laid by the mason and the contractor can determine if a profit is being made. Second, enough materials must be ordered beforehand to keep the masons working continuously. Almost all masonry jobs operate on the principle of recording the amount of masonry work done. This indicates to the foreman and contractor if any changes or improvements must be made to maintain a profit on the job.

Progress Meetings

On large construction jobs, such as commercial buildings where there are many different subcontractors working at one time, *progress meetings* are usually held once a week. The meetings are not necessarily long, but allow time for the superintendent to meet with the foreman or representatives of all of the subcontractors working on the job.

The purpose of a progress meeting is to discuss or work out any problems relating to keeping the job on schedule. The mason foreman must attend the meeting and discuss any problems which concern the masonry work. Progress meetings are usually indicated on the job specifications, or there is an understanding between the general contractor and all subcontractors when the contracts are awarded. If it is deemed necessary, the architect can also attend. A good progress meeting is very important to the successful completion of any large structure.

Follow The Chain Of Command On The Job

If a problem arises concerning the correct action to take in a particular situation, or if a mistake is found in the plans, the competent mason foreman will follow the proper chain of command before making a decision. Unless otherwise stated, the foreman contacts the job superintendent first. If the job superintendent cannot help or wants a change made that is not in the contract, then the foreman should contact their employer before making the change. The contractor may give the foreman the authority to get in touch with the architect for an answer to the problem. As a rule, however, this is done through the contractor's office. If the contractor agrees to a change in the original contract or specifications, a letter from the contractor or architect should be sent to the mason foreman relieving the foreman of any responsibility in relation to the change.

Before making any changes on the job, an authorization letter or note should be obtained and signed by the person desiring the change. This is extremely important, because if it is not done, the foreman may be held responsible for any changes. The letter from the architect authorizing changes is known as a *change order* and is essential for the foreman to have before starting any extra work.

Reporting Errors In Estimating Discovered On The Job

When ordering materials for use on the job, if an error is discovered according to the estimate the contractor had previously made, it should be reported immediately to the contractor's office. An example of this is when ordering concrete masonry lintels for over openings. Although it is the job of the estimator to take off the correct number, thickness, and length of all lintels (either steel or concrete masonry) from the plans, it is usually left up to the mason foreman to actually order the materials for delivery when they are needed. There is usually a *take off sheet* on all jobs against which the order can be checked. The estimator who works in the main office, figure 34-5, can check back to determine where the mistake occurred and see if it will affect the profit of the job.

Keeping Up With New Materials And New Methods Of Work

The foreman should be aware of new developments in the trade, such as new materials and advanced methods. To keep up with these advances, the foreman can attend evening seminars or special classes sponsored by manufacturers, architects, or others involved in the trade. It is a good experience for the foreman to attend building material shows and conventions with people who have similar interests in the construction business. A subscription to a trade magazine can also be a valuable aid.

Concerned contractors generally reimburse (repay) a conscientious mason foreman for the expenses involved in retraining or upgrading, if this was discussed earlier between the foreman and contractor. These experiences also help prepare the motivated foreman for any future promotions.

SAFETY RESPONSIBILITY OF THE FOREMAN

Safety is one of the most important responsibilities of a mason foreman. An employer is required by law to provide a safe working environment. Since the foreman represents the

Fig. 34-5 Estimator checking the plans in the contractor's office.

contractor, it is the foreman's function to see that the job is performed according to the law.

The *occupational safety and health standards* are standards which require the use of practices, means, methods, operations, or processes that provide safe and healthful places of employment. The law places the major burden of performance directly on the employer (or contractor in this case). The contractor is responsible not only at the beginning of the job, but at all times during the job when workers are there, from start to finish. No part of this responsibility can be shifted at any time from the employer to the worker.

It is the foreman's responsibility to establish and carry out good safety practices on the job. Therefore, since safety is a condition of work, it is important that the foreman is constantly checking for safe working practices and possible hazards that may be present.

Hard Hats

An important safety aid that protects the worker from bumps or blows on the head is the *hard hat*. If the work area is specified as a "hard hat area," figure 34-6, a sign should be posted so all the workers know that a hard hat must be worn.

The foreman should inspect the hard hats to be sure they are approved models, figure 34-7, and meet required standards. Light, plastic "bump hats" that resemble hard hats are a poor substitute and do not conform to standards. The foreman should remind the workers that they must wear the hard hats when working. In turn, the foreman should also wear a hard hat to set a good example.

Fig. 34-6 Construction job with a sign warning all workers that hard hats must be worn on the job.

Scaffolding

Before workers are situated on newly erected scaffolding, it is the job of the mason foreman to check that the scaffolding is safe. Nails protruding from the scaffolding, broken boards, and other hazards (figure 34-8) should not be present because they can cause accidents. Scaffolding planks should be laid on the scaffold frames to provide a sound footing which cannot slip or move. Any large holes should be boarded over and nailed securely. Many workers have been injured by stepping through holes left when the scaffold was built.

Masonry materials are stacked on the work area in a safe manner so they cannot be upset or accidentally tipped over. Equipment and machinery should have the proper guards in place to protect the worker. All of these items should be checked by the foreman.

Conducting Safety Inspections

Periodical safety inspections should be conducted by the foreman in addition to constant observation of job conditions. Workers should be encouraged to report any unsafe conditions, and quick action should be taken to correct problems.

Workers must be aware that the foreman is concerned about their safety. An occasional brief safety meeting in the morning just before beginning work can make the foreman aware of the worker's comments on job conditions. Concern shown for the workers usually results in better safety attitudes.

There are so many areas of safety in the masonry construction business that it would be difficult to mention them all here. Therefore, only some of the more important ones have been discussed.

Fig. 34-7 Mason wearing an approved hard hat while working.

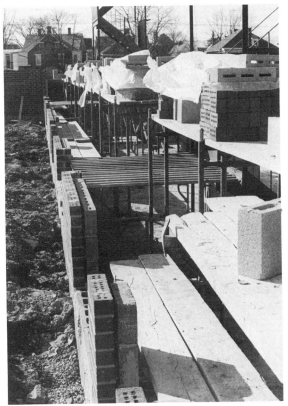

Fig. 34-8 The scaffolding shown here is unsafe to work on because of the numerous holes and obstructions present. All holes should be boarded over and the metal strap projecting out should be removed.

Injuries

All injuries should be reported to the foreman immediately. There are two reasons for this; prompt action and first aid may prevent the injury from becoming more serious, and workman's compensation insurance requires that injuries be reported as soon as possible after they occur. The foreman must make sure that quick action is taken in the event of an injury, including the filing of the proper papers and forms reporting the accident. If the accident is not reported, the employer may be subject to a fine.

Fig. 34-9 Mason foreman directing work on the job.

As a rule, neither worker nor foreman should try to treat the injured person if the injury is serious. Improper treatment can be harmful. It also exposes the foreman or contractor to a possible lawsuit. In the case of severe injuries, qualified help should be called and the injured worker transported to a doctor or hospital.

CONCLUSION

The role of a mason foreman is varied and requires a vast amount of working knowledge of the masonry trade. A foreman must also show concern for the safety of the workers and should encourage positive attitudes while directing the work, figure 34-9.

In addition to all of the other job responsibilities of a mason foreman, it is the task of the foreman to ensure that the job is finished on schedule and that a reasonable profit is made for the employer. A mason foreman is usually someone who began as an apprentice, worked up to journeyman, and then became foreman. To become a foreman usually requires years of experience and the ability to manage and get along well with people.

ACHIEVEMENT REVIEW

Answer the following questions relating to the foreman running a masonry job. Review the unit if necessary.

1. What is the single most important responsibility the foreman has to the employer?
2. List four important qualifications a mason foreman must have to be successful in managing a job.
3. Why is the foreman's attitude considered so important when running a job and dealing with workers?
4. Why is it so important for a mason foreman to cooperate and get along with the other tradespeople and contractors working on the job?

5. List and discuss three areas of responsibility for a mason foreman on a typical job.
6. The foreman must operate a safe job because the employer is liable under law. Describe three areas of safety discussed in this unit.

When the achievement review is completed, and the instructor has checked it, discuss the unit in class with other students.

SUMMARY, SECTION 10

- Planning and managing the job efficiently requires that the foreman keep up with advances made in equipment and materials along with providing careful supervision.
- The cost of labor on a masonry job makes up more than half of the total cost of the job.
- The ratio of mason to labor depends on the type of work being done and how difficult it is to supply the mason with materials. It is very important, however, to always have enough labor to keep the mason supplied with materials.
- An efficient laborer is one who is well trained and equipped with the tools necessary to do the job.
- A mason's laborer or tender should be treated as an equal by the mason if good relations are to exist.
- Masonry materials should be stocked in the immediate area of the mason, where they are readily available. They should be stacked on boards and covered to keep dry until ready to be laid in the wall.
- The area where the mason is going to be working should be level and free from hazards.
- It is a good practice to lay out the first course of masonry units before locating a group of masons on the wall. The cuts can also be made ahead, once the wall is laid out.
- A story pole and gauge rod should be made before any of the masonry work is built.
- The corner mason should be given a head start on the corner before a crew of masons start on the wall.
- Metal door bucks should be set in place and braced before the masonry work is begun.
- Metal corner poles, replacing the regular masonry corner, have been a giant step forward in increasing the mason's productivity.
- The mason's productivity can be improved by spacing the masons evenly on the wall, elevating mortar boards or pans, and stacking the materials in such a manner that they are readily available to the mason.

- Mortar must be supplied to the mason at all times.
- The mason foreman should plan ahead so that the laborers have mortar mixed by starting time and tempered when the masons return from lunch.
- The standard mortar joint thickness of 3/8 inch should be used whenever possible for laying masonry units.
- The type of scaffolding used should allow the mason the easiest access to the wall. The type depends on the design and height of the structure.
- It is more productive to build a masonry wall when facing toward the finished wall.
- Productivity increases if the mason is protected from bad weather.
- The process of laying masonry units in mortar can be improved by studying the movements involved. This is known as motion study.
- The mason foreman plays an important part in management-labor relations, being a representative of the contractor.
- A good mason foreman shows qualities of leadership such as being a self-starter, being fair to all of the workers, being able to meet the production schedule, having knowledge of the trade, and having a good attitude.
- The foreman should study the plans and specifications at home before starting a job.
- A foreman should always try to start and end the job on time.
- Inside work should be saved for rainy or bad days if possible.
- A good mason foreman knows the capabilities of the workers and places them to suit their best individual potential, in order to achieve maximum productivity.
- Clean drinking water should be supplied to the workers on the job.
- The mason foreman must cooperate with other tradespeople on the job if the job is to run smoothly.
- One of the important jobs of the mason foreman is to make sure that story poles and bench marks are available.
- Accurate records such as the daily work count, time book, and accident reports must be kept by the foreman.
- The foreman should attend all progress meetings.
- When there is a problem on the job, the mason foreman should follow the proper chain of command in reporting and solving the problem.
- It is the duty of the mason foreman to immediately report any errors discovered in estimating to the main office of the contractor.
- A responsible foreman should be aware of new developments in masonry materials and equipment.

- The safety of the workers is very important. Since the foreman represents the contractor on the job, it is the foreman's responsibility to see that safe working conditions are maintained in accordance with all state and federal laws.
- Regular safety inspections should be made by the foreman in addition to constant observation of job conditions as the job progresses.
- All injuries should be reported to the proper authorities and medical care should be arranged for the injured worker.
- Experience is the best teacher. This is the reason the mason foreman is usually someone who has worked up through the ranks to a position of leadership. Once there, good foremen will constantly upgrade themselves to stay on top.

SUMMARY ACHIEVEMENT REVIEW, SECTION 10

Complete each of the following statements referring to information found in Section 10.

1. More than half of the cost of an average masonry job is spent on _____.

2. A poor attitude shown by the foreman toward the mason usually results in a drop in the mason's _____.

3. The first course of masonry units is laid out ahead of the crew to establish the _____.

4. Before starting to build any of the masonry walls on the job, the foreman should mark off all heights and coursing on a _____.

5. The metal door frames that are set in place to be walled in by the mason are called _____.

6. The traditional masonry corner has been replaced on many jobs by the _____.

7. The even spacing of masons and materials on the wall greatly reduces the mason's _____.

8. Mortar pans or boards should be dampened with water before filling with mortar to prevent _____.

9. The process of adding water to mortar that has been lost through evaporation is called _____.

10. The most productive size mortar joint to use in masonry work is _____.

11. The most popular type of scaffolding used on an average job is _____.

12. Regardless of the type of scaffolding used, it must meet strict _____ _____ regulations.

13. The practice of laying bricks by standing on one side of the wall and reaching over it is known as _____.

14. The mason foreman represents the _____ on the job.

15. When dealing with workers, a good mason foreman always considers their feelings. How a person displays feelings is known as _____.

16. A good foreman prepares for a new job by studying the _____ and _____ beforehand.

17. In warm weather it is the responsibility of the mason foreman to provide _____ for the workers.

18. The mason foreman should cooperate with all of the other trades working on the job because cooperation affects the _____.

19. The term *making the count* refers to the _____.

20. If there is a problem on the job that the foreman cannot answer, the first person the foreman should consult is the _____.

21. If changes from the original plans and specifications must be made on the job, there should be a letter sent to the masonry contractor requesting approval of the change. This letter is called a _____.

22. Under law, it is the foreman's responsibility to make sure that the job is run in a _____ manner, free from _____.

GLOSSARY

abutment — the skewback of an arch and the masonry which supports it.

anchors — metal ties that have a bend on each end; used for tying stone and backing masonry together by building in with mortar.

arch — a section of masonry work that spans an opening and supports not only its own weight, but also the weight of the masonry work above it.

arch axis — the median line of the arch ring.

ash dump — flat, metal door with a flange built flush with the inner hearth of a fireplace; used for dumping ashes into the ash pit.

ashlar stonemasonry — stone that is cut to the dimensions shown on the shop drawings. It is cut, dressed, and finished at the mill to precise job requirements, and shipped to the job as a finished product. Ashlar stone is set in a constant mortar joint thickness, and is laid in a definite repeated pattern or in a varied bond called random ashlar.

ash pit — area directly below the inner hearth of a fireplace where ashes from the fire are collected by means of the ash dump.

ASTM (American Society for Testing Materials) — an organization that sets standards for building materials.

backfilling — the process of filling earth in around a foundation.

barbecue — a raised masonry structure for the smoking, drying, or preparing of food; commonly used to describe a small outdoor fireplace without a chimney.

basket weave bond — brick masonry pattern bond in which the bricks are alternated in a vertical and horizontal position every 8 inches.

batter — the upward and backward slope of a masonry wall (such as a retaining wall) from the face side. It is built to withstand the pressure of the earth fill behind the wall.

batter boards — right-angle corner posts and boards, erected on the job to hold and preserve the actual wall lines of the structure during the period of excavation and construction of the foundation.

bearing wall — a masonry wall that supports a load other than its own, such as joists, concrete floors, or roofs.

Bedford limestone — limestone that is quarried in Bedford, Indiana.

belt course — a course of stone or brick that completely encircles a building at one level for ornamental purposes.

bench mark — a definite mark put on some stationary object on the job site, to be used as a reference point in determining heights for a structure. This mark is based on the elevation of the ground above or below sea level.

bituminous — used to describe materials that are derived from coal tar, petroleum, and so forth.

bond beam — course of masonry units inserted with steel reinforcing rods and filled solidly with mortar; may serve as a lintel or reinforcement beam to strengthen a wall.

bonded arch — an arch that is composed of alternating bricks that are bonded together, such as a jack arch that is 1 1/2 bricks high.

bonded jack arch — a jack arch that is higher than one brick and is bonded by alternating the mortar joints so they are staggered in appearance. It is the strongest type of jack arch.

brick buggy — a cart with wheels that can be powered by a mounted gasoline engine, with or without hydraulic controls. It is used to transport masonry units to the mason.

brick veneer — the brick facing laid around an existing frame building. It supports only its own weight and is not considered a load-bearing wall. Brick veneer is also called brick casing.

building code — rules and regulations governing the construction of buildings; established by local, state, and federal agencies.

building line — the main line used to lay out a building; usually the front wall of the structure. The building line is generally taken from the property line.

building paper — paper applied to walls or floors as a cushion and dampproofer in the construction of a building. Thirty-pound felt paper is the most popular type to apply to framing walls before laying the brick veneer.

bush hammer — a heavy, steel forged hammer with a toothed head either on one end or both ends. It is used for dressing soft stone such as limestone or sandstone. The toothed head grinds the stone to a fine edge without cracking or splitting the stone.

calcium chloride — chemical that is added to water for the mixing of mortar. It should not exceed more than 2 percent of the portland cement by weight, and should not be used in mortar that contains metal wall ties because it can cause corrosion.

camber — the relatively small rise of a jack arch.

casting — method of fabrication of masonry panels involving the combining of masonry units and grout in a form. This can be done either in a vertical or horizontal position.

caulking — the process of filling in cracks or cavities around windows, doors, or expansion joints, using an elastic material usually from a caulking gun.

caustic soda — a chemical that is used for removing old paint from masonry work. It is also called sodium hydroxide or lye.

ceramic glazed brick — brick that is made from clay and shale, fireclay or a mixture of both, and glazed on the face before firing in a kiln. Glazed bricks are highly desirable where a brick with a very dense, water-repellent finish is specified.

chain of command — a series of executive positions, in order of authority.

cheeks — the sides of the porch wall that extend above the porch.

chill factor (or windchill) — the temperature plus the wind speed. The windchill makes the air seem much colder than the thermometer indicates.

chimney pad — footing for a chimney.

cleanout door — metal door installed in the masonry work below a flue at the bottom of the chimney. It is used for cleaning out soot from the chimney.

closure — the last masonry unit (such as the last brick) laid in a wall or step. A closure unit is usually laid in the center of a wall.

CMU — abbreviation for concrete masonry unit.

colonial ring-around fireplace — fireplace that has only the facing of bricks stacked up the jambs with an exposed soldier course over the top.

common jack arch — a jack arch that is only one soldier brick in height.

constant cross-section arch — an arch that has a constant depth and thickness throughout the span.

correlated — to be mutually related to each other or to work together efficiently. The laying of six standard courses of brick together, backed up with 2 courses of concrete block, would be level with each other, or correlated.

cove — the angled, thickened coat of portland cement mortar troweled on at the bottom of the outside of the foundation wall; used to turn water away.

cove base — specially made glazed tile with a projected curved lip at the bottom which is made for the finished floor to fit against.

crawl space foundation — foundation built deep enough to allow space for plumbing pipes and electrical wiring; usually a minimum of 24 inches from the foundation floor to the bottom of the floor joists.

creepers — bricks in the wall next to the arch which are cut to fit the curvatures of the outer edges of the arch ring.

crown — the apex (center) of the arch ring. In symmetrical arches the crown is at midspan.

cube — a banded stack of bricks or a pallet of concrete blocks.

curtain wall — a non-load-bearing masonry wall used to enclose a building. It is tied to the structural frame by means of angle iron clips (or some other type of tie), bolted or welded to the framework. Curtain wall construction is very popular in prefabricated masonry construction.

damper — metal device installed over the firebox in a fireplace to control the draft from the firebox into the chimney flue.

dead load — the weight of the superstructure such as floors, roofs, and walls. Dead load is constant, steady, and unmoving.

density — hardness of an object.

depth (of an arch) — the dimension which is perpendicular to the tangent of the axis. The depth of a jack arch is its greatest vertical dimension.

diamond bond — brick masonry pattern bond in which the bricks are arranged to form a series of diamond designs.

door bucks — the wood or metal subframe that a door is hung on.

dowel pin — a metal pin, usually made of brass or galvanized metal, used in stonework. The dowel pin is inserted in the top of a masonry wall with the end projected to fit into the dowel hole in the coping stone.

draft — the current of air created by the variation in pressure resulting from differences in weight between hot gases in the flue and the cooler air outside the chimney.

drain tile — terra-cotta or concrete pipe that is laid on a bed of crushed stone to drain water away from a masonry structure.

dry bonding — laying out masonry units without mortar to establish the bond for a wall.

dummy flue — flue lining installed at the top of the chimney that is not hooked up to any heat source. It is used only to appear as a flue and balance out a chimney that has only two flues but a large empty space between them. Dummy flues should be filled with mortar on the side to prevent birds or insects from building in them.

efflorescence — the whitish deposit found on the exposed face of masonry units in a wall. It is caused by the leaching of soluble salts from within the wall.

end construction — the vertical coring of the cells of a structural clay tile.

engineered brick masonry — load-bearing brick masonry walls that are constructed following a careful analysis of the structure and based on generally accepted analysis procedures. The brick walls are built to the story height and the floor is installed bearing on the wall. The next story is then built and the process is repeated until the structure is complete. No steel or concrete frame is required.

epoxy mortar — mortar used for pointing up head and bed joints in glazed tile work; consists of a resin, hardener, and powder. The finish of epoxy mortar is very hard and durable, and recommended for a joint finish where a sanitary condition exists.

expansion joint — elastic joint of rubber or fiber used in masonry to allow the movement of the masonry to prevent cracking from expansion and contraction.

extrados — the convex curve which bounds the upper limits of the arch.

face hammer — a heavy hammer weighing 4 to 6 pounds; used for dressing and breaking stone for masonry work. It has a tapered edge on one end and is also sometimes called a stone axe or stone hammer.

fill — a coarser grade of earth used to fill in and around foundations or walls.

firebox — area of the fireplace built of firebricks or bricks where the fire is built.

fixed arch — an arch that has a skewback which has a fixed position and inclination. Plain masonry arches are fixed arches.

flagstone — a hard flat-faced stone usually no more than 1 1/2 inches thick. It is laid either in mortar or sand for walks, patios, and floors. Flagstone can be laid in a squared section bond or a random pattern with irregular pieces. Several kinds are slate, New York, and fieldstone.

flash set — the rapid setting of portland cement caused by adding water that is too hot (usually in excess of 160 degrees F).

floor joists — wood or metal supports that span from wall to wall, usually 2 inches x 10 inches or 2 inches x 12 inches; used in the construction of a building to support the floor.

flue lining — fireclay or fireproof terra-cotta, used to line the flue of a chimney. It may be round or rectangular in shape, and is usually 2 feet long.

footboard — a wood board that is laid on concrete blocks or bricks and is used by the mason to stand on when setting stone or laying brick. Sometimes called a hopping board, a footboard is the simplest type of scaffolding.

footing — the enlarged base, usually made of concrete, that the foundation rests on. Its function is to support the building and distribute the weight over a greater area.

formed footing — concrete footings that are placed into a form on the job.

foundation — walls of a building that are built below grade (or below the first floor joists) and support the structure that will rest on it.

Franklin stove — cast iron stove resembling a fireplace; invented by Benjamin Franklin.

frost line — the depth of the earth where it is likely that frost will occur. The frost line varies in different places according to the temperatures in each particular area.

fullback — concrete block that has all of the cell webs cut out except the last one that holds the block together.

garden wall — a masonry wall that surrounds a property or divides the grounds into sections for gardens or secluded areas.

gauge rod (or sill stick) — shortened version of the story pole usually about 4 feet high. It is used for building the corner to the sill height.

gin pole — a wood pole or metal pole that is held in position by guy wires and cable. The end of the pole has a block and tackle or hook attached on a cable for the lifting of stones with a winch.

glass wool — insulation made of glass fibers.

Gothic arch (or pointed arch) — an arch with a rather high rise, with sides consisting of arcs of circles, the centers of which are at the level of the springing line. The Gothic arch is often referred to as a crop, equilateral, or lancet arch, depending upon whether the spacings of the centers are less than, equal to, or more than the clear span.

grade — the ground level around a building in relation to the building. When the earth is on a slope it is said to have a rising or falling grade.

grade beam — concrete reinforced beam that is placed just below the finish grade line for the purpose of supporting a masonry wall.

grading — the process of filling around a building with soil so the grade slopes downward and away from the building to carry the water away.

grout — very thin, proportioned mortar which is placed between two walls for reinforcement or liquid concrete that is placed in the center of a reinforced masonry wall; consists of portland cement, lime, and aggregates.

hacking — the process of unloading bricks from a truck by manual labor using brick tongs.

half-high block — a rip block that is cut in half lengthwise.

halfback — concrete block that has the inside of two cells cut out.

hawk — a square platform made of wood or metal, hand-held using a vertical handle in the center. It is used to carry plaster or mortar.

header brick — brick laid lengthwise in the wall with its smallest end facing out; used for corbeling in the smoke chamber.

herringbone bond — pattern bond used in brickwork in which bricks in alternate courses are laid on an angle in opposite directions. It presents a zigzag, diagonal pattern.

high-bond mortar — portland cement mortar that has a special additive for extra strength. High-bond mortar makes it possible to build masonry panels in 4-inch wythe thicknesses, in a factory, and ship to the job site to be erected.

high-rise building — a trade term describing a building that is built more than three stories high.

Indiana limestone — limestone quarried in the state of Indiana.

inner hearth — the flat bottom part of the firebox that the fire is built on; usually built of special firebrick.

intrados — the concave curve which bounds the lower limits of an arch. The difference between the soffit and the intrados is that the intrados is linear, while the soffit is a surface — in this case the bottom of the arch.

island fireplace — fireplace built in the center of the room that is open on more than one side.

jack arch — a flat arch capable of carrying a load, but should be supported on steel angle irons if the openings are over 2 feet long. It is the weakest of all arches.

jamb — vertical side of an opening, such as the side of a window or door.

jig — a form, template, or guide used for installing special forms or shapes in woodwork, masonry, or metal.

keystone — the wedge-shaped piece (stone or brick) at the top center of an arch which locks together the other pieces that form the arch ring.

lewis pins — steel pins that are inserted in a predrilled hole in a stone, on an angle of approximately 60 degrees, in an opposite direction to each other. The pair of pins are connected by a chain to a large center ring to which a rope or cable is attached, for lifting a stone to the wall. Lewis pins are usually from 5/8 inch to 1 inch in diameter.

limestone trim — term describing Indiana limestone used for coping, sills, belt courses, and trim on a building.

live load — load applied to a structure other than the weight of the structure itself; examples are snow on the roof, furniture, people, and any moving load.

lowboy — a low-built type of trailer used to transport equipment and heavy vehicles such as bulldozers.

lug sills — limestone sill set in a brick or masonry wall, extending beyond the widths of the openings, having masonry jambs set on the end or lug of the sill. Lug sills cannot be removed unless the masonry jamb is cut free.

making the count — the counting and adding together of the number of masonry units laid by the mason in one day. This count is necessary to discover if enough work is being laid by the mason to keep the work on schedule.

mantelpiece — the projection of wood or masonry materials that forms a shelf on the front of a fireplace; usually very ornate.

masonry saw — saw powered by an electrical motor used to cut masonry units, either wet or dry.

membrane — thin sheets or layers of tar paper or cloth used with tar against the foundation wall to waterproof the wall.

metal heat-circulating fireplace — prebuilt metal fabricated form fireplace unit, with hollow walls and vented openings to allow the passage of heat.

modular — masonry work laid out on a 4-inch grid called the module.

modular bricks — bricks made to fit the 4-inch grid system of measurement.

modular coordination — based on the 4-inch grid, this method applies the principles of mass production to reduce the cost of construction work. Different types of construction materials can be built together with no cutting or alteration. The plans are drawn on a modular scale and the building materials are also manufactured on a 4-inch grid.

module — the unit of measure used as a standard for modular construction; a module is 4 inches.

mortared paving — the laying of masonry paving in mortar on a concrete base.

mortarless paving — the laying of masonry paving without the use of mortar, generally on a crushed stone or sand base.

multicentered arch — an arch whose curve consists of several arcs of circles which are usually tangent at their intersections.

multiple-opening damper — type of damper used in a multiple-opening fireplace. It is much larger than a standard damper.

multiple-opening fireplace — fireplace that is open on more than one side.

multiwythe construction — masonry wall built of at least three thicknesses of masonry units in width.

nominal — a dimension greater than a specified masonry dimension because of an amount allowed for the thickness of a mortar joint not exceeding 1/2 inch.

nonmodular bricks — bricks that do not fit into the 4-inch modular grid system of measurement.

outer hearth — the part of the hearth that is built out in front of the fireplace; can be flush or raised.

pallet — a portable platform for handling, storing, or moving materials.

parging — the application of portland cement mortar to the outside of the foundation wall to prevent water from entering.

perch — unit of measure for stonework. A perch of stone is 16 feet 6 inches long, 1 foot high, and 1 foot 6 inches thick. This equals 24 3/4 cubic feet of stone. In some localities 16 1/2 cubic feet or 22 cubic feet are used as a perch.

pinch bar — a small crowbar of metal that is made with a hooked end. Pinch bars are used for prying objects or pulling nails out of wood.

pitching chisel — a heavy-duty chisel with a beveled edge, available in different widths, and used for cutting and dressing the face of building stone.

plain chisel — chisel built to withstand rugged use in stonecutting. It has a beveled blade, 1 1/4 inch to 2 3/4 inches wide. This chisel is used for all-purpose cutting and trimming of building stone.

platform — the part of the porch that is the landing surface at the top of brick steps. It can also be considered a flat surface usually higher than the floor or surrounding area.

plumb — exactly vertical or true.

point chisel — chisel used for removing small knobs of stone that project off a stone face, that must be removed without a lot of cutting. The point chisel is drawn to a point on the end.

prefabricated masonry — masonry wall panels that are built in a factory-type environment using high-bond portland cement mortar with special additives. The design is based on rational engineering standards used by the mason. After the masonry panel is made, it is shipped to the job site by truck and installed into place with the use of cranes.

prefaced concrete masonry units — lightweight concrete blocks that have had a thermoplastic resin finish applied to the face, and have been fired in a kiln to make the finish and block inseparable.

prehydration — the process of premixing water with pointing mortar, using only enough water so that the mortar will form into a damp ball when squeezed with the hand. After the mortar has set for one hour, more water is added to bring the mortar to the consistency needed for pointing in mortar joints with a tool.

quarterback — concrete block that has the inside of two cells cut out.

radial center point — the point where the radius line begins; used to lay out a jack arch.

raised flat joint — a method of tooling the mortar joints by forming a smooth, raised flat joint. This is very popular in ashlar stonework.

raised hearth — masonry hearth built above the finished floor level.

retaining walls — any wall built to hold back or support a bank of earth.

rip block — concrete block that is less than full size in height. It is generally used as a starting piece.

rise (of an arch) — the rise of a minor arch is the maximum height of arch soffit above the level of its springing line. The rise of a major parabolic arch is the maximum height of the arch axis above its springing line.

riser — the distance in step construction from one tread to the next tread; also the distance an object rises in height in construction work.

rock cut face — term used to describe building stone that is cut with the pitching chisel leaving the face in a roughly squared shape.

rolling bead joint — a method of tooling mortar joints with a concave striking tool that forms a rolled projected mortar joint. This type of joint is often pointed with a white portland cement mortar for design purposes.

Roman arch — the name given a semicircular arch because of its use by the Romans in buildings and bridges.

roof buck — scaffold frames made of lumber, built to fit the slope of a roof. They are usually constructed by the carpenter or mason on the job.

rubblework — stonemasonry in which the stone is laid either as it comes from the quarry (irregular in size) or roughly squared with a chisel and hammer by the mason on the job.

runway — a wood (or metal) platform that extends from the outside of a structure to the inside at floor level. It is used to support material handling equipment such as wheelbarrows.

salamander — type of metal stack heater that burns fuel oil in the bottom and radiates heat from the chimney stack pipe. It is used for heating areas where masonry or construction work is being done.

Salem limestone — select limestone that is quarried in Indiana. Salem limestone lies directly under the overburden which is composed of a limestone-dolomite formation. Salem limestone is characterized by a fairly uniform texture, color, and strength.

scratch coat — roughening the first coat of cement mortar parging with a stiff brush or wire lath to permit a better bond between the coats.

sedimentary stone — stone formed over a long period of time by the sediments of minerals and other natural materials.

segmental arch — an arch that has a circular curve which is less than a full semicircle.

semicircular arch — an arch that has a full semicircular curve; commonly called a Roman arch.

serpentine wall — masonry unit wall (usually bricks) built on an alternating curve from an established centerline. The unique feature of a serpentine wall is that it is usually only 4 inches thick and resists movement due to its curving design.

setting derrick — a wood-framed derrick used in the setting of stone. It is mounted on wheels and can be moved from place to place.

shooting points — establishing level marks around a foundation with the aid of a builder's level or transit.

shop drawings — the drawings showing details of construction and identification of the stone by numbers; supplied by the stone manufacturer or mill.

shove joint — method of forming a solid mortar joint by spreading the mortar on the brick and flattening out the mortar with a trowel instead of furrowing with the trowel point as is usually done. The brick is then shoved into the mortar, forming a solid joint.

side construction — the horizontal coring of the cells in a structural clay tile.

silicones — chemical dampproofer that comes in a liquid form for either brushing or spraying on masonry walls to make them water-repellent. They are effective in preventing absorption of water because they change the contact angles between the water and the walls of the capillary pores in the wall.

sill high — height of the masonry wall where the window sills will be installed.

6-8-10 method — formula used to square the corners of a structure as follows: the base squared plus the rise squared equals the hypotenuse squared. This forms a right triangle.

skewback — inclined (sloping) masonry units built into the wall, which the beginning arch bricks rest against at each end of an opening, such as over a window or door.

slab — the side thickness of a concrete block used where a thin facing block is needed.

slicker — metal tool used to strike a flat joint, such as in the hearth of a fireplace.

slings — heavy canvas, fiber, or leather straps used for lifting stones to the wall. The sling is attached to a cable and lifted by a derrick or crane.

slip sill — stone sill made the same size as an opening (such as a window or door) minus a mortar head joint on each end.

smoke chamber — the area of a fireplace that extends from the level of the damper up to the first flue lining. The slopes of all sides should be equal.

smoke shelf — brick ledge or shelf, laid level with the bottom edge of the damper and back of the firebox. It serves to deflect the air currents that come down the chimney back up the opposite side of the flue.

smoke test — test conducted by the mason after the fireplace is completed, consisting of building a fire at the bottom of the flue and then covering the top of the flue tightly. The escape of smoke into any other flue or into the building indicates a leak that must be made tight.

soffit — the undersurface or bottom of an arch.

soft joints — the joints between panels of prefabricated masonry in curtain wall construction. Panels are supported by the structural frame and joints are either pointed with mortar or filled with an elastic sealant.

soldier stack bond — stack bond in which all of the masonry units are laid in a soldier position.

spall — small chip of masonry materials such as brick or concrete block.

span — the horizontal distance between abutments (masonry jambs on each side of an opening). For minor arches, the clear span of the opening is used. For a major parabolic arch, the span is the distance between the ends of the arch axis.

splay — an inclined surface or angle such as the angled sides of a firebox in a fireplace.

springing line — for minor arches, the line where the skewback cuts or intersects the soffit. For major parabolic arches, the intersection of the arch axis with the skewback.

starter piece — a masonry unit or glazed tile cut specially for use at the corner to establish the proper bond lap.

stepped footing — concrete footings built in the shape of a set of steps. They are used because of a sharp drop in the grade or the presence of soft ground. Stepped footings should be graduated in increments of eight inches to work courses of concrete blocks.

stocking of materials — placing masonry materials in the work area for the masons to use in their work.

stonemasonry — the art of cutting and building stone walls either with mortar or without mortar.

stone setter — term applied to a mason who specializes in setting large, precut stone. This type of stonework concentrates more on the skills of handling and setting large stone, rather than the dressing or cutting of stone.

story high — the height of the masonry story from the floor line to the height where the wood plates will be set for the ceiling joists.

story pole — wood pole approximately 3/4 inch x 2 inches x 10 feet on which all of the various heights and individual courses of masonry units are laid off. The pole is used as a gauge to build the masonry corners.

stretcher stack bond — stack bond in which all of the masonry units are laid in a soldier position.

striking hammer — a heavy steel hammer used in settling stone into place or driving stakes. The striking hammer weighs from 2 to 4 pounds and is also called a mash hammer.

structural clay tile — a fired masonry unit consisting of light burning, de-aired fireclay used in the construction of masonry walls. It can be either glazed or unglazed depending on the job requirements.

take-off sheet — the list made by the estimator when estimating the materials for a job.

termite shield — metal shield, made of copper, aluminum sheet, or similar material, installed on top of foundation walls to prevent the upward movement of termites into the floors of a building.

thermal movement — expansion and contraction of masonry caused by changes in temperature.

thimble — a round terra-cotta or fireclay insert which fits into a chimney to receive the furnace pipe and form a fireproof connection into the masonry wall.

throat — the part of the fireplace that narrows at the top of the firebox and which the gas smoke passes through. If a damper is set, it becomes the throat.
thrust (of an arch) — the tendency of an arch to spread to the sides under a load.
tile shapes — various shapes of structural tile that are made to adapt to special needs in the construction of structural clay tile walls.
tooth chisel — a steel chisel in which the blade is shaped to form a series of teeth; used for dressing stone.
topping — finishing layer of mortar on concrete.
topping out — the process of laying the last section of the masonry wall to the finished height of the building.
tread — the flat part of the step that a person steps on.
tree well — masonry wall, usually round in construction, that holds back the earth from the roots of the tree after the grade has been raised.
trench footing — concrete footings that are placed in an earthen trench.
trimmer arch — masonry arch used in fireplace construction to support the rough hearth or the rough opening over the firebox.
tuck-pointing — the process of refilling mortar in mortar joints using a slicker.
tudor arch — a pointed, four-center arch of medium rise-to-span ratio.
viaduct — a long bridge built of a series of masonry spans supported on piers or towers. The Romans used viaducts with arches over the piers to transport water from the mountains to the cities.
voussoir — one of the wedge-shaped masonry units which form the arch ring, such as a brick in a jack arch.
wall-area method — method of estimating masonry, consisting of multiplying known quantities of materials required per square foot by the net wall area (gross areas less areas of all openings).
washing down — process of cleaning masonry work with acid and water.
water table — a masonry unit or stone that forms a ledge usually at the top of a story of a building or ground level. Water tables provide an ornamental effect; the slope of the top drains water away.
weep holes — small holes left in the mortar bed joints just above the grade line to allow water to come out. Usual spacing for bricks is every third head joint.
working to one's hand — working a left or right-handed mason to the best advantage off of the corner. A right-handed mason is more at ease working from the left end of a wall towards the center, and a left-handed person is more comfortable and productive working from the right end of the wall to the center.
working overhand — the process of building a masonry wall by working only from one side, although the wall may be tooled on both sides.
wythe — wall built of only one masonry unit in thickness.

ACKNOWLEDGMENTS

The author extends special thanks to the Brick Institute of America, for providing technical assistance and special materials throughout the preparation of this text.

Reviewers

Alan H. Yorkdale, P.E., Director of Engineering and Research, Brick Institute of America
Matthew Meizinger, Instructor, Albany County Board of Cooperative Educational Services
Thomas Redmond, P.E., National Concrete Masonry Association
William F. Roark, Director of Manpower Development, Brick Institute of America
Dr. Donald Maley, Professor and Chairman, Department of Industrial Education, University of Maryland

Dr. Kenneth Stough, Department of Industrial Education, University of Maryland
Kenneth A. Gutschick, National Lime Association
William Hotaling, Jr., Hotaling Associates
Richard Siebert, Instructor, Associated Builders and Contractors
Dorothy Kender, Executive Vice President, Building Stone Institute
Carl Deimling, Vice President, Stark Ceramics, Inc.
Wayne Smith, Executive Secretary, Facing Tile Institute

Contributions by Delmar Staff

Publications Director — Alan N. Knofla
Supervising Editor — Mark W. Huth
Source Editor — Judith E. Barrow

Copy Editors — Noel Mick
Rosanne F. Budrakey
Photography Editor — Sherry Patnode
Editorial Assistant — Virginia Styczynski

Other Help Was Provided By

Carl P. Bongiovanni
Harold Staley
Cleon R. Stull, Sr.
Tedd Godbee
Robert Sheckles and Sons
Frederick Contractors, Inc.
Sherwood Mackenzie
Helen Corgill
Duane Eaton
Aubrey Markus Lyles
Hubert Schneider
Larry Huston
Indiana Limestone Company, Inc.

Vestal Manufacturing Company
Patricia Marshall
The Burns and Russell Company
Charles W. Chamberlain, III
AMSPEC
Fleming Devices, Inc.
Hanley Company
PCM, Division of Koehring Company
Vetovitz Masonry Systems
Aeroil Products Company, Inc.
Parks Tool Company
Norton Construction Products

Tom Shoemaker
Steve Firme
Stover Brothers
Modular Masonry Inc.
Lefty Kreh
Irving Swope
R.W. Otterson
Joe Thomas
Donald Kreh
Students, Thomas Johnson High School, Frederick, MD
The Masonry Instructors of Maryland

The author wishes to express special thanks to his wife, who typed and proofread the original manuscript; and his son, who helped with the photography.

Contributions To Classroom Testing

The instructional material in this text was classroom tested at Middletown High School, Middletown, MD; and Western Vocational Technical Center, Catonsville, MD.

Contributions Of Illustrations

Clipper Saw Company — figures 6-1, 6-2, 6-13
Vestal Manufacturing Company — figures 8-2, 8-3, 8-5, 8-10, 9-2, 9-3, 9-4, 9-5, 9-6, 14-4, 14-5, 14-6, 14-10
Brick Institute of America — figures 8-4, 9-1, 10-8, 10-9, 10-12, 10-14, 10-15, 10-16, 10-17, 10-18, 10-19, 11-4, 11-8, 12-8, 13-5, 13-6, 13-17, 13-18, 14-8, 14-9, 15-4, 15-8, 16-1, 16-2, 16-3, 16-4, 16-5, 16-6, 20-3, 20-5, 20-6, 21-2, 22-7, 28-3, 28-7, 29-4, 29-5, 30-1, 30-2
Dur-O-Wal — figure 13-3
National Concrete Masonry Association — figures 13-14, 13-15, 16-7, 16-8, 16-9, 16-10, 16-11, 16-12, 16-13
Facing Tile Institute — figures 17-3, 17-5, 17-6, 17-7, 17-8, 17-9

Burns and Russell Company — figures 19-1, 19-2, 19-3, 19-4, 19-5, 19-6
Indiana Limestone Company — figures 24-1, 24-2, 24-13
Goldblatt Tool Company — figures 24-10, 25-2, 25-3
Masonry Specialty Company — figures 24-11, 28-1, 33-6, 33-7
Norton Company — figure 32-8
Fleming Devices, Inc. — figure 20-2
Park Tool Company — figure 32-9
PCM, Division of Koehring Company — figures 32-6, 32-11, 32-12
AMSPEC, Inc. — figures 31-6, 31-7, 31-8, 31-9, 31-10, 31-11, 31-12, 31-13, 31-14, 31-15, 31-16
Vetovitz Masonry Systems — figure 32-13

INDEX

Abrasive blade, masonry saw, 68, 69
Accelerators, cold weather mortars, 392
Access to foundation, 9
Accidents, 460
Accuracy, in estimating, 227
Advances in masonry equipment, 423-431
Advantages, modular coordinated system, 225
    prefabricated masonry panels, 419
    prefaced blocks, 275
Air ducts, metal heat-circulating fireplaces, 127
American Standard Association, modular coordination, 216
Anchor bolts, 42, 328
Angle cutting, masonry saw, 74
Angle irons, 40, 417
Antifreezes, cold weather mortar, 392
Arch form, semicircular arch, 289-290
    curvature, 291
    marking off arch bricks, 292
    removing, 295-296
    setting in place, 291
Arches, classes, 282
    defined, 281
    history, 281
    jack arch, 300-303
    parts, 284-286
    present day uses, 287
    terminology, 282, 286
    two-rowlock semicircular arch, project, 297-299
    types, 282, 283
Areaways, foundation walls, 41
Ash dump, fireplace, 94, 95
Ashlar stonework, 315, 318
    raised flat joint, 344, 345
Attitude of foreman, 451
Automated plant fabrication of masonry panels, 414

Backfilling, 51, 57
    retaining walls, 193
    waterproofing foundation, 57
Barbecue, brick, 204
    building, 199
    project, 209
Base materials, brick paving or flooring, 144-145
Basement, waterproofing, 50
Basement fireplaces, 80
Basement foundation, 8
Basket weave bond, 140, 141
Basket weave brick garden wall, 169, 170
Basket weave pattern in panel wall, project, 174-175
Batter boards, 3-4
    depth of foundation, 8
    locating wall lines on, 7
    project, 12-14
Battered stone wall, 184, 185
Beam pockets, 40
Bearing walls, 43
Belt courses, 325
Belt-type conveyors, 428

Bench marks, 2
    foreman's responsibility, 455
Bituminous primer, 51, 373
    waterproofing, 54
Block bond, 252
Block foundation wall, story pole for, 24
Blocks, corner, 35
    first course, 35, 440
    placing materials, 36
Blockwork at grade, 38
Bolts, anchor, 42, 328
Bond beam, 74
Bonded arch with soldier course, 293
Bonded jack arch, 300, 301
    project, 306-308
Bonds, brick paving patterns, 139-141
    concrete block foundation, 33
    estimating tables, 229
    fireplaces, 92, 106
    glazed tile, 252
    modular walls, 223
Boundary cut, limestone, 321
Bracing, foundation walls, 57
Breast or setting derrick, 327
Brick and concrete retaining wall, project, 195-197
Brick arches, bonded, 293
    laying arch bricks on form, 293
    marking on form, 292
    rowlock or header arches, 293
Brick barbecue, plan for, 204
Brick bearing masonry walls, 400-408
    comparison with steel and concrete frame construction, 402
Brick buggies, 429
Brick courses, story pole, 25
Brick fireplace and chimney, project, 114-118
Brick flooring and paving, 138-149
    cleaning, 148
    concrete base, 143
    drainage, 142
    edging, 142
    estimating materials, 149
    installation methods, 141
    interior, 146-148
    mortar, 143
    mortared paving, 142
    patterns and bonds, 139-141
    reinforced, 148
    units, sizes and shapes, 138
Brick incinerator fireplace, 207
Brick Institute of America, building codes, 401
Brick masonry, engineered, 400-408
    estimating, 228
Brick paving units, 138
    estimating, 149
    laying in running bond, project, 151
Brick porch and steps, building project, 162-165
Brick retaining walls, 187

473

Brick steps and porch, 154-161
Brick stretcher course, 157
Brick units, types of, 403
Brick veneer, over existing structures, 375-382
Brick wall, repointing, project, 365
    story pole for, 24
Brick windowsill, 3
Bricks, ceramic glazed facing bricks, 255-257
    chimney, 83
    garden walls, 166
    laying on concrete base, 143
    modular, equivalent factors, 232
        estimating, 229
        sizes, 216
    nominal dimensions, 217
    patios, 134
    prefabricated masonry panels, 411
    standard, dimensions, 221
Brickwork, removing old paint, 356
    starting at grade, 38
Bucks, 361
Builder's level, 4
Building codes, 2, 401
Building lines, 2
    batter boards, 3-4
Building steps, 159-161
Bullnose corner tile, 251
Bush hammer, 329, 338, 339
Buttering, 303

Calcium chloride in mortar, 392
Cantilever wall, counterfort type, 186
    retaining wall, 181, 182, 183, 184
Capping patio wall, 135
Caps, glazed tile, 246
    masonry retaining wall, 189
Care, story pole, 29
    tools, 20, 75
Carving limestone, 322
Cast iron damper, 102
Casting, prefabricated masonry panels, 414
Caulking, brick veneer construction, 380
Cement bags, care, 36
Cementitious materials, cold weather mortar mix, 392
Center partition bearing walls, 43
Ceramic glazed facing bricks, 255-257
Ceramic glazed structural clay tile, 244
Chain of command on job, 456
Change order, 456
Cheeks, porch, 154
Chill factor, protection against, 394
Chimney flues, dummy, 85
    linings, 84
    old, cleaning, 363
    two-flue, project, 87
Chimney footings, 16
Chimney pad, 16
Chimney tops, rebuilding, 362
    repairing, 361
Chimneys, 83-85
    clearance around wood, 84
    finishing, 85
    flashing, 85
    footings, 84
    masonry saw cutting, 55-76
    more than one fireplace, 108-110
    outdoor fireplace, 206
    project, 114-118
    section view, 95
    solid masonry, 83
Chisels, stonemasonry, 339
Circular and running bond, 140
Circular retaining wall, 191
Circulation of heat, metal heat-circulating fireplace, 124
Clay tile, *see*
    Glazed structural tile
    Structural clay tile
Cleaning, brick floors, 148
    fireplace, 360
    glazed brick wall, 268
    glazed tile, 260, 265, 266
    limestone, 333
    masonry work, outdoor fireplace, 208
    rubble stonemasonry, 347
Clipped header, 101
Closure brick, 160
Closure units, special, 223
Cobblestones, walks or patios, 347
Coffee breaks, 454
Cold weather, construction in, 388-397
    accelerators and antifreezes, 392
    heating sand, 390
    heating water, 391
    mortar, 390, 392
    preparation of units, 393
    protecting work and workers, 394, 447
    sound base, 393
    storage of materials, 389
    summary of practices, 397
Collar joint, 228
Colors, glazed facing bricks, 256
    glazed structural tile, 245
    limestone, 323
    prefaced masonry units, 273
    stone, 314
Columns, footings for, 16
Common jack arch, 300, 301
    project, 304-306
Concrete, curing time, 20
    curling, 369
    footings, 16, 20
Concrete base, laying bricks on, 143
Concrete block, 1
    cutting with masonry saw, project, 77-78
    electrical mortar spreaders, 430
    lintels, 40
    partition wall, 43
    slab, 73
Concrete block and brick foundation, project, 45
Concrete block foundation, 33-44
    story pole for, project, 30-32
    waterproofing, 50-58
Concrete frame construction, comparison with brick bearing walls, 402
Concrete masonry units, estimating, 231-239
    patios, 134
    retaining walls, 187
Concrete retaining wall, 187
    reinforcing, 190
Concrete saws, 428
Construction, arch, 282
    factors affecting, 1
    footings, 15-22
    metal heat-circulating fireplace, 125-126
    prefabricated masonry panels, 412-416
    retaining wall, 189
Conveyors, 428
Convex joints, 346
Coping, garden walls, 167
    limestone, 325
    patio walls, 135
Corbeling, metal heat-circulating fireplaces, 127
    smoke chamber, 104, 106
Coring glazed tile, 251
Corner batter board, construction, 3
Corner blocks, laying in mortar, 35
Corner mason, responsibility, 28
Corner pole, 38, 442
Corners, concrete block foundation, 33-34
    glazed tile, 263
Costs, labor versus materials, 438
Counterfort wall, 186
Coursed ashlar, 316
Coursed rubble stonework, 341
Courses, story pole, 25
Coursing, glazed tile, 253-255
Cove base units, glazed tile, 248, 262-263
    prefaced masonry units, 272, 273

Covering work, cold weather, 393
Cracks, repairing, 366-369
Cranes, advances in, 424
Crawl space, foundation, 8
Curvature, semicircular arch, 291
Cuts, angle, 74
 around electrical boxes, 72
 bond beam 74
 masonry saw, 70-72
 prefaced masonry units, 275
 rips, 73
 slabs, 73
Cutting, flagstone, 348
 glazed tile and bricks, 268
 limestone, 321-334
 pitching chisel, 342
 squared rubble stone, 341
Cutting machines, hydraulic, 429
Cutting methods, masonry saw, 68-69

Damper, fireplace, 82, 97, 102, 103
 multiple-opening fireplaces, 120
Dampproofing, 371
Dead load, 1
Definitions, glossary, 465-472
Depth, footings, 15-16
 foundations, 8
Design, factors determining, 1
 garden walls, 169-172
 patios, 135
Design series, prefaced masonry units, 272
Diamond blade, masonry saw, 68, 69
Diamond bond, garden wall, 171
 panel wall, project, 176-178
Dimensions, nominal, 217, 220
 See also Size
Door buck frames, 441
Door heads, story pole, 25
Doors, lintels, 40
 modular construction, 233-237
Dowels, 328
Draft, fireplace, 81-83
 multiple-opening fireplaces, 120
Drain tiles, around footings, 55-56
 retaining walls, 188-189
Drainage, brick paving, 142
 patio walls, 136
Drinking water, 454
Dry bonding, fireplaces, 92, 106
Dry cutting, masonry saw, 69
Ducts, metal heat-circulating fireplaces, 126-127
Dummy flues, 85

Economy, metal heat-circulating fireplaces, 124
Edging, paving or walls, 141, 142
Efficiency, job planning practices, 438-448
Efflorescence stain, 138, 167
 removal, 369
Electrical boxes, cutting around, 72
Elevations, concrete block foundation, 31
Elevators, advances in, 425
Encased columns, 369
Enclosed scaffolding, 395, 396
Enclosing structure, 394
End construction, glazed tile, 251
Engineered brick masonry, 400-408
Epoxy mortar, glazed tile, 260
Equipment, access to foundation, 9
 advances in, 423-431
 limestone setting, 326
Erecting prefabricated masonry panels, 416-419
Errors in estimating, reporting, 456
Estimating, modular masonry units, 227-239
 outdoor fireplace, 206
 reporting errors in, 456
 stone, 336
Estimating tables, brick paver units, 149
 modular coordination, 228-239
Excavating foundation, 9

Expansion cracks, 366
 repairing, 367
Expansion joints, brick paving or flooring, 144
 glazed brick laying, 267
 retaining walls, 189
Eye protection, 68, 100

Face hammer, 338, 339
Facing, fireplace, 106-108
Facing work, 446
Failure, retaining walls, 179
Familiarity with job, 452
Federal Occupational Safety and Health Act, 424
Fieldstone, 316, 317
Fill, around tree, 192
 retaining wall, 181
Finished outer hearth, 94
Firebox, 99
 outdoor fireplaces, 205, 206
 section view, 101
 walls, 100
Firebrick, hearth, 98
Fireplaces, 79-83
 basement, 80
 brick, project, 114-118
 building fire in, 83
 cleaning, 360
 construction principles, 96
 design and construction, 90-111
 footings, 16
 heat from, 83
 history, 79
 mantels, 79
 masonry saw cutting, 66-76
 metal heat-circulating, 123-127
 modern, 80
 multiple-opening, 119-123
 outdoor, 198-208
 random rubble stone, 344
 repairing and rebuilding, 359
 workmanship, 81-83
First course of block, 35
Flagstone, 347-349
Flashing, brick veneer, 379
 chimneys, 85
Flooring, interior, 146-148
Flue linings, chimneys, 84
 fireplace, 105-106
 metal heat-circulating fireplaces, 127
Flue size, multiple-opening fireplaces, 123
Flues, chimney, cleaning, 363
 chimney with more than one fireplace, 108, 109, 110
 dummy, 85
 metal heat-circulating fireplaces, 125
Footings, 15-22
 brick veneer, 376
 chimneys, 84
 concrete block foundation, 33
 concrete for, 16, 20
 corners, 33, 34
 curing time, 21
 defined, 15
 formed, 17, 18, 19
 garden walls, 167
 low retaining walls, 188
 outdoor fireplaces, 202
 porch and steps, 154
 retaining walls, 181
 size, 15
 steel reinforcement, 16
 steps, 156
 types, 17-19
Foreman, 450-460
 attitudes, 451
 coffee breaks, 454
 drinking water, 454
 everyday problems, 453
 familiarity with job, 452
 general responsibilities, 454-457

Foreman (continued)
  lifting heavy materials, 454
  punctuality, 453
  qualifications, 451
  safety responsibilities, 457
  working to one's hand, 453
Forklift tractors, 427
Formed footing, 17, 18, 19
Formula, squaring foundation, 4-7
Foundation walls, anchor bolts, 42
  areaways, 41
  center partition, 43
  cracks and repairing, 368
  grade, 38-40
  lintels and beam pockets, 40
  porches, 41
  solid joist bearing, 42
  termite shields, 43
  waterproofing, 50-58
Foundations, 1-9
  access to, 9
  batter boards, 3
    project, 12-14
  brick veneer, 376
  building lines, 2
  changing grade line, project, 45
  concrete block, 33-34
  crawl space, 8
  depth, 8
  excavating, 9
  fireplace, 91
  garden walls, 167
  laying out, 4-7
  loads, 1
  locating lines, 7
  outdoor fireplaces, 202
  parging, project, 59
  patio walls, 135
  porch and steps, 154
  units for constructing, 1
Framing around openings, brick veneer, 380
Franklin back fireplaces, 79
Front facing, fireplace, 106-108
Frost line, 8
Fullback cuts, 72
Functional series, prefaced masonry units, 272

Garden wall, construction, 166-172
Gauge rod, 27, 441
Gin pole derrick, 326, 327
Glass wool insulation, multiple-opening fireplaces, 121
Glazed and prefaced masonry units, 244-276
  glazed brick, 255-257, 266-268
  prefaced concrete masonry units, 270-276
  structural clay tile, 244-255, 259-266
Glazed facing bricks, 255-257
  cutting, 268
  laying, 266-268
Glazed structural tile, 244-255
  bonds, 252
  care before laying, 259
  colors, 245
  coring and scoring, 251
  coursing, 253-255
  dimensions, 221
  grades and types, 245
  laying, 259-266
  metal reinforcement, 253
  non-absorbent, 255
  prefabricated masonry, 412, 415
  properties, 245
  shapes, 250
  sizes, 250
  standard units and shapes, 246-249
Glossary, 465-472
  arches, 285-286
Gothic arches, 282, 283, 284, 287
Grades, glazed tile, 245-250
  limestone, 323

  starting brickwork, 38
  stepping up, 40
  walls of outdoor fireplace, 203
Gravity retaining wall, 183
Grids, location of masonry walls, 220
  masonry units, 222
  modular design, 217-219
  module, 216
  window details, 218
Grilles, metal heat-circulating fireplaces, 126
Grills, outdoor fireplaces, 205
Grout, 143
  prefabricated masonry panels, 411
  retaining walls, 187
Groutings, high-lift and low-lift, 190

Half-high rip, 73
Half-lap bond, glazed tile, 252
Halfback cut, 71
Hammers, 338
Hand method of building, prefabricated masonry panels, 412-414
Hand tools, 329
Handling care, prefaced masonry units, 274
Hard hats, 458
Hawk, 358
Head joints, cut limestone, 332
Head protection, 458
Header course rings, 293
Hearths, fireplace, 93, 94, 98
  inner, 94, 95
  laying, 98
  metal heat-circulating fireplaces, 126
  raised, 98
  rebuilding, 360
Heat chamber, metal heat-circulating fireplace, 124
Heat circulation, metal heat-circulating fireplaces, 123-127
Heat radiation, fireplace, 83
Heating sand, 390
Heating system, protection during rebuilding of chimney top, 362
Heating water, 391
Height, brick step, 157
  checking, 27
  concrete masonry courses by, 234
  fear of, 453
  gauge rod, 27
Herringbone weave bond, 140, 141
  garden wall, 170, 171
High-lift grouting, 190
High-rise buildings, 400-401
  cranes used on, 424
Hog in the wall, 25
Hoists, for limestone, 326
Horizontal coursing tables, glazed tile, 254
Hydrated lime, 392
Hydraulic-equipped material trucks, 426
Hydraulic masonry cutting machines, 429
Hydraulic-operated machines, 427
Hydrostatic pressure, 50, 57

Incinerator fireplace, 206, 207
Indiana limestone, 321
  grades, 323
  mortar, 326
Injuries, 460
Inner hearth, 94, 95
Inspections, safety, 459
Installation, brick paving or flooring, 141
  interior floor, 146
  metal heat-circulating fireplaces, 126, 127
  prefabricated masonry panels, 417-419
  prefaced masonry units, 274
Insulation, brick veneering construction, 381
  metal heat-circulating fireplaces, 126
  multiple-opening fireplaces, 121
Interior brick flooring, 146-148
Interiors, limestone for, 326
Irregular squared rubble work, 341
Island fireplaces, 120

Jack arch, 283, 286, 287
    construction, 300-303
    projects, 304-308
Job planning and practices, 438-448
Joints, *see* Mortar joints

Keystone, bonded jack arch, 301

Large stones, setting, 316
Laying brick paving in running bond, project, 151
Laying cobblestones, 347
Laying flagstone, 348
Laying glazed brick, 266-268
Laying glazed tile, 259-266
Laying hearth, 98
Laying out foundation, 4-7
Laying out steps, 158
Laying rubble stone, 339-340
    squared rubble stone, 341-344
Laying structural tile, 259-266
Laying tile to the line, 265
Layout, increasing job production, 440
Lead, glazed tile, 263
Leadership on job, 450
    attitude, 451
    coffee breaks, 454
    drinking water, 454
    everyday problems, 453
    familiarity with job, 452
    general responsibilities, 454-457
    lifting heavy materials, 454
    punctuality, 453
    qualifications, 451
    safety responsibility, 457
    working to one's hand, 453
Leveling batter board, 4
Lewis pins, 328
Lifting devices, advances in, 427
Lifting heavy materials, 454
Lime, cold weather mix, 392
Limestone, cut, 321-334
Lintels, concrete block, 40
    fireplace opening, 107
    fireplaces, 79
    metal heat-circulating fireplaces, 127
    porch walls, 42
Live load, 2
Loads, 1-2
Location, fireplace, 90
    outdoor fireplaces, 199
Low-lift grouting, 190
Low retaining wall, footing, 188
Lug sills, 324

Major arches, 282
Making the count, 455
Mantels, fireplaces, 79, 107
Mesh hammer, 338, 339
Mason, cold weather protection, 394, 447
    responsibility for marks on story pole, 28, 455
    spacing, on job, 443
Masonry arches, classes, 282
    terminology, 282-286
    types, 283-284
    uses today, 287
Masonry, alteration, repair and maintenance, 355-364
    protecting in cold weather, 393, 394-397
    removing old paint, 356
    stonemasonry, 312-319
Masonry caps, 189
Masonry construction, cold weather, 388-397
    engineered brick masonry, 400-408
    prefabricated masonry panels, 410-420
Masonry equipment, advances in, 423-431
Masonry layout, nominal dimensions, 220
Masonry saws, 66-76, 428
    cutting concrete block, project, 77
    using to best advantage, 75
Masonry silicones, 371
Masonry units, 1
    modular coordination, 224
    size and weight, 445
Masonry walls, cracks, repairing, 366-370
    protection while under construction, 381
    See also Walls
Material trucks, hydraulic-equipped, 426
Materials, access to foundation, 9
    brick paving, estimated, 149
    chimneys, 83
    cold weather storage, 389
    concrete masonry mortar, 239
    engineered brick masonry, 406
    garden walls, 166
    interior floors, 146
    job productivity, 439-443
    modular, estimating, 227-239
    new, 457
    outdoor fireplaces, 199
    patio floors, 144-145
    placing, 36
    porch and steps, 155
    retaining walls, 187
    spacing, 443
    traffic lanes for movement, 37
Mayan stonework, 312, 313
Measurement, story pole, 24, 455
Membrane, 51
Metal corner pole, 38
Metal heat-circulating fireplaces, 123-127
Metal wire joint reinforcement, retaining walls, 181
Minor arches, 282, 288
    terminology, 286
Miters, glazed tile, 246
Modern fireplaces, 80
Modular coordination, 215-225
Modular masonry materials, estimating, 227-239
Modular size flue linings, 105
Module, 216
Moisture, waterproofing, 50-58, 189, 372
Mortar, brick paver units, 149
    brick veneer, 377
    cold weather preparation, 390, 392
    efficiency in handling, 444
    engineered brick masonry, 404, 405
    flagstone work, 348
    garden walls, 166
    glazed bricks, 267
    glazed tile, 259
    jack arch, 302
    limestone, 326
    modular construction, 237-239
    outdoor fireplaces, 200
    outdoor paving, 143
    patio walls, 135
    prefabricated masonry panels, 411
    removing from joints, 357
    repointing work, 358
    retaining walls, 187
    rubble stone, 340
    stone, laying, 338
    waterproofing, 51
Mortar boards, checking, 443
    spacing, 443
Mortar cove, 52
Mortar grout, 143
Mortar joints, arches, 294, 295
    brick arches, 292
    cut limestone, 331
    grid location, 222
    modular design, 217-219
    prefabricated masonry panels, 419
    prefaced masonry units, 275
    raised flat, 344
    raked, 345
    removing old mortar, 357
    rolling bead, 346
    rubble stone, 340
    steps, 160
    stonework, 344-346

Mortar joints (continued)
　　striking, glazed tile, 264
　　thickness, 220, 445
　　waterproofing, 372
Mortar mixer operator, 445
Mortar spreaders, electrically operated, 430
Mortared paving, 142
Mortarless paving, 141
　　bricks, 144-145, 147
　　estimating table, 149
Multicentered arch, 283
Multiple-opening fireplaces, 119-123
Multistory wall bearing, brick construction, 400-408
Multiwythe construction, 228
　　concrete masonry units for, 236

National Oceanographic and Atmosphere Administration, bench marks, 2
New developments in trade, 457
Noise-absorbent glazed tile, 255
Nominal dimensions, 139
　　importance, 217
　　understanding, 220
Nonmodular size flue linings, 105
Nonstaining mortar, 326

Offset cracks, 369
Offsetting fireplace, 91
One cell out cut, 71
Oolitic limestone, 321
Openings, fireplace, 98
　　modular construction, 233-237
OSHA, safety standards, 424, 458
Outdoor fireplaces, 198-208
Outdoor living trends, 133
Outdoor structures, 133-172
　　fireplaces, 198-208
　　garden wall construction, 166-172
　　patios and brick paving, 133-149
　　porches and steps, 154-161
　　retaining walls, 179-193
Outer hearth, 94
Ovens, outdoor fireplaces, 200-202

Paint, old, removing from brickwork, 356
Panel wall, basket weave pattern, project, 174-175
　　diamond bond pattern, project, 176-178
Parabolic arch, 283
Parapet walls, cracks, 369
Parging, after first three courses, 55
　　back of arch, 294
　　foundation, project, 59
　　planter walls, 138
　　Portland cement mortar and tar, 51
　　waterproofing, 52-54
Partition wall, 43
Patching, cut limestone, 334
Patio walls, building, 136-138
Patios, cobblestones for, 347
Patios and brick paving, 133-149
Pattern bonds for paving, 139-141
Patterns, garden walls, 169-172
Paving, patterns and bonds, 139-141
Piers, footings for, 16, 17
　　garden walls, 169
Pilasters, garden wall, 168, 169
　　retaining walls, 185
Pitching chisel, 339
　　cutting with, 342
Plain-bladed cutting chisel, 339
Planning and managing job, 438-460
　　job planning and practices to improve productivity, 438-448
　　leadership on job - mason foreman, 450-460
　　familiarity with, 452
　　framing, 95
　　154, 156

Plumb bob, 17
　　corners on footing, 33
　　veneer work, 377
Plumb bond, 252
Plumb line, stonework, 336
Plumbing glazed tile, 264
Plywood arch form, 289-290
Point chisel, 339
Pointing, cut limestone, 332
　　flagstone, 348
　　mortar, 358
　　stonework, 359
Poker type damper, 102, 103
Pole derrick, 327
Porch cheek, 161
Porch foundations, 41
Porches and steps, 154-161
Portland cement mortar, waterproofing, 51
Portland cement paints, 372
Poured joint, 333
Powered wheelbarrows, 429
Prefabricated masonry panels, 410-420
Prefabricated metal stairs, 405
Prefaced concrete masonry units, 270-276
Prescription method, concrete, 17
Problems, dealing with, 453
Productivity, job planning and practices to improve, 438-448
Progress meetings, 456
Projects
　　barbecue, 209
　　basket weave pattern in panel wall, 174-175
　　bonded jack arch, 306-308
　　brick and concrete retaining wall, 195-197
　　brick fireplace and chimney, 114-118
　　brick paving in running bond, 151
　　brick porch and steps, 162-165
　　brick wall, repointing, 365
　　common jack arch, 304-306
　　concrete block and brick foundation on changing grade line, 45
　　cutting concrete block with masonry saw, 77-78
　　diamond bond pattern in panel wall, 176-178
　　jack arches, 304-308
　　laying out batter boards for foundation, 12-14
　　parging a foundation, 59
　　story pole for concrete block foundation, 30-32
　　two-flue chimney, 87-89
　　two-rowlock semicircular arch, 297-299
Proportioning fireplace openings, 98
Protection, cut limestone work, 333
Protection from weather, 447
Punctuality on job, 453

Qualifications, foreman, 451
Quarterback cut, 71

Radial center point, 291
Rain spouts, retaining wall, 181
Raised flat joint, 344
Raised hearth, 95, 108
Raked joints, 345
Random ashlar, 315
Random rubble stonemasonry, 344
Random rubble stonework, 339-340
Rebuilding chimney top, 362-364
Rebuilding fireplaces, 359
Records, foreman's responsibility, 455
Reinforced brick floors, 148
Reinforced concrete masonry retaining wall, table, 191
Reinforced cutting blades, should not be used, 69
Reinforced masonry cantilever wall, 183, 184
Reinforcing retaining wall, 180
　　concrete, 190
　　metal wire joint, 181
Relationships with other trades on job, 454
Repairing, chimney tops, 361
　　cracks, 366-369
　　cut limestone, 334
Repointing brick wall, project, 365

Index

Repointing mortar joints, 357
Repointing work, mortar for, 358
Responsibilities, general foreman, 454-457
Restoration work, 355-364
Retaining wall, 179-193
    brick and concrete building project, 195-197
Right-angle corner batter board, 3
Right-triangle method, foundations, 4-5
"Ring around" fireplace, 80
Rips, 73
Rise, stairs, 154
Riser and tread, brick steps, 156-158
Rock cut face, 343
Rolling bead joints, 346
Roman arch, 281, 282
Rotary control damper, 102, 103
Rough hearth, installing, 93-94
Rough opening, hearth, 94, 97
    allowance for raised hearth, 98
Rough rubble stonework, 314
Roughly squared rubble stone, 314, 315
Rowlock arches, 293, 294
Rowlock bricks, 157, 158, 159
Rubble stone, laying, 339-340
Rubble stonemasonry, cleaning, 347
Rubblework, 314
Running bond pattern, 140, 141
    laying brick paving in, project, 151
Runways, 37

Safety, OSHA standards, 424
    responsibility of foreman, 457-460
Safety inspections, 459
Safety practices, masonry saw, 69-70
Sand, heating, 390
    moisture in, 228
Saws, concrete, 428
Scaffolding, 36
    chimney tops repairing, 361
    effect on productivity, 446
    enclosure, 395, 396
    engineered brick masonry, 405
    safety responsibilities, 459
Scheduling, 403
Scored series, prefaced masonry units, 272
Scoring, glazed tile, 251
SCR acoustile, 255, 265
Screen blocks, 134
Sedimentary stone, 314
Segmental arch, 283, 286, 287
Semicircular arch, 281, 282, 283, 287
    construction, 289-296
    two-rowlock, project, 297-299
Serpentine wall, 172
Setting practices, cut limestone, 330
Setting tools, 326-329
Settlement cracks, repairing, 368
Shapes, glazed tile, 246-250
Shipment, limestone, 322
Shop drawings, cut limestone, 326
Silicones, dampproofing, 371
Sill high, 29
Sill stick, 27
Sills, glazed tile, 246
Sills and coping, limestone, 323-325
Single-wythe walls, 228, 229
6-8-10 method, squaring foundation, 4-5
Size, fireplaces, 90, 96
    footings, 15-16
    glazed facing brick, 256
    glazed tile, 250, 251
    metal heat-circulating fireplaces, 125
    nominal dimensions, 217, 220
    prefaced masonry units, 272
    retaining wall, 181
Skewbacks, 301
Slabs, 73
Sledge hammer, 338
Slicker, 330

Slings, 327
Slip sills, 323, 324
Smoke chamber, 103-106
    flue lining, 106
Smoke shelf, fireplace, 82, 101
Smoke test, rebuilding chimney, 363
Soft joint, 419
Soldier course, 142, 293
Solid joist bearing, 42
Solid masonry, chimneys, 83
Spacing of masons and materials, 443
Specifications, familiarity with, 452
Springing line, 291
Squared rubble stone, 341-344, 345
Squaring foundation, 4-7
Stack bond pattern, glazed tile, 252, 253
Stack bond weave, 140
Stairs, prefabricated metal, 405
Standardization, modular coordination, 215-225
Starter pieces, bond for modular walls, 223
    glazed tile, 261
Steel frame construction, comparison with brick bearing
        walls, 402
Steel ovens, outdoor fireplaces, 200, 201
Steel reinforcement, footings, 16, 17
Steel reinforcement rods, 187
Stepped footings, 19
Stepped masonry wall, 185, 186
Steps and porch, 154-161
    project, 162-165
Stone, properties of, 313-314
Stone barbecue, 200
Stone cutting and laying, 336-349
Stone porch and steps, 156
Stone retaining walls, 187
    battered, 184, 185
Stone setter, 313
Stone walls, 316
Stonecutting bench, 341, 342
Stonemasonry, development, 281-287, 312-319
    applied arch terminology, 286
    arches, 282-286
Stonework, face jointings and finishes, 317
    pointing, 359
Stoop, 154
Storage of materials in cold weather, 389
Story high, 19
Story pole, 24-29, 441
    foreman's responsibility, 455
    nominal dimensions, 220
    project, 30-32
Straight brick garden wall, 168
Strength, masonry work, 314
Stretcher and header brick steps, 161
Stretcher and rowlock brick steps, 161
Stretcher course, 157
Stretcher stack bond, glazed tile, 252, 253
Stretchers, glazed tile, 246
Structural Clay Products Institute, 401
Structural clay tile, 244
    history, 244
    manufacturing, 244

Tar, waterproofing, 51
Tar compound, applying over parging, 54
Tarpaulin, 394
Tender labor, 438-439
Terminology, arch, 282, 285-286
    glossary, 465-472
Termite shields, 43
Thermal movements, 367, 369
Three cells out cut, 72
3-4-5 method, squaring foundation, 6, 7
Throat, fireplace, 102
Thrust of arch, 282
Time, punctuality, 453
Tooling glazed brick mortar joints, 267
Tools, care, 20
    hand, to cut limestone, 329

Tools (continued)
　　masonry saw, 75
　　setting, limestone, 326-329
　　stonemason, 338-339
Tooth chisel, 330
Topping out, 29
Topsoil, 9
Tractors, 427
Tradespeople, relationship, 454
Traffic lanes for material movement, 37
Transit level, 34
Travelers, limestone equipment, 322
Tread, steps, 154, 156, 158
Tree wells, 191
Trees, fill around, 192
　　lowering grade, 192
　　retaining walls around, 191
Trench footings, 17, 18
Trim, limestone, 325
Trimmer arch, 93
Trucks, hydraulic-equipped, 426
Tuck-point, 357, 358
Tudor arch, 283
Two-against-one stonework, 315
Two cells out cut, 71
Two-flue chimneys, 81-83
Two-(or more) flue chimneys, 83-85
Two-rowlock semicircular arch, project, 297-299
Type M mortar, 51, 135
　　outdoor paving, 143
Types, glazed facing bricks, 255
　　glazed tile, 245-250
　　prefaced masonry units, 272

Veneer, brick, 375-382
Vertical coursing table, glazed tile, 254
Viaducts, 281

Walking buggies, 406
Walks, cobblestone, 347
Wall-area method, brick estimating, 228
Wall lines, locating on batter board, 7

Walls, basket weave pattern in, project, 174-175
　　brick and concrete retaining wall, project, 195-197
　　brick veneer, 379
　　building, 38
　　center partition, 43
　　diamond bond pattern, project, 176-178
　　firebox, 100
　　fireplace, 99
　　glazed brick, procedures for laying, 267-268
　　glazed facing bricks, 255
　　grid location of mortar joints, 222
　　joining patio walls to house, 137
　　outdoor fireplace, 203, 204
　　retaining, 179-193
　　tying in arch, 295
　　waterproofing interior faces, 372
Washing down, fireplace, 108
Water for drinking, 454
Water pressure, foundation walls, 50
Water tables, belt courses, 325
Waterfall, 138
Waterproofing, foundation wall, 50-58
　　interior wall faces, 372
　　retaining wall, 189
Weather, protection from, 447
Web cuts, 71-72
Wet cutting, masonry saw, 68, 69
Wheelbarrows, gasoline engine, 429
Windbreak, 384
Windows, lintels, 40
　　modular construction, 233-237
　　nominal dimensions, 218
Windowsills, bench mark, 3
　　limestone, 323-328
　　story pole, 25
Wood arch forms, semicircular arch, 289, 290
　　curvature, 291
　　setting in place, 291
　　removing, 295
Working overhand, 446
Working to one's hand, 453
Worm, damper, 103

LEEDS COLLEGE OF BUILDING LIBRARY
NORTH STREET
LEEDS LS2 7QF
Tel. 0532 430765